GRAINS OF SAND

Martin Buckley began his journalistic career as a sub-editor in Bombay. Later he joined the BBC, where his work ranged from reports on *From Our Own Correspondent* to a documentary about Tantra. His column *Notes From a Nomad* appeared in the *Daily Telegraph*, and he has presented travel documentaries for Discovery Channel.

Martin Buckley

GRAINS OF SAND

VINTAGE

Published by Vintage 2001

2 4 6 8 10 9 7 5 3 1

First published in Great Britain by
Hutchinson 2000

Vintage
Random House, 20 Vauxhall Bridge Road,
London SW1V 2SA

Random House Australia (Pty) Limited
20 Alfred Street, Milsons Point, Sydney
New South Wales 2061, Australia

Random House New Zealand Limited
18 Poland Road, Glenfield,
Auckland 10, New Zealand

Random House (Pty) Limited
Endulini, 5A Jubilee Road, Parktown 2193,
South Africa

The Random House Group Limited Reg. No. 954009
www.randomhouse.co.uk

A CIP catalogue record for this book
is available from the British Library

ISBN 0 09 927735 2

Papers used by Random House are natural, recyclable products made from wood grown in sustainable forests. The manufacturing processes conform to the environmental regulations of the country of origin

Printed and bound in Great Britain by
Cox & Wyman Ltd, Reading, Berkshire

CONTENTS

PREFACE   x

MAPS   xi

Part I: Into Africa
1   Dinner With Buffet   3
2   Vast Eternities   8
3   White Men Drive Jeeps   15
4   Out of Eden   19
5   The Bibliophobe   26
6   It's Good to Kill   33
7   Faya and Water   40

Part II: Tibesti
8   Vast Eternities   47
9   *L'homme moyen sensuel*   52
10   The Waste Land   62
11   Gentian Violet   70
12   The Crater of Salt   74
13   The Sky at Night   80
14   A Photograph   85

Part III: To the Atlantic
15   Days and Nights in N'Djamena   93
16   The Government Inspector   98
17   Hausseman   104
18   An Interlude   110
19   Sahel Taxi   115
20   Agadez   118
21   Desert Storm   129
22   Hasheesheen   134
23   Of Mere Being   142
24   Sand   147

25   To Timbuctoo   161
26   Altared State   176

Part IV: A Thirstland
27   Nothingnessland   185
28   Solitary Pleasures   190
29   *Deutschland Über Alles*   194
30   The Romantic Belgian   200
31   Bushmen and Bushwomen   208

Part V: Between the Andes and the Ocean
32   The Longest Country in the World   221
33   Ice Cold in Anto   226
34   Copper Country   229
35   Blood and Sand   233
36   To Peru   238
37   The Nazca Lines   241
38   Señor Grau   245
39   Señor Grau's Great-Grandfather   249

Part VI: Sonora
40   In Which I Plumb the Depths   255
41   Salvador and the Meat Market   262
42   Snakes, Teddy Bears and Desert Rats   266

Part VII: Americana, Australiana
43   On the Road   273
44   Koyaanisqatsi   277
45   Abbey Country   283
46   Indian Country   287
47   High and Lows   297
48   Death Valley   302
49   Red Centre   306
50   Diesel Rock   313

Part VIII: From the Gobi to the Empty Quarter
51   *Le désert, c'est l'avenir*   327
52   Desert is Too Hot   332

CONTENTS

53   The Desert Blooms   339
54   Heaven's Gate   347
55   Sand City   354
56   Silk Cut   360
57   Kashgar   368
58   Karakorum Highway   375
59   Pakistan   381
60   The Great Indian Desert   387
61   Civilization   393
62   Hell is Other People   399
63   Iran   402
64   Happiness   409
65   Tehran   416
66   The Empty Quarter   422

Part IX: The Christian Desert
67   The Desert Fathers   431
68   A Grain of Sand   438

ACKNOWLEDGEMENTS   445

To my wife, Penny
And to the memory of my father

To see a World in a Grain of Sand,
And a Heaven in a Wild Flower,
Hold Infinity in the palm of your hand,
And Eternity in an hour.

William Blake, '*Auguries of Innocence*'

And yonder all before us lie
Deserts of vast eternity.

Andrew Marvell, '*To His Coy Mistress*'

He told me that his book
was called the Book of Sand,
because neither the book nor the sand
has any beginning or end.

Jorge Luis Borges, '*The Book of Sand*'

PREFACE

This is the story of a circumnavigation of the world via the deserts, the two rings of sand and rock that would – were it not for the oceans – completely circle our globe. The deserts have always impeded travel, though our ancestors were more adept at crossing them than we often imagine. I wanted to see the deserts not as an obstacle, but as a place in their own right, a unity. I started out simply liking desert landscapes, and ended up feeling completely at home in them – perhaps more at home than I do anywhere else.

I could not afford, and did not want, to mount an expedition; it was my ambition to travel like the local people, and I did many thousands of miles by bus, jeep, truck, train and – occasionally – camel or mule. I made the journey in three main legs of around six months each, interrupted by periods at home for rest and recuperation. There were also several supplementary trips. But the essence of a near-continuous overland journey, mostly east-west, through the deserts of the world, is accurately described. Politics and discretion have dictated the changing of a few people's names.

I talk at greatest length – almost half the book – about the Sahara, because in many ways it is the quintessential desert. Travel there is still so difficult and dangerous that one is forced to go at the pace the terrain allows, and human relationships are slow-building and profound.

Occasionally I get a touch mystical, and sometimes I comment disapprovingly on the material obsessions of the Western world. I hope readers who do not share these perspectives will be tolerant: anyone who spends time alone and far from the West, and in particular with poor desert dwellers, comes to see the virtues of a simpler life.

You may wonder why I should have wanted to make such an arduous journey. Yet I think we all live with a sense that somewhere inside us there is a wilderness, an area of unexplored anxieties, hatreds, desires – but also, of spaciousness and calm.

Everyone has his own desert.

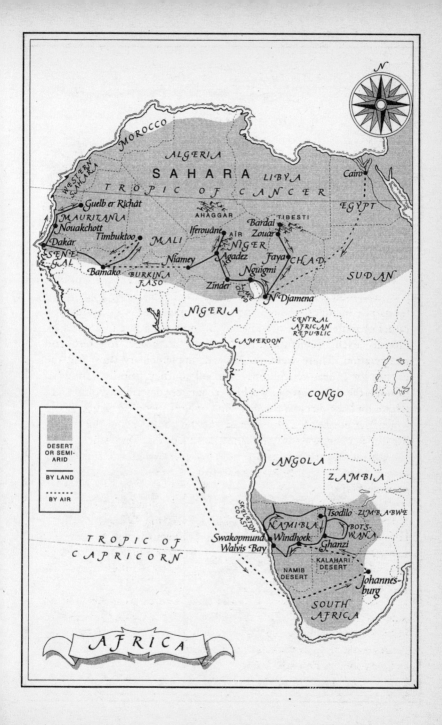

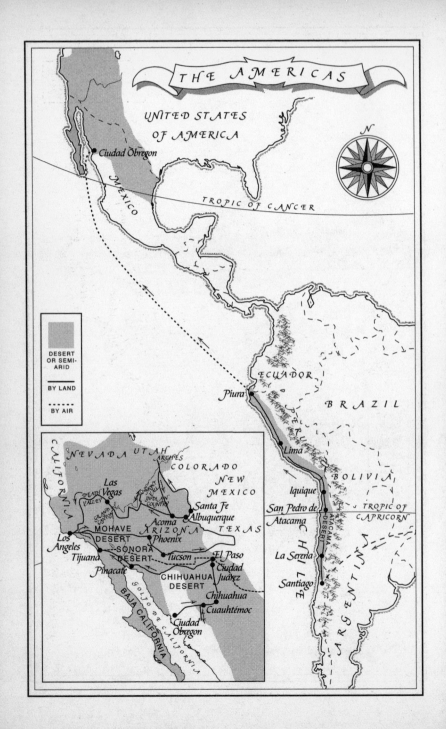

THE AMERICAS

UNITED STATES
OF AMERICA

Ciudad Obregon

MEXICO

TROPIC OF CANCER

N

ECUADOR

Piura

BRAZIL

Lima

BOLIVIA

Iquique

San Pedro de
Atacama

TROPIC OF CAPRICORN

La Serena

Santiago

ARGENTINA

DESERT
OR SEMI-
ARID

BY LAND

BY AIR

NEVADA UTAH
ARCHES
COLORADO
NEW
MEXICO

CALIFORNIA

Las Vegas
DEATH VALLEY
GRAND CANYON
INDIAN COUNTRY

Santa Fe
Albuquerque

Acoma
ARIZONA

MOHAVE
DESERT

Phoenix

TEXAS

Los
Angeles

SONORA
DESERT

Tucson

El Paso

Tijuana

Ciudad
Juárez

Pinacate

CHIHUAHUA
DESERT

Chihuahua

BAJA CALIFORNIA

Cuauhtémoc

GOLFO DE CALIFORNIA

Ciudad
Obregon

CHILE

ATACAMA DESERT

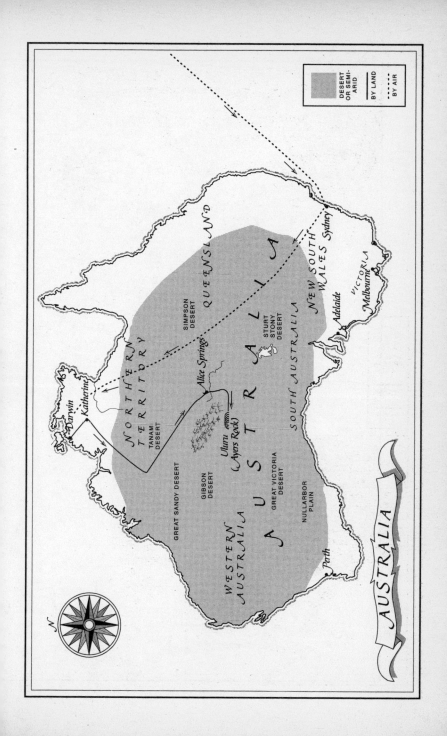

DESERT OR SEMI-ARID
BY LAND
BY AIR

QUEENSLAND

NORTHERN TERRITORY

SIMPSON DESERT

Alice Springs

Darwin

Katherine

TANAM DESERT

Uluru (Ayers Rock)

GREAT SANDY DESERT

GIBSON DESERT

GREAT VICTORIA DESERT

WESTERN AUSTRALIA

NULLARBOR PLAIN

Perth

AUSTRALIA

STURT STONY DESERT

SOUTH AUSTRALIA

NEW SOUTH WALES

Sydney

VICTORIA

Melbourne

Adelaide

AUSTRALIA

N

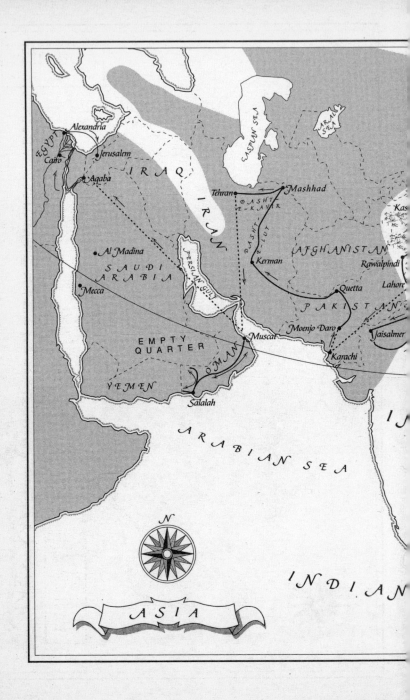

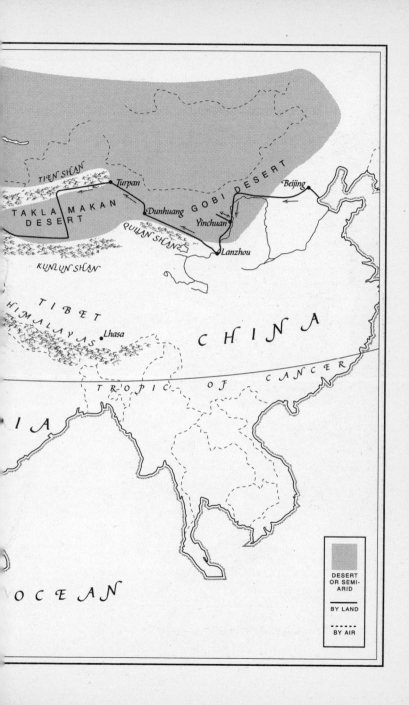

TIEN SHAN

Turpan

TAKLA MAKAN
DESERT

Dunhuang

GOBI DESERT

Beijing

Yinchuan

QILAN SHAN

KUNLUN SHAN

Lanzhou

TIBET

HIMALAYAS

Lhasa

CHINA

TROPIC OF CANCER

IA

OCEAN

DESERT
OR SEMI-
ARID

BY LAND

------ BY AIR

PART I

Into Africa

CHAPTER ONE

Dinner with Buffet

'PERHAPS YOU'D CARE TO join me for dinner,' Bernard Buffet had said. 'I have a guest who knows all about the desert. He used to live in Tibesti, near the Libyan border. Come early, around six. Don't worry if I'm late, *la patronne* will look after you. *Elle s'appelle Mari-Jo.*'

As the evening settled, the Yamaha and I puttered north along the banks of the broad River Chari. The Buffet residence was in a well-irrigated security compound just outside the Chadian capital of N'Djamena, overlooking the river. *La patronne* herself met me at the door.

She was young and black, while M. Buffet was old and white – nothing too original there – but she seemed more like the professional escorts who hung out around N'Djamena's hotel pools than your conventional ex-pat wife. She was pretty, stocky, with a bulbous backside imprisoned in black lycra. On the top half of her body she wore a loose, thin cotton smock, and nothing else. It was 6 p.m., and Mari-Jo was plastered.

She was not alone. Her friend Michèle was a more petite woman, in a pillar-box-red dress with a frilly neck. She had the most extraordinary hair I had ever seen. It had been bleached and dyed strawberry-blond, then divided into concave sheets. It seemed like an attempt to recreate the Sydney Opera House in hair. I tried not to stare, but I need not have worried: they both carried on as though I wasn't there.

'It's a cat's head that keeps evil spirits away,' said Michèle.

'No,' said Mari-Jo with drunken pedantry, 'it's the head of a *dog*.'

'But if they give you a cat's head to eat, you *have* to eat it,' insisted Michèle.

Mari-Jo noticed that the television wasn't on, and jabbed the remote control at the screen. Heads of state were seen arriving in aeroplanes and press-conferencing in huge gilt chairs. One of the men was squat and wore a snot-green suit.

'God, he's ugly!' said Mari-Jo.

'I can't stand him,' said Michèle, 'he's so *black*!'

They began to giggle helplessly, collapsing sideways on the sofa.

When Michèle went to the bathroom, Mari-Jo whispered confidentially that her friend was waiting for *le patron*. 'A colleague of *le patron* has been sleeping with her for the last three days. He was short of cash, so he sent her here for *le patron* to give her a cheque.'

I nodded, understanding nothing. The doorbell rang.

It was *le patron* with his two guests, both employed, like Buffet himself, by French NGOs. The first was François, a man in his mid-forties, tall and curly-haired with a pocked, sulky face. He shook my hand curtly and walked to the bar, filling a tumbler with ice and inverting a bottle of Red Label until whisky sloshed around the rim. He took a swig, and greeted Michèle warmly, putting his hand around her waist.

The second man, addressed by Buffet as *M. le docteur,* was rounder, softer and older. Buffet handed him a triple gin and tonic, and *M. le docteur* sank into the sofa and asked Mari-Jo how she was.

'Tipsy, I'd say,' said Buffet, deftly plucking the gin bottle from Mari-Jo's reach just as she leaned forward to take it. She shot him a dangerous look, which he ignored. 'M. Buckley here is interested in deserts,' he told the Doctor. 'He wants to go to your Tibesti.'

'Really?' The Doctor said.

'You know the area?' I asked.

'I don't know if I *know* it. I was based in the desert for a while.'

'Tibesti,' announced François, turning his attention from Michèle, 'is a fucking shit-hole.'

I stared at him.

'They're racists in the north,' added Michèle.

'M. Buckley, there's a tension, you see,' the Doctor told me, 'between northern and southern Chad, between Arab, Moslem North Africa and the black, mostly Christian, south.'

'This country,' said Buffet, 'is black-controlled. But not many miles north of here you cross a fault-line which separates Black Africa from the Arab world. The *toubous*, the northern tribes of the desert,

are a constant thorn in the south's political flesh. They have to be . . .
conciliated.'

'Or else', continued the Doctor, 'they may change allegiance –
which they're dying to do. You have to remember that when this
country was at war with Libya in the eighties, many northern
tribesmen fought for Gaddafi. He promised them their own state.
They're still unstable.'

'And they hate the blacks,' put in Michèle.

'They're racists,' said François.

'*You're* the racists,' said Mari-Jo, with a sulky pout.

A silence descended, and all eyes turned to her. Clutching her glass
in one hand, she gestured with the other at the television and grinned
uneasily. 'Well, haven't you just elected all these fascists?'

The French election results had been on satellite TV all day, showing
the far right making gains. Buffet laughed. 'Ah, *Madame la directrice* does
have a point. My friends, denunciations of racism sound a little hollow
on French lips today!' He changed the subject, lecturing us for ten
minutes on the far right and left in post-war French politics. Even in
mid-flow, his beady eye never let any glass get more than half empty –
unless it was Mari-Jo's. I stared at this tall, slightly paunchy man with
his Gauloises voice and long, tapering, nicotined fingers with the
unclipped fingernails of a man who inhabits his mind, not his body.
What path had brought him to a life in Chad with Mari-Jo?

As the evening wore on, the tone of the conversation steadily
dropped. Certainly no one wanted to talk about the desert. The
teasing of the French by the two women – members of formerly
subjugated races – became more shrill, as though they were taking
advantage of intimacy to tease their actual masters, as though the
subjugation had never ceased.

'Why', Michèle asked rhetorically, 'are the former British colonies
all better off than the French ones? I've been to Kenya, Nigeria –
they're rich compared with here. And another thing – the English
actually *left* their colonies. *Mais vous êtes encore là!*'

François looked surly. 'I'll tell you why the former English colonies
are wealthier,' he said. 'Because the English were more effective
oppressors of Africa than the French. Because they grabbed the
wealthiest bits for their colonies before the French got here! And
now, of course, the Americans want it *all*.' He tottered, and lost his

grip on Michèle's waist. '*Je vais pisser*,' he slurred, putting down his whisky tumbler heavily and making for the toilets.

'*Alors, pissons*,' added the Doctor aimiably, getting up to join him.

'Don't take anything personally, M. Buckley,' Buffet said. 'I encourage people who come here to be themselves. People need to . . . unwind, no? It's all between friends. Take nothing to heart.' He gave me an avuncular wink, then cast an eye around the room. '*Madame la patronne*, the time has come to eat!'

We gathered at the table. Two bottles of Bordeaux were uncorked, and Mari-Jo began to bring hors d'oeuvres from the kitchen, where Buffet's chef had been silently labouring. Buffet took up a carving knife and sharpened it vigorously. 'M. Buckley, I think you'll find this an excellent *saucisson*.' Taking a wrinkled, red, seven-inch salami, he placed it on the chopping board and brought down the newly keen blade.

'Gentlemen, this may make your eyes water; I recommend that you avert them!'

François was telling the Doctor an anecdote about a stewardess in the plane in which he had flown from Paris, including a reference to her 'arse'. Mari-Jo leapt up. 'Why do you mention her arse? You're just trying to find a way to insult me!' And she flounced into the kitchen, the crumpled tail of her smock lying on top of her cantilevered buttocks as if on a shelf.

'I wasn't insulting her,' slurred François. 'Silly bitch.'

The Doctor went to appease Mari-Jo, but as they returned, his arm around her shoulder, she shook him off and pouted, 'You don't love me any more either, since I've put on weight.'

'Well, I don't think you love *me* any more,' the Doctor mumbled self-pityingly, slumping into his chair.

The cook brought out the main course, *filets* sliced finely like wafers of blood-red onyx. With the help of a third bottle of Bordeaux, spirits seemed to lift. The chef departed and the evening wore on, Buffet benignly deadpan in the midst of prods, pinches and bottom slaps. As dessert was served, there was an electricity cut. Amidst much business with lights, torches and candles there were sharp reports from slapped flesh, and Mari-Jo's high-pitched, gleeful squeals. Suddenly the lights came on. Mari-Jo had wine splashed down her left side, a crimson stain over her breast, the nipple stiffening. 'I warned you,' she muttered, and threw a glass of mineral water at François. As he twisted to avoid it, she started away from the table. 'You could have hurt my breast!' she said angrily to the Doctor. He looked bemused.

Mari-Jo was swaying in the middle of the room. François and the Doctor were staring at her dopily. She raised her arms above her head. 'Undress me,' she ordered.

'M. Buckley,' said Buffet hurriedly, 'you shouldn't leave it too late to get back. It's really not safe without a car, you know, after dark.'

We went out onto the veranda. Beyond the trees, beyond a tall wire fence, the river rippled in the moonlight.

'I'm going to miss this,' said Buffet. 'I mean, not just this place – Africa.'

'Are you leaving?' I asked.

'Oh yes.' He sighed heavily. 'Back to France. Back to my family, my wife. She returned several years ago.'

'Ah.'

'Yes . . . Well, M. Buckley, thanks to your presence none of us has said a word about work this evening. Very rare and very agreeable. But we haven't been much use to you about the desert, have we? Well, this may be helpful. On my one visit to Tibesti there was an Italian woman I met at the oasis of Faya, married to a local chap. She was really something, you know, not at *all* bad-looking, and so elegant, in the middle of all that dust.' His hand sketched an appreciative curve. '*Soignée, quoi.* Go and ask the Italian Consul about her.'

CHAPTER TWO

Vast Eternities

THREE MONTHS EARLIER, in the last days of winter, the Western Desert of Egypt had been brilliant and crisp. I was travelling with Penny, the woman I would soon marry – though neither of us had spoken of it yet.

We were visiting the Valley of the Kings, that tract of desert bedrock riddled with shafts intricately decorated with gold and indigo hieroglyphs. Dead kings, whose subjects worshipped them as gods, were entombed here, launched on mythic journeys into the afterlife.

I had especially wanted to visit the Ramasseum. Ramses II had raised colossal statues in his likeness all over Egypt, and not without reason: his military exploits made Egypt wealthy, and he was seen by his people as a great king. Yet Ramses has become a symbol of human vanity.

The felled torso on the west bank of the Nile is the size of a bull elephant; a broken finger is a yard long. The complete statue weighed 900 tons, and loomed over a funerary temple of incredible grandeur, designed to impress on Ramses's subjects his ineluctability in life and sure and certain resurrection after death. A thousand years ago, undermined by the flooding Nile, the statue toppled. It crashed through the southern gateway of the temple, demolishing walls and pillared roofs, and shattered when it struck the temple floor. Fragments still lie around in the dust, among the weeds and drink cans, granite of a tenderly mottled, fleshy pink. How are the mighty fallen.

It was the ancient Greek historian Diodorus who misread Ramses's name as 'Ozymandias', inspiring Percy Shelley's well-known sonnet:

> *I met a traveller from an antique land*
> *Who said: Two vast and trunkless legs of stone*
> *Stand in the desert . . . Near them on the sand,*

Half sunk, a shattered visage lies, whose frown,
And wrinkled lip, and sneer of cold command
Tell that its sculptor well those passions read
Which yet survive, stamped on those lifeless things,
The hand which mocked them, and the heart that fed.
On the pedestal these words appear:
'My name is Ozymandias, king of kings:
Look upon my works, ye Mighty, and despair!'
Nothing beside remains. Round the decay
Of that colossal wreck, boundless and bare
The lone and level sands stretch far away.

It is a romantic image, the temple's antique grandeur reminding one that every eighteenth-century country house had to have a ruined folly; a moral image, the sneering tyrant Ozymandias belittled. We sense that *all* human works must come to naught; yet there is something strangely soothing about those lone and level sands closing finally like a spring tide over everything, presaging cleanliness and the restoration of natural order.

The river boats were becoming easier to see, thanks to the reflections of their navigation lights in the darkening water. As dusk settled on the Valley of the Kings and the small ferry chugged over the Nile, Penny and I were looking at each other intently. The statue of Ramses had moved us. It lay on the edge of the desert, which reaches south and west for over 3 million square miles – an area big enough to swallow the US or Australia whole. And I was shortly to set off on a journey across that desert, the sort of journey you didn't undertake lightly.

But we were both happy. I thought Penny's expression radiant, expressing love and imagining love into the future. I sensed that her thoughts, like mine, were turning to marriage, and I talked excitedly – anything to prevent her from voicing those thoughts. I wanted to voice them myself, but the moment was not right.

A few nights later, however, on another boat on the same river, I asked her to marry me and she accepted.

It was Penny who drew my attention to the hieroglyph that meant 'the land beyond the green fields' – the desert. It is a triple horizontal wave, three ripples that might have been suggestive of water but were intended to evoke sand. Water is denoted by three choppier zig-zags, while the desert waves gently undulate. Now we were edging slowly along a tall granite wall, a densely carved white cliff, searching for a good example of the hieroglyph. 'This is like walking along the Thames at low tide, looking for pottery and clay pipes,' said Penny. 'Somehow you filter out all the wrong-shaped pebbles and flotsam, everything that isn't what you're searching for – ah, look!'

We moved closer and examined the hieroglyph: uneroded, well modelled, shadowed by the brittle midday light.

For the ancient Egyptians, the desert that yawned behind the west bank – the desert into which I was headed – was the land of Death. I asked Penny, 'Do you think they really believed in life after death?'

'Well, they tried hard enough to get there,' she replied. 'Didn't crossing the Nile symbolize a journey towards immortality? I don't think the elaborate coffins and boats and food for the journey were merely about display and status.'

'But does that prove that they really believed any of it? I find myself wondering if there's *ever* been a time when the whole mass of humanity truly believed in resurrection.'

'I think most people believed, or at least wanted to believe, that they'd carry on existing. I think they still do.'

'Do you believe?'

'I don't think something recognizably me will *literally* be reborn after I die. But something goes on, a sort of energy – and perhaps, who knows, my DNA. That kind of rebirth is all around us in plants, rivers, everything.'

Of course, marriage can be considered as a sort of rebirth; two individuals reborn as one.

We were eating dinner in a restaurant close to the Nile.

'When we were searching for the hieroglyphs,' Penny told me, 'I realized that for me, the desert was just as abstract as that hieroglyph; three sine waves, a symbolic notion, not a reality. I was looking at the map of the Sahara you're going to cross, and it's just a huge blank. No roads, hardly any landmarks. I can't imagine what it's like to cross nothing. Even travelling through the Egyptian desert, on a single narrow road, I look to the west where I know you're going, and I just see a blurry nothingness. I know that just because you'll have no contact with me it doesn't mean you'll really be nowhere and meeting no one. But from my point of view, with no phone or post, it will be as though you've put on a cloak of invisibility made out of the blank sand on that map. I do find myself wondering if you will come out on the other side and let the cloak fall.'

As the sun broke from the horizon, a Coptic church clanged its solitary bell. It was lost among the amplified calls of muezzins all over Luxor. The locomotive wheezed, but remained motionless.

There was a rumour of a new railway line from Luxor, crossing the Western Desert to the oasis of Kharga. It was so new that few people in Luxor had heard of it. Even at the railway station there was an element of doubt. Eventually an official explained that the train was doing trial runs. The line would not officially open for another six months – but a train was leaving today.

The people of Kharga already knew about the train. It made a short cut through the desert, while the road wound painfully around it. Most of the passengers were farmers and their families loaded with bundles and bales and chickens in cardboard boxes, bound for the pioneering new settlements in the desert. Between Luxor and Cairo a string of oases loops out into one of the driest deserts in the world. But this arid landscape is being transformed beyond recognition. The Egyptian government is constructing an irrigation scheme to rival the Nile itself – the New Valley.

'When do we go?' I asked the conductor.

'When the train fills up,' he replied.

At length, the train pulled away, trailing the glossy banks of the Nile. There was no buffet car, and all around us giant picnic breakfasts were unpacked. Boisterous children vied to stuff Penny and me with falafel and fried chicken. The train turned into the desert. The sun rose, and the day grew warmer. One by one, we dozed.

I felt the hot air on my skin and smelt the spicy scent of the desert before I opened my eyes. Then I was dazzled by a bleached vista of dunes rolling like a heavy sea. The train ploughed on into the abstract, strangely soothing landscape of the Western Desert.

By 11 a.m. the carriages were ovens. Sand gusted through the windows, and I had to keep wiping it out of my mouth and brushing a thick golden crust off my sunglasses. In the early afternoon we reached Kharga's brand-new station, a white-domed pile rising from the dunes like an antique ruin.

At the heart of the New Valley scheme, Kharga is no longer a sleepy oasis, but a concrete New Town bustling with farmers and fertilizer shops. We checked into the Oasis Hotel, a Stalinist government block. We were the only guests. A sign on the door warned that 'Breakfast is Compulsory.'

That night we walked through palm groves to the ruined Coptic necropolis of Al-Baqawat, on the northern edge of town. The sun was daubing the domed mausoleums red as we ducked into their dark interiors. The site watchmen ran their torch beams over 1,500-year-old murals of Adam and Eve, Noah, Jonah emerging from the whale's belly, Jesus and St Paul.

In the fourth century this eremitic outpost of Christianity provided refuge for St Athanasius, who had been banished for arguing that Jesus was not of a lesser substance than God. (Dissidence has always sprung from the desert, from Jewish and Christian Monotheism against pagan Rome to Monatism, from Donatism to Mohammedanism.) Eventually Athanasius won the theological day, and the three-in-one doctrine of Father, Son and Holy Ghost was born. Influenced by the clean, rigorous lives of the monks here, Athanasius wrote a biography of the original desert father, St Anthony, which laid the foundations of the monastic movement. Ever since, Christian monks have seen their vocation as a journey into a metaphoric desert, the same desert where Jesus went to be tempted by the devil.

We walked back down to the main road in darkness. A donkey cart clattered out of the gloom, and its drivers bade us climb aboard. We rode back into Kharga on a bed of just-cut grass.

After the modernist concrete of central Kharga, we half-expected every oasis in the Western Desert to have been reconstructed. But at midday we reached an oasis of cinematic splendour, a palm-ringed pool flanked by a great cliff of pink rock slowly dissolving into a skirt of sand. Dakhla is a dozen interlinked villages with walled gardens, ponds of ice-blue water and lush, irrigated fields. Occasional sand dunes loom over the trees like golden domes. In one place they had overwhelmed a row of telegraph poles, so that only the crowns were clear of sand, and the wires trailed on the surface of the dunes.

The administrative centre of Dakhla would like to be modern. Its dusty blocks of flats stand forlornly among empty plots. One day soon, the dreams of Egypt's state architects will be realized, and Dakhla will resemble suburban Bucharest. Throughout the developing world, architects ape mediocre modernist architecture instead of looking to their vernacular forms for inspiration. And the ruined Dakhla village of Al-Qasr should serve as inspiration. Narrow alleys twist round crumbling mud-walled houses. Bats flit in the shadows. Set into the walls are minute shuttered windows. Door lintels are carved with Koranic inscriptions in intricate Kufic script. This is the organic architecture of the desert, lovelier and cooler than concrete.

The day we reached Dakhla, a team of government engineers also arrived to survey it for an airport. Every oasis, they told us, must have one, as the government was anxious to open the desert up to tourism. At night, the plain around Dakhla glowed with the lights of tube-wells. Great canals are being cut into the desert, and the prehistoric groundwaters located and sucked to the surface. In ten years, these sand-locked oases will have been absorbed into a broad, cultivated prairie.

The road continued north. We stopped for the night in the White Desert, where the wind has carved huge mushroom forms from the brilliant chalk of a prehistoric seabed. Long after midnight, we walked by the light of the moon through a surreally pure landscape.

The next day we reached Farafra, the most remote oasis in the Western Desert. We put our bags in the government rest house and toured the gardens that cover the western slopes of the oasis. These shady, walled gardens, alive with the sound of birdsong and the water that trickled through irrigation channels, were havens of tranquility. I dozed, and woke up to see Penny talking to an old man. He invited us into his garden, taking us on a tour of green beans and date palms. The sun began to set in veils of yellow over sandstone mountains in the west.

Back in town we met Hassan, the Bedouin proprietor of the Nice Times Restaurant. When we told him where we were staying, he wrinkled his nose. 'The government guest house is a sewer, but it's the only hotel here.' He pointed to the new hotel rising on the other side of the town square. 'The man who is building that is going to make a fortune.'

Hassan insisted that we move into his house, taking the grand matrimonial bed, while he and his wife and daughter slept on the living-room floor. After dinner, he drove us out to a hot spring. A second hotel was planned for the hill nearby, and the desert scrub was covered with markers where irrigation ditches and fields would come. We asked Hassan, settled member of a nomadic people, how he felt about the rapid transformation of the desert.

'I hate it. They're planning a new road to run due east from here, you know, across the desert, to the Nile. We won't be cut off any more. And many thousands of people are going to be brought in from the slums of northern Egypt, resettled here. It's a social revolution. There will be crime, a change in morals, people won't know each other any more. This will stop being an oasis . . .'

He whooped and plunged into the pool. Hot, sulphurous water gushed from a fat agricultural pipe. As the three of us lay in the warm, pitch-black water staring up at the stars, Hassan said, 'I love this pool. I wonder how much longer it will be here.'

Penny and I decided that we would marry in the desert, somewhere in the western Sahara when I had completed my crossing. But where exactly? Mali? Mauritania? Morocco?

CHAPTER THREE

White Men Drive Jeeps

THE JET SLID OVER a hazy Sahara. Occasionally I made out a frozen yellow ocean, its waves etched with fine lines of black sand as symmetrical and intricate as an Escher drawing. Algeria? Libya? Refused a visa for Libya, I was crossing the Sahara Desert north-south – seas of sand, archipelagos of jagged rock – making for Chad.

Somewhere in that fawn haze was my first Saharan destination, the huge but little-known region of Tibesti, close to the Libya–Chad border and part of the mountain chain that bestrides the central Sahara from Algeria to Egypt. There were remote oases in Tibesti, vast volcanic craters in desolate mountains, and drawings scratched or painted on rock 5,000 years ago when the Sahara was green. War had kept travellers out of the region for thirty years. I was determined to be one of the first to return.

I had found it impossible to obtain up-to-date information about Tibesti. Unlike the American deserts, spanned by effortless freeways, there is no paved road into the central Sahara, no railway track, only a handful of military airstrips. I was not mounting an expedition with gleaming Land Rovers; I wanted to cross the world's deserts by the same forms of transport as the people who lived in them. How did the locals get into the central Sahara in mid-summer? I would soon discover that unless they could possibly help it, they did *not*.

The desert routes I had planned took in many of the world's most physically punishing environments, though in some places the impediments were human ones – politics and warfare. Now I was committed to the journey; that fawn haze was going to be my life for many, many months ahead. I gazed into its undifferentiated vagueness and wondered if I was up to it.

We reached N'Djamena at 3 a.m. The terminal had the resonant, hollow feel of the early hours: queues of dopey mid-sleep-cycle passengers, prodigiously yawning customs officials. Outside, beyond a couple of street lamps, was a wall of black: the plane had been eight hours late and the taxi drivers had given up and gone home. I cadged a lift with Patrick, a young French diplomat whose friends had come to collect him. They drove break-neck through unlit streets, their headlights scouring the whitewashed walls of colonial villas.

As I stepped into the Hotel Chari, a sick-looking young girl, a prostitute, scuttled up to me.

'*Monsieur, avez-vous besoin de moi?*'

'*Non, merci.*'

I woke late next morning, went to the pool-side restaurant and ordered an omelette. Women were lounging by the pool, beautiful women with black skin and Day-Glo bikinis, gossiping and giggling and rubbing oil into their thighs. They greeted me with professional warmth. I ate the omelette and felt a sudden wave of giddiness. I spent the rest of the day in bed, vomiting and reading *Crime and Punishment*.

N'Djamena is one of the world's more modest capitals. A colonial French creation, it resembles a provincial Mediterranean town circa 1925. The main street is flanked by shady trees and colonnaded white villas. A dusty bookshop sells *Paris Match* and *Le Monde Diplomatique,* perused at the adjoining *café-pâtisserie* by well-creased expatriates savouring the odours of butter and coffee. Nearby are the post office, the ministries, the TV station with its red-and-white pylon, the law court with its pavement vendors selling fountain pens and sunglasses. Cars obediently wait at the city's only traffic light.

Heading south along the Avenue Charles de Gaulle you pass the barrack-like modernist cathedral, then the actual barracks, then the walled presidential palace. The avenue climaxes at the aggressively spiked *Grande Mosquée*, spearhead of Islam, a gift from Saudi Arabia. Around it is N'Djamena's bazaar, with its Koran sellers, henna painters, carpet vendors, *qat* vendors, whisky shops, shoemakers, bordellos. Under a spreading tree, colonizing an entire pavement, I found an ageing Arab motorbike repairman willing to rent me a moped.

I Yamaha'd until I found the Métropole, the cheapest hotel in N'Djamena that was not a brothel – though the dividing line was somewhat blurred. My room was a gloomy cell with an eighteen-inch window that clamped shut like a valve on a bathysphere.

I set about the business of penetrating the desert beyond N'Djamena's deceptive tree-cover. I learned that a Frenchman, an expert on the desert north of Chad, lived and worked in N'Djamena. He turned out to be on leave in France. No one at the university had actually been to Tibesti, but everybody knew someone who had. Tracking down any of these useful individuals proved impossible.

At the Ministry of Tourism, an official explained to me that a trip into the desert was not a foregone conclusion but a privilege. I would need a visa extension, plus travel and photography permits signed by the Minister – if he could find the time. Naturally all this would cost a lot of money.

After two days of official stonewalling, I made for the French Embassy to find the young diplomat, Patrick, who had given me a suave assurance of his assistance in any possible way. When I finally got past the revolving steel drum of a front door and stood before him, dusty and perspiring, Patrick shot me a look of high-caste horror and produced the address of a French non-governmental agency. 'I am sure M. Buffet here will be able to help you,' he said.

Bernard Buffet was a tall, stooping man with a beaked nose, slightly prominent teeth in a smallish mouth, and slicked-back hair. He had impeccable, rather patrician manners, a slow-motion grace and charm, the distracted air of an intellectual. So I wanted to get to Tibesti? Well, my dear fellow, there are no commercial flights. The only way is to fly in a French military aircraft; you have to apply to the French Interior Ministry through your embassy in Paris.

'No,' I said, 'I want to go overland.'

'*Overland*? Can you afford to hire a jeep?'

'Er, no.'

'Well, there's no public transport, *mon vieux* . . . The only alternative is to travel by *gros porteur*. But I don't know anyone who's done it.'

I had heard of the big, six-wheel drive trucks that ply their way across the sands of the Sahara, successors to the great camel caravans of the past.

Buffet lit a cigarette. He chain-smoked unfiltered Gauloises, which he held between tapering nicotine-yellowed fingers. 'They run between N'Djamena and Libya,' he continued. 'A lot of it is smuggling. And you go in the back, on top of the cargo. That, my friend, is how people get into the Sahara. Not that many people do at this time of year. Rather hot and dusty. But if you have the money, you can sit next to the driver.'

'And where do you find these trucks?'

'There's a sort of parking area they leave from – look, my driver can run you over and make some enquiries. By the way, I'm having some people who know a thing or two about the desert over to dinner this evening. Perhaps you'd care to drop by?'

Buffet's driver took me to a flattened wasteland on the northern edge of town where a single lorry stood, a monstrous, bulbous object at least twenty-five years old, painted a dull military green and sporting an implausible quantity of wheels. When we called out, a sleepy face appeared in the window of the cabin, and the driver came down to meet me.

'Are you going to Tibesti?' I asked.

'Yes . . .' he said, guardedly.

'When?'

'When I have enough cargo and enough passengers. Five days, ten . . .'

'And how long does it take?'

'It depends on how heavy the cargo is, and where the passengers are going. A week, ten days . . .'

'How much do you charge?'

His eyes narrowed craftily; he was not a good liar. The figure he named was many thousands of CFAs – the currency of the Coopération Financière en Afrique – an outrageous amount.

'That's too expensive!'

He shrugged. 'That's the price.'

As we drove away, I asked Buffet's driver if he thought the price was reasonable. 'Of course not,' he said, 'but you're a white man.' As he dropped me back at Buffet's office he watched in astonishment as I unchained the moped. 'But white men . . . ' he murmured, making it clear that I was letting down my race, 'white men drive Jeeps.'

CHAPTER FOUR

Out of Eden

*J'ai toujours aimé le désert. On s'assoit sur une dune
de sable. On ne voit rien. On n'entend rien. Et
cependant quelque chose rayonne en silence . . .*

Antoine de Saint-Exupéry

A T THE DINNER PARTY *chez* Buffet I had learned little about the
desert, but had come away with the suggestion that I make
contact with the Italian consul. The next afternoon I sat on the lawn
of the Consulate with Hermana Delacroix, Honorary Consul. I gave
her Buffet's description of the *soignée* Italian in Faya and asked if she
could help. She gave a discreet smile. 'I think the woman you are
looking for might have been me. I left Faya two years ago.'

Yves Delacroix soon arrived, a tall, impossibly thin man of fifty-
something summers, pencil-moustached, military and brisk. His
mother had been Chadian, his father a French soldier. A ten-year
veteran of the Libyan war, several times wounded, Yves was now an
official in the Ministry of the Interior. He had met Hermana a decade
earlier when she was working for the Red Cross and he was fighting
Libyans. 'He was,' Hermana told me with a smile, '*so* handsome in his
uniform.'

The Delacroix home-cum-Consulate was a large villa in the
diplomatic quarter, flying the *tricolore* on a pole in the garden. The first
thing I saw among the roses and bougainvillaea was a baby gazelle
delicately evading a naked toddler – Chuti, the youngest Delacroix. I
was given an iced beer, and we sat on the veranda and watched the
sky redden. 'M. Delacroix,' I said, 'you live in Eden.'

'Eden?'

'Babies, gazelles, flowers.'

'The damn gazelle is *eating* all the flowers. When it gets bigger I'm going to cut its throat. Gazelle meat is excellent, you know.'

The next night, I was invited back for drinks. Hermana asked where I was staying, and when I told her she looked appalled. 'But it's a, er . . .'

'No, it isn't,' I said. 'A little seedy, perhaps, but not actually a, er . . .'

She turned to Yves. 'Martin is going to stay with us.'

The Delacroix's veranda began to fill with friends and acquaintances – pilots, military attachés, doctors, aid workers. It felt like some TV chat show, 'N'Djamena Tonight'. Yves was interested in all that was going on, and had the connections to make things happen. When he reached the improbable conclusion that my book might be good for Chad's tourist effort, he made a few phone calls. A senior official from the visa office immediately dropped by to tell me that my papers would be ready the next morning. The next day the pesky Ministry of Tourism official had become my devoted servant, and I received a prompt audience with the Minister, who assured me of his personal commitment to the happy outcome of my journey north.

The only problem that remained was how to get there.

It was Hermana who had the thought. 'Who's that priest who looks after the Mass in Faya?' she asked Yves.

'Bessita.'

'The blacks in the north are all Christians,' she told me, 'mostly government workers from the south – civil servants, soldiers. There are churches, but they have no priest. So Father Bessita drives to Tibesti two or three times a year to celebrate Mass for them. You never know, he may be going soon, and maybe he can offer you a lift.'

As I puttered into the yard of the Convent of the Sacred Heart, *Père* Bessita was standing erect and motionless, dressed in a teepee of a shirt screen-printed with a reproduction of the Last Supper. He was a bear of a man, more than six feet tall and barrel-shaped, with shovels for hands. Dotted around this colossus were some pygmies of normal height – two local schoolboys, an oriental nun and a white French priest with a florid, beaky face. A big Toyota stood nearby, infested with aerials, roof-racks, ladders, spare wheels and jerrycans.

The priest listened to my request in silence, regarding me through narrow eyes that gave nothing away.

'As a matter of fact, I'm about to do a trip north. I'm leaving the day after tomorrow.'

My heart leapt. '*Mon Père,* do you think you—'

'Unfortunately I already have two other passengers, both *blanc,* and both fee-paying.' Sorry and all that, he said, manifestly not sorry.

I went straight to the long-distance lorry park. The truck that stood empty had gone; a different truck and driver were in place. 'He said it would take him a week to load up!' I spluttered.

'It depends,' the present driver told me. 'He had a good offer, so he didn't have to wait to fill up with passengers. He just went. It's like that sometimes.'

'When do you expect to go?'

'Who knows? A week, ten days . . .'

'And how much would you charge to let me go with you?'

He narrowed his eyes craftily and named an outrageous sum.

'The hell with it,' I thought. I would spend the next day by the Novotel pool. Give it a couple of days, then go back to the lorry park to negotiate.

In the pool, tanned French soldiers and their African girlfriends were making a sort of non-penetrative aqueous love. I thought of my own love, far away under a hazier English sun. It would be almost half a year before I saw her again, to be married in a still-undecided, West African location. I felt lonely, and the journey I had ahead of me seemed unimaginable. I sat with my pale body under a sunshade and sulkily read *Crime and Punishment.*

When I returned to the Delacroix's, Hermana had a message for me: Ring Bessita at once.

'Someone has dropped out,' Father Bessita told me. 'Do you still want to come?'

'Yes!'

'You understand that I need you to make contributions to expenses?'

'Naturally, Father.'

'Excellent.' He sounded as relieved as I was. 'Be here at seven a.m.'

That night, Yves took me to stock up on black-market French army rations, openly on sale throughout N'Djamena. The French must be the only army in the world to provide their combat troops with *pâté de foie gras.*

'Take care in Tibesti,' he told me. 'There are unexploded mines all over the place. Couple of years ago I had to go in and get a Frenchman who'd blown his foot off – could have blown my own foot off. Saved his life; he never even sent me a postcard. Anyway, don't go off on any solitary walks.'

I arrived at the convent early the next morning to find preparations anything but complete. Bessita was issuing orders, dressed in a fawn military uniform with scarlet epaulettes and beret. Waiting by the jeep was the other *blanc*, Sylvestre, arms folded, legs well apart, surveying events with an air of authority. He shook my hand with a steely grip.

There were three other passengers. One was a religious colleague of Bessita's who would be coming only half way. Loading the Land Cruiser was Romain, Bessita's driver, who had a quick, wide grin and an air of fanatical devotion to his tasks; slender and sinewy, he was working up a sweat stowing bags and tensioning straps. The third man was Bessita's nephew, Thomas, a muscular eighteen-year-old hefting duffel bags like cabers.

Not to be outdone, I busied myself filling some jerrycans with water, but *Père* Bessita suggested I leave the packing to Romain and Thomas. I sat down and opened my book. I had already finished the Dostoevsky, and started on Graham Greene's African-set novel of Catholic anguish, *The Heart of the Matter*.

The sun filtering through the leaves was soon uncomfortably hot. At last Bessita climbed into the driver's seat and his colleague took his place beside him. The remaining four of us squeezed onto two small bench-seats facing each other at the back of the jeep, behind bundles, bags and a stinking oil drum of diesel fuel.

Summer's fickle rainstorms had begun, and every fold in the landscape was a sudden steaming pond. Bony cattle cropped sun-bright spikes of new grass. Tree-cover diminished and the body of the land grew visible, a rippling plain of pale brown earth. Rainfall engenders chemical processes, darkening the soil, breaking up rock, making humus, washing out salts; but desert soils are salty, devoid of organic contents, white.

N'Djamena lies on the northern edge of the Sahel, a semi-desert purgatory between Central Africa and the Sahara, a broad savannah of low-growing grasses dotted with thorny trees. In the twentieth century the felling of these trees brought soil erosion and the southward march of the Sahara. In the 1970s and 1980s, droughts exterminated most crops and livestock, bringing famine. Today, farmers still cultivate groundnuts and millet, and pasture scrawny livestock, but their survival is tenuous.

A hundred miles from N'Djamena the tarmac abruptly became a dirt track. We stopped for some food in an exhausted, dust-caked town. In the shade of a wall an open-air butcher hacked at purple, fly-speckled beef. There were urchins all around; even the waiters in the gloomy restaurant were little boys with prematurely solemn faces. Another watched us from the doorway, his face immobile. He carried a bundle of possessions on a stick over his shoulder like Dick Whittington, and his forehead was bound with a dirty red cloth. Romain mimed pulling the rag down over his mouth. '*Ça commence déjà, le désert,*' he said. The desert is starting already.

The first desert I ever went into was Christ's; I was eight. My family was living in Istanbul, and my father had proudly taken delivery of a Peugeot station wagon shipped direct from Paris. Finned and fashionable in bronze metallic paintwork, it exuded competence. My parents decided to make that year's family holiday a drive to the Holy Land.

The journey took us into the Negev Desert, where the gaudy pages of my *Illustrated Children's Bible* sprang to life. We bobbed and giggled in the buoyant Dead Sea, and knelt on the sandy banks of the River Jordan, bottling holy water for the folks back home. I remember being lifted onto a saddle with exaggerated care by an unshaven man in a shabby European suit, his eyes bright with the uninhibited love of children that is so common in the Middle East. The donkey picked its way into a hollow, honey-stoned wadi where rock seemed to have melted like the ramparts of a sand-castle breached by the sea. It was a hot, windy day. My mother's knee-length summer skirt billowed, my father seemed solid in his Fred Perry shirt and black wrap-around Carl Zeiss sunglasses. Tanned and healthy, a family from an ad in *Life* magazine, we did not belong to this harsh landscape.

Before I began the journey in this book, I had travelled to deserts on several continents. I was attracted to the stoicism of their inhabitants and the elemental, uncluttered beauty of the sands. I remember lying on the skin of a 400-foot sand dune, watching the stars take their places in the sky. I felt wrapped in an intense and timeless calm. As the air grew cool and the surface of the dune grew chilly, I burrowed into the sand, which held heat to a depth of about a foot. Later I slowly slid, trod sand, swam to the foot of the dune, and when I stood up I was heavy with the weight of sand in my pockets and shoes. I was drunk with sand.

The shape of a sand dune is so particular; the three-dimensional sickel that swells at first, then narrows to a delicately sculptural edge. There is a mesmeric purity, a liquidity. You plunge your hand into a dune's skin – through the surface tension – and penetrate a universe of sand, a Big Bang of glittering grains of rock; lift up your hand and watch sand flow through your fingers in a golden wave, returning to seamless unity with the dune. Sand dunes seem to be made from a billion identical sand atoms, yet they have their hierarchies. Their component grains vary in colour from black to bright red, and find different levels according to their mass; they take different forms, given by desert-dwellers beautiful names such as *barchan* and *gassis*; they are in a constant state of motion, of migration and transformation. Sand dunes are complex, and elementally, ontologically, simple.

I was sitting in the bath when it hit me. I had been planning a desert holiday, poring over a soggy map. Maps usually camouflage the deserts, making mountains brown and lowlands green, but this was a map of physical environments. Binding this green planet were bands of golden desert. I saw that you could almost circumnavigate the earth without ever leaving the desert. The route would be a ragged one, winding across two hemispheres and making leaps across the oceans, but a route it was.

It struck me for the first time that the desert lands are giant strata, a reality interleaved within our own. They cannot be ignored, only turned away from, as so many traders and soldiers have been forced to do over the centuries, or, these days, crossed in one of those long-haul flights where you gaze for hours at a dazing panorama of sepia, sun-beaten earth.

The tropics, for all their humid image, are mainly dry. The Tropic of Cancer etches through a giant swathe of deserts from North America to Africa, the Near East and Asia, while the Tropic of Capricorn crosses deserts in South America, Southern Africa and

Australia. Common sense suggests that the planet's driest regions should be at the equator – but common sense is wrong. The rotating earth deflects hot equatorial air to the north and south, where it sucks moisture out of the atmosphere – and makes deserts – before looping back to the equator and plunging down as rain. A zone of equatorial super-fertility is thus sandwiched between rings of sand. In my bath, I re-imagined these obstacles as a highway. Look at a map and you will see what I saw: an arid track around the earth.

Having that vision of a sandy circumnavigation made me feel almost compelled to realize the journey. There were other factors; I had spent nearly a decade working too hard for an organization whose ideals were being destroyed by gangsters called 'managers'. I felt that in the rush to keep up with the flow of so-called 'ideas', my inner life was in peril. I was ready to go into a sort of Christly wilderness of my own. In the desert I would have time and space, and the chance to hold up a mirror to my inner world. I would be sure not to like what I saw – and that was the point.

There was also the approach of my fortieth year, which was joked about by my friends as a standard mid-life crisis. To be pushing forty unmarried and childless was unnatural, they told me. I was selfish, unrealistic, a Peter Pan. Some saw this long journey as another symptom of the complaint, others as a last spasm before I reclined into the warm bath of domesticity.

As I prepared for the journey, I began to be excited by the possibility of a book that might convey something of the staggering scale and diversity of desert lands. I wanted to do something to reverse their bad press. In French, the word *désert* does not have the negative connotations it has in English, but is closer to our 'wilderness'. It has some of the overtones we associate with the American mystics' sense of their land, of a spaciousness, a purity, an as yet unspoilt landscape against which to set the exploration of human possibilities. That was the desert I was looking for.

CHAPTER FIVE

The Bibliophobe

The camel does not see his hump.
Arab proverb

ROMAIN HAD BEEN RIGHT. Beyond Massakori the road turned to dust. The Land Cruiser bounded along the rutted track with its hind axle bucking like a donkey. Our pelvises pounded on the hard bench-seats; to avoid being thrown around we needed to crucify ourselves with feet and arms taut. But the journey was just beginning and conversation was fresh. Like prisoners in a paddy wagon, we took each other's measure.

Bessita's young nephew, Thomas, had a permanent grin and a glib tongue. Everything was a joke or an improvised song, he was cheerful and mocking, with an adolescent's boastfulness. He was a boxer, a local champion, and he wore the sleeves of his fashionable American T-shirt ostentatiously peeled back over lumpy biceps. He had short, strong fingers with chewed nails, and a boxer's neckless, shaven head. There was something about him of the playground bully.

Easiest to bully was Romain, fifteen years his elder, but his uncle's employee. Romain was friendly, but cautious and deferential. He knew that Thomas was the Father's nephew and that we were rich whites. Romain was not without status: *he* was *Père* Bessita's driver, and it niggled him: why was the Father insisting on driving himself? But Bessita seemed happy in the driver's seat, chatting to his friend. There was air-conditioning to the front seats only, and occasional teasing zephyrs of cool air wafted our way.

The other passenger was the Frenchman Sylvestre. He was in his early twenties, of medium height, with brown curly hair and bright, dark eyes. In the curly depths of an untrimmed beard one glimpsed small, sharp teeth. He was doing his *service militaire*, the option to replace a year's army service with two years' social service, in his case with an NGO sinking wells in the rural south.

As the afternoon wore on, there was a steady increase in temperature, and gradually conversation ceased. At every bump we were tossed into the air, to land with a painful thud, and we began to peer into the haze ahead, dreading the next bump and silently cursing Bessita for failing to slow down or avoid it. The Land Cruiser was gaol and instrument of torture combined. I christened it the Iron Maiden.

In our trail of dust we left small herds of goats and cattle and shepherd boys who looked after us incuriously, unable to imagine themselves in a jeep bustling towards the horizon. We overtook open-backed pick-ups crammed with blank-faced people, whom I imagined looking neither to the future or the past, but into an unalterable present.

The hours passed slowly, and I tried to read. I had to brace myself extra hard with two feet and just one hand, and the book would rotate and pitch as the rear wheels tossed me about. But I persisted.

'You never stop,' said Sylvestre, with an unfriendly grin.

'Stop what?'

'Reading. Don't you ever stop?'

'Only when I'm interrupted,' I said, with an equally false smile. After all, I was minding my own business. Conversation had grown so thin that I could hardly be accused of unsociability. 'Why?' I asked. 'Don't you like to read?'

'No,' he said flatly. 'Reading complicates things unnecessarily.'

'What, reading a novel?'

'All these bookshops full of *books*, all the so-called experts, academics who spend their whole lives studying nothing whatsoever, and these journalists who constantly *go on* about everything. They're all pretending they've got something incredibly significant to say, but does it change anything? No. Does life get better as a result? No. Can most of them prune a tree or build a wall, or do anything useful whatsoever? *No*. I'm happier in the bush.'

I smiled at him. 'Surely we need writers, Sylvestre. They're story-tellers, entertainers, you find them everywhere – including the bush.

Civilization may be out of balance in all sorts of ways, but you're not going to improve it by abolishing reading.'

'Well I would. I'd get them all out in a field, working with their hands. Then they might learn a thing or two, actually have something to say about the real world.'

Suddenly I felt angry. Oh yes, I thought, Pol Pot had the right idea. Soft-handed writers – shove 'em in a mass grave.

Sylvestre was getting to me.

After that, Sylvestre and I addressed most of our remarks to the Africans and not to each other. Romain and Thomas were quick to sense our division. Miniature eddies of aggression were whirling around the cabin, the start of a pecking order. But inscrutable, unlieutenanted Bessita was the boss. And he sat up front, chatting to his friend or listening to Radio France Internationale, its political soundbites drifting back on wisps of air-conditioning.

Night fell, and long-tailed rats and a racoony fox darted across the briefly lit proscenium of headlights. Two hours after dark, we realized we were lost. Bessita consulted his map and concluded that he had missed a right fork, and we were now heading north instead of north-east. We could double back and waste several hours more, or turn due east into the bush and try and pick up the track. Bessita turned off this trail with its well-worn tramlines, and the Iron Maiden leapt over fallen palm logs, crashed through thorn bushes and slithered in loose sand. God only knew where we were. Bessita, I realized, was navigating by the moon, the pale creature jigging above our heads. I suggested that I might come up front with my compass. He hesitated. He always hesitated before responding to a question. 'All right,' he said.

It was a turning point. I could sense the waves of envy surging from behind me as I slithered up the social hierarchy into the front cab and sat beside Bessita with my pen-torch pointed at my compass, issuing directions.

Sometimes a dim light swam in the distance and we headed for it, but the light always vanished. Were these mirages, or cautious villagers damping their lamps at the sound of an engine? Civil war had only recently ended, and there were bandits hereabouts, known to be demobbed soldiers.

As the Iron Maiden wallowed through a slough of sand, we saw a clutch of round huts, and men with oil lamps warily approached.

'Which way to Moussoro?' asked Bessita.

They pointed at a hoof-marked trail. 'It'll bring you to the road eventually,' they said.

It did, and we regained the tracery of ruts that was the road to Libya. I looked at Bessita's tired eyes and suggested that I might take over for a while. After long deliberation, he accepted.

We reached Moussoro in the early hours, and drove its sandy unlit streets trying to find *Père* Bessita's religious compound. Here at last it was, a double door in a wall, and sleepy figures coming out to undo padlocks and back us in. We unrolled nylon mats and slept.

'In Chad in those days, the sound of gunfire was so constant that in the end we'd say, "We're used to the guns now!" And we'd dance to them!' Romain strutted and waved his arms, 'Bom, bom, baba bom-bom.'

It was afternoon and we were together inside the compound walls, in the shade of an acacia tree. The old Arab who kept the place for Bessita was wandering around with a plate of meat, and his grandsons were kicking a ball in the dust. His son sat in the shade of a room opposite, playing a single-stringed instrument that made a plangent twang. Bessita was away on Church business, and Thomas and Sylvestre had gone to the bazaar.

'Aeee, Martin!' Romain laughed as he fell back into his wooden chair, then shook his head. 'He wanted the whole of the north, you know, that bastard Gaddafi. All this land, the desert, where the Arabs live and there are hardly any blacks – he wanted the lot. Maybe there's oil down there, maybe that's why he came . . .'

Romain paused. I said nothing, and after a few moments he went on, 'Gaddafi fought with mines, you know. Some days when I was driving a truck, the bodies were so thick on the ground you had to drive over them to advance. Or other days, on foot, you had to climb over them. I've seen forty men killed by mines, just like that. The Libyans would lay them, then retreat, and you'd hear them calling out, 'If you were men, you'd attack!' So we'd get up and go for them, and – bang. They used to lay a mine and put a new hammer on top of it, or some money, or a pistol. You'd go to pick it up, and – bang. But we learnt all their little tricks.

'He dropped acid, too, from planes. They were always up there, his planes. And suddenly these clouds of stuff would drop out of the skies . . .

'But we threw the bastard out. The Libyans ran like rats, you know. Not far from here we surrounded a huge encampment of arms, planes, tanks. Thousands of prisoners. It was a disaster for Gaddafi. He threw in the towel after that, he negotiated. But we should have chased him to Tripoli, I'm telling you. Next time, we will . . .' He slapped my thigh and grinned. 'But we saw him off, eh, Martin? He won't be back in a hurry.'

He relaxed in his chair and stared across at the compound walls. There was a long pause. The boys played football in the dust. From the door opposite came the sound of an instrument plucked.

Romain turned to me. 'I was in the army for thirteen years,' he said. 'War brings you into close contact with people, Martin. You share everything, you go back for each other if you're wounded, you tell each other everything – maybe death is coming, why hold back? You drink, you take drugs, you have women . . . You kill people together, and you see your friends killed. Extremes. That's why some people even get a kick out of war . . .'

He smiled. 'It's good to kill, Martin. Killing a man teaches you a lot. About life, I mean.'

The next day, as we set off from Moussoro, Sylvestre climbed into the driver's seat. He had obviously had a quiet word with Bessita. Romain's dark eyes clouded with emotion, but after two hours he was given his first spell behind the wheel – and he messed it up. He rode the clutch badly, making Bessita roar at him, 'Have you ever driven before? My God, you'll burn out the clutch!' Thomas whispered that Romain had never driven for Bessita before, he was on trial. The clutch screeched, and Bessita impatiently ordered Romain to the back. He slunk there like a scolded dog.

Twenty miles from Moussoro we came upon the spectacle of a trans-Saharan truck like those I had seen in the lorry park in N'Djamena, but marooned, jacked-up, resembling the husk of some giant dead insect. Scores of jerrycans, plastic bottles, plastic teapots – anything that could hold water – dangled on cords.

Bessita pulled up twenty yards off, and a woman ran towards us.

'*Mon Père, mon Père* – I know you from N'Djamena, Father, I'm a Christian. Please help me!'

Bessita listened impassively as her story poured out. Four days ago at dawn two jeeps had forced the truck to stop. Three armed men had ordered the passengers off one by one and searched them, then ransacked the truck. She had lost her gold earrings and her money – everything she owned. Two hours later the truck's drive-train had given out, and the driver had set out to hitch back to N'Djamena. They had been waiting ever since. She was desperate to get to Faya for an entrance exam at the college.

'Please, Father, it's in two days' time. It's my only chance, I've lost everything, I'm desperate. Please help me!' She began to sob.

'Get in,' said Bessita.

Joy and incredulity swept over her face. In seconds, she had gathered up her bundle of clothes and scrambled into the overcrowded jeep, where she sat hugging her bundle, gazing about in disbelief.

'What's your name?' I asked.

'Estelle.' She fondly eyed the back of Bessita's head. 'The Father is a good man. Maybe I'm going to make it after all.'

'What's the entrance exam for?'

'The college at Faya. My application has already been accepted, but I have to sit a written entrance exam. It's a formality, but I have to be there.'

'What will you be studying?'

'To be a student teacher.'

'There's a teacher-training college in Faya?' I had thought of it as a remote oasis.

'There's a college, not a very good one. But what can we do, those of us *sans avantages*? They say Faya is a dump – imagine, studying so far from your family, and to return home once a year you face *this*. Nobody wants to live there, so it's easier to get admission in Faya than at any other college in Chad.'

'The students are black people from the south?'

She looked doubtful. 'I think so . . .'

Bessita interrupted. 'There are plenty of southerners in Faya,' he called back over his shoulder, 'mostly soldiers, civil servants, their families. The government has to maintain facilities there, or the locals will complain. But there isn't enough local demand to make it viable, so students attend from the south.'

Estelle shrugged. Politics didn't interest her. 'All I know is, I have to get there. I've done a few things in my life, but now I need to be able to support myself. This is my chance.' She stared back through the rear window. The pall of our dust had long erased the broken-down truck. 'My God,' she said, *'les gens ici sont malheureux.'* The phrase seemed redolent of ancient suffering. 'They told us this happens all the time, you know. Sometimes they kill people. *We* were lucky . . .'

Estelle tore at the sliding side-window and vomited. She leaned weakly on her bundle of clothing, and every few minutes retched again. 'I can't travel in the backs of cars,' she mumbled. I called forward to Bessita and said Estelle would have to travel up front. Sylvestre, sitting next to him, flashed me a look of hatred.

I pondered why Sylvestre and I were so uncomfortable together. Our shared Europeaness made us like some too-long married couple who took almost any amount of irritation from others, but whose tolerance for one another was nil.

As Estelle went forward and Sylvestre climbed in through the back doors, I picked up my book.

Romain grinned at me. 'You read too much, M. Martin.'

This was all Sylvestre needed. 'Books, newspapers — especially newspapers — they're all full of shit.'

I looked at him. 'So how do you learn anything, Sylvestre?'

'I . . . talk to people. I walk, I hunt, I fish. I don't need *books* to learn. I've left all that behind.'

'Left it behind?'

'Yup. For good.' Sylvestre, bearded hunter-fisherman, Bringer of Appropriate Technology, at home among the un-welled Africans.

'You believe you can, do you? Make such a change in yourself? Wait till you get back to France. It's your culture, and you can't adopt illiteracy as a pose back there. The Noble Savage is all very well, but it's not so easy to isolate yourself back home. Once you're there it'll get its hooks into you again.'

'No,' said Sylvestre, noing and knowing better.

The Man of Action was about to have his day.

CHAPTER SIX

It's Good to Kill

W E DROVE INTO A HAMLET, a cluster of circular huts where the children who ran out to greet us had their hair shaved Mohican-style. Piles of logs stood at the roadside. The villagers made a living by cutting down the few remaining trees on the fringes of the Sahel, speeding the desert's advance.

A young woman approached us, wrapped in rags, and – to my amazement in an Islamic village – bare-chested. Someone said she was a madwoman. She pointed at my finger. 'How much?' she said. I was wearing no ring, and everyone began to laugh. 'His finger,' she insisted. 'How much?'

I wondered how she had lost her wits. Perhaps she was a casualty of the wars that had flailed this region for thirty years.

Beyond this last human settlement lay steppe. We bowled across a porridgy plain under a celluloid-blue sky. This was where Sahel and Sahara merged. The jeep slowed, and I realized that everyone was looking attentively ahead.

'What is it?' I asked.

Romain turned. 'We might find some gazelle here.'

Hunting. How, I wondered vaguely, would they kill it? Did Bessita carry a gun?

We saw no gazelle, but for once the motion of the Iron Maiden was less than torturous and I began to feel drowsy. At the precise moment that I dropped off, Romain slapped me on the knee.

'*Y'a une gazelle!*'

The jeep left the rutted track and plunged across the desert, pursuing not one but two gazelles, slender creatures pale brown against the pale brown sand. They loped elegantly ahead of us at a steady twenty miles an hour, and Bessita followed. After three or four minutes, amazingly, a third gazelle joined them.

Bessita slowly increased his speed. We watched, excited. The gazelles lightly danced away from us, as though this were play. Bessita kept up a running commentary over his shoulder. 'I have to keep my distance. If I go too fast, they'll dart off in another direction faster than I can turn, and I'll lose them. I have to stay close, like this, but no closer. I have to exhaust them.'

'You need a rifle,' said Sylvestre. 'In the south I have a .22, it would be perfect for this.'

Bessita said enthusiastically, 'I'd *love* a .22! Will you sell it to me before you leave Chad?'

Perhaps he edged too close, and in a moment two of the creatures had peeled away and were far behind us, and we were following just one gazelle. But it was the largest. Romain was watching closely, 'A female,' he said.

Bessita nudged the gazelle in a long curve back to our path, and we recrossed our previous tracks. 'The trick,' he said, 'is to get it running in the direction you want to go in.'

Long minutes passed, and the chase lost its excitement. We pursued the gazelle for half an hour at a steady twenty-five mph, nudging it north-east. Its endurance was astounding.

I had read about this motorized form of hunting, that, combined with the wide availability of automatic rifles, had decimated deer populations throughout the Sahara. 'This is sick,' I muttered.

Sylvestre turned around. 'It's too late to stop now,' he said. 'The gazelle can't survive after a chase like this – its whole system is overstressed, its heart is damaged. It would die anyway.'

I didn't want to be the po-faced foreign know-all. I watched the lovely creature, loping steadily. She still had stamina.

And then she stopped, side-stepped, and darted back the way she had come. 'Aha, she's weakening!' said Bessita. Half an hour earlier this sort of strategy would have saved the animal, but now she was tired.

As Bessita made a wide turn, Romain clambered out of the rear door and hung on to the ladder. Bessita heard the door bang and called back, 'What's going on?'

'It's Romain,' said Thomas, with a note of envy. *He* wanted to be in on the kill. Meanwhile Sylvestre had unsheathed an African dagger, a gleaming triangle of ancient design.

'That looks like a good knife,' said Thomas.

'It's very good,' said Sylvestre, fondly. 'It's rare to find such good steel in Chad.'

As we bore down on the gazelle she side-stepped again, forcing Bessita into another turn. The animal paused for breath, and as we passed I could see her flanks heaving. Then Romain was away, and we watched him dive. With a graceful bound the gazelle escaped him, and everyone laughed. Romain was lean and fast, running with an athlete's gait; the loping antelope and pursuing man made a fine, atavistic sight. On foot, though, a man is outclassed. I imagined a ring of ancient men, using skill and cunning to catch their prey, and often failing. We had diesel power.

At an area of tall, resinous bushes with lime-green leaves, dotted twenty or thirty feet apart, the gazelle disappeared. The jeep bore down, and she broke from a bush and raced towards another. '*Ah, la méchante!*' Bessita cried gleefully, '*Ah, la villaine!* Keep your eyes on her!'

We zig-zagged across the savannah, looping or turning in tight angles. God knew where Romain had got to, and nobody cared. We'll have to retrace our tracks to find him, I thought. Now it was Thomas who had climbed on the roof, and was hanging on for dear life each time the jeep perilously cornered. The gazelle was visibly weakening. She took shelter behind a small shrub, and through the feathery leaves we could see her sandy flanks. The Toyota thundered in, spraying sand, the animal leapt and stumbled, and suddenly there was Romain leaping too. The gazelle escaped him, bounding to the next bush, and Bessita circled, trying to drive her back. She bounded and stumbled again, and Romain was hurling himself forward, catching a hind leg, whooping, grinning, hanging on to the twisting animal, trying to gather all its desperate writhing into his arms. Bessita braked hard, and as Thomas jumped down Sylvestre tumbled through the back door and was upon the gazelle with the long blade of his dagger, Romain holding it down with a hand around its head and another on its legs and Sylvestre expertly opening the creature's throat. Ah. *Jouissance.* The slit yawned like a snake's mouth, and spurted purple juice. The gazelle quivered. Blood spurted, spurted, then trickled, soaking into the fawn sand. When I lay my hands on the creature's pale, silken flanks its heart had stopped beating. The body was obscenely hot.

Breathless and proud, Romain held up his dripping trophy, blood and sand matting its flanks, blood on his shirt. Sylvestre and Thomas grinned. Bessita oddly showed no interest in the triumph – oddly, because the chase had excited him. He stood, looking absently into the distance. Then he looked at his wristwatch. 'Tie the gazelle on the roof-rack,' he said.

The oasis of Koro-Toro lies on the shore of the Erg du Djourab, a sea of dunes that must be crossed to go north. In the Libyan war it had been a military stronghold, and was still surrounded by minefields. To avoid them we needed to make a long detour into the sand dunes, and so we needed a guide. A scattering of stunted government-built block-houses had grown up near a well outside the minefield. Bessita spoke to an old man sitting in the shade of a wall. Without ado he got to his feet, told a younger man to bring him a bundle from inside the house, and climbed into the front passenger seat.

In the early evening the land was the creamy colour of my untanned flesh. There were no dunes, not even ripples, only an outflung plain. Beneath it there must be ragged rocks, covered by sand the way an undersea ridge is covered by water, an ocean of sand seeping into every cranny and finally smoothed flat by the wind. This was land you could sink into, and drown in its dark, white depths. In 1960 part of the skull of an early female hominid was discovered in the sandstone here, and named *Tchadanthropus uxoris* by her finder. She had protruding jaws, small canines, broad cheekbones, a large face, a big brain – indicators of intelligence. She died between 200,000 and 500,000 years ago.

As evening came on the sky grew briefly and unsentimentally pink, and darkness fell. We drove along a highway built from memory, the milestones heaps of camel bones. Nothing could live here permanently, but these camels had died en route to the butchers' shops of Libya, a thousand miles hence. The weaker animals collapsed or were slaughtered for food, and lay, tanning, until the wind and sand wore away the leather and revealed bones as white as toothpaste.

I began to make out domed silhouettes in the moonlight, then the old guide was indicating a turn. With a grunt, our Iron Maiden breasted the first sand dunes of the Erg du Djourab.

I climbed out through the side window and onto the roof, putting my hand in the darkness on the soft corpse of the gazelle. Its blood had spread over the plywood roof-rack and set hard in a broad mahogany stain. I shone my torch into the gazelle's eyes: lenses lifeless as glass.

It was a landscape like pictures remembered from a childhood book: moonlight-silvered hills, rounded as pillows, hills of pure sand. As we ploughed across them I clung to the roof-rack as if to the handrails of a pitching boat. It was preferable to being down in the nauseating cabin with its stink of diesel. There were bound bedrolls on the roof-rack; I lay back and looked at the sky, now half-obscured by light-veined cloud, and became so drowsy that there was a danger of falling asleep and being hurled off – it might be hours before anyone noticed that I was gone. I worked my limbs tightly under the retaining ropes until I was strapped down like Gulliver, then slept.

It was late when we stopped to camp. The Englishman was given the task of making tea. While the water boiled, I held the torch for Romain and Sylvestre to butcher the gazelle. Using the lovely dagger, they expertly unzipped the gazelle, peeled off its skin and pulled out its stinking guts and torn stomach that spilled half-digested greens on the sand. These marbled sacks of refuse were carried off into the night by Romain and dumped, where they fed God knows what, perhaps the sun. The two men broke the gazelle open like a great Christmas cracker, and Romain grabbed handfuls of spongy crimson flesh, happily letting the dagger fall, crack! crack! like a cleaver, hacking into ribs and spinal chord and legs.

My excellent English-style tea was universally condemned as undrinkable. The situation was remedied by Estelle, who poured a further half-cup of tea and half-pound of sugar into the saucepan and stewed it to the colour of creosote, whereupon it was universally acclaimed as just what the doctor ordered.

Estelle had automatically slipped into the feminine role of cook, chopping onions and chillies and bringing rice to the boil. Red cubes of butchered gazelle were brought to her, and she braised the crimson flesh into grey.

We ate, laid out our bedrolls, and slept.

At dawn, I climbed a hill of sand that overlooked a dried lake-bed.

The physicists say that solid objects are really billions of frantic grains. Solid-seeming sand dunes open when you step on them, and close around your foot the way they have closed over so much else. I imagined that I was walking over the bones of camels, pottery shards from vanished cities, corpses of the war-dead. Here and there were shards of grey, slate-like compacted mud, remnants of the vast lake that only a few thousand years ago covered much of the Sahara between here and Egypt.

Back at the camp, Estelle washed the dishes with no apparent concern for the amount of water she was using, and Thomas washed himself, splashing water everywhere. Bessita said nothing. Did he have an extra supply stowed away? As far as I could tell, all we had left were twenty litres, five gallons. I carefully filled my own water bottle and tucked it behind the seat.

Sylvestre began the day at the wheel, and within a minute had run us deep into sand. Out with the shovels and sand-ladders – the aluminium grids used when a vehicle is stuck to give it purchase.

Romain, experienced desert driver, took over, and at once ran us even deeper into sand. More manic leaping and heaving. Some people, I reflected, seem to enjoy this sort of thing. A problem involving bodies and masses, forces and leverage, spatial awareness.

'Why did the guide bring us this way?' I asked Bessita. 'Why leave the tracks we were following last night?'

'I suppose he thought it was a short-cut,' Bessita replied through gritted teeth. 'But now he's obviously lost.

'*I'll drive*,' he told Romain impatiently, and climbed behind the wheel – promptly burying us in sand a third time.

Our guide was standing to one side in his dirty djellaba, flapping his hands. '*Mon Père*,' called Romain, 'the old man says we have to go back the way we came.'

'I could have told *him* that,' retorted Bessita.

'Is the sand too soft?' I asked him.

'Yes, I think the wind has piled it up recently here, and it isn't packed enough yet to take the weight of the jeep. We'll just have to reverse back along our tracks.'

As we shovelled sand away from the wheels, two men appeared at the top of a dune to the north and slowly descended its steep face towards us. I noticed Romain stiffen, but as they approached we could tell that they were harmless. They were skeletal. Their truck had broken down a kilometre away, they said, and they had been here two weeks trying to repair it. One man addressed Bessita hesitantly. 'Sir, is there any possibility of one of us getting a lift . . . ?'

'No, I'm already overloaded.'

They had not expected any other answer. 'Then, sir . . . perhaps some water and food?'

Bessita told Romain to find them some tinned fish and bread, and ten litres of water. The sudden change in the two men was palpable, a surrender of tension so acute they both seemed physically to contract. Grins of gratitude loosened their faces. As Romain handed them the food, one of them looked at me and gave a high-pitched laugh. 'What are you doing here *writing*,' he asked, 'in the middle of the desert?'

I had seized on this break in the proceedings to catch up on my diary. Given their predicament, and the fact that we ourselves were enmired in sand, I must have seemed like some kind of deranged clerk.

Thomas was improvising a song. 'Martin can't find a moment to write his notes,' he sang, 'but he should just relax. He'll have time to write everything there is to know about Chad before we ever get out of here . . .'

I grinned, and put away the notebook. We all lent our weight to shifting the jeep.

CHAPTER SEVEN

Faya and Water

In the desert you celebrate nothing but water
Michael Ondaatje, *The English Patient*

THE IRON MAIDEN SHUDDERED and lurched, the air temperature inside her cabin something over 110 degrees Fahrenheit (the manufacturers of my pocket thermometer had not thought it necessary to extend the calibration beyond 110). The only remaining water was a few dribbles in some plastic bottles, and my three-pint canteen, which was full. Faya was still several hours off.

The desert surface was a cycle of splintering rock, gravel and gentle gradients of loose sand that constantly threatened to swallow the wheels and maroon us. You had to pick up speed on the gravel then speed over the sands, jerking the steering wheel to the left and right to prevent the front wheels from burrowing straight down. Several times we had shuddered to a standstill, wheels spinning and the stink of burning rubber poisoning the air. Even more unpleasant was the motion produced by the stretches of serrated rock, a spine-juddering vibration that dislocated everything inside the jeep, and threw us on top of each other.

We slewed to another standstill. The murdered gazelle was taking vengeance on Romain; he had severe diarrhoea. Now he seized the opportunity of an impromptu halt and crouched in the sand, clutching his belly.

Slithering in pools of mingled sweat, unbearably hot, we had been getting increasingly irritable with each other. As we prepared to restart, I told Bessita I was going up on the roof.

'But it's nearly midday.'

'It's OK.'

He shrugged. Let the English boy learn the hard way.

I climbed into the white glare and squatted on the scorching nylon tarpaulin. The metal bars of the roof-rack were too hot to touch, and I hung onto the nylon ropes binding the cargo. Coarse, glassy filaments bit into my palms. I could feel the sun burning the bare flesh of my forearms and feet; despite my hat, my head throbbed; my mouth was dry. I was aware that in Saharan summer midday heat, with almost no water in the vehicle, I was inviting rapid dehydration. At least I had those three pints in my canteen down below.

The power of the desert sun is awesome. Uwe George's book *In the Deserts of this Earth* tells the story of a couple of tourists who had set out in summer to drive the twenty miles between two oases, and made it without difficulty. Emboldened by the ease of the journey, they set off on the return journey without refilling their water bottles. Halfway back they ran out of fuel. They decided that the woman would remain sitting in the shade of the vehicle, while the man went on foot to bring help. He returned just five hours later, and she was still sitting there. The sun had moved around, and she was no longer in the shade. She had dehydrated to death.

I began to feel ill, and knew that it was the sun. But if I went back down too quickly I'd lose face. Supplementing the hat, I pulled my shirt over my head to provide a further layer of protection from the sun – a sunburned body seemed the lesser evil. My hands were raw from gripping the chafing nylon. I daydreamed about the water I had stowed down below.

At last Bessita stopped the jeep to let some air out of the overheated tyres, and I hopped down as casually as I could. I stretched my stiff limbs, climbed inside and asked Thomas to pass me my water bottle. He grinned and shook it. It rattled. 'It was all the water we had,' he said. 'But we saved you some.' A wave of fury surged through me, and I was aware that Sylvestre and Romain were watching me with interest.

'Pass it over,' I snapped.

I emptied the bottle in two gulps. It did not begin to quench my thirst. It was taking me a moment to adjust to the fact that the water I had been dreaming about – *my* water – had been shared by the group. But I had filled the bottle that morning from the common jerrycan, so what made it *mine*? The fact that *I*, that I . . . Nothing made it mine.

Ever since we left N'Djamena it had troubled me that we were travelling with so little water. The received wisdom is that in the desert you carry a minimum of twelve pints – six litres – per person per day. White desert-wallah wisdom, not something you find in the Holy Koran. We seven should have been carrying 250 pints for a three-day journey. But we had set off with barely 100 pints, 50 litres – two-fifths of the scientific recommendation.

I felt angry with Bessita for getting us into this position. True, we had acquired a passenger along the way, and several times had been obliged by the rules of the desert, or common compassion, to give water to the passengers of broken-down vehicles. But why had we failed to refill in Koro-Toro when we had the chance? Why no system of rationing? How could Bessita let people be so, so—? Never, I told myself, *never again* will I let myself be put in this position by another man.

We rattled on for three more hours. Staring through the dust-smeared glass, I felt as helpless as a fly trapped behind a window pane on a summer afternoon. I could only think of Faya – and water.

We were driving through a rocky gully, and suddenly through a crack in the rock wall I glimpsed a plain beneath us with a cluster of black dots.

'Trees,' I said.

Thomas, Romain and Sylvestre span round to see where I was looking. Now, nothing was visible but rock.

'Bullshit,' said Thomas.

'I just saw trees,' I said.

'Stop talking shit.'

Five minutes later the jeep rounded a bend and we all saw a great natural depression, as though a disc of yellow sand miles wide had descended into the earth, leaving behind this crater-like rim. Silence fell over us as we devoured the sight of hundreds of date palms: they meant water. As we reached the base of the cliffs we passed the first palms, stunted, ragged, caked with dust so that they were khaki instead of green, but evidence of water nonetheless, and nothing could diminish our enthusiasm for them – not the fact that the track into Faya was a slow wallow through almost undriveable sand, not

42

that the southern outskirts of the town were a battlefield littered with the torn hulks of tanks, not even our first impressions of Faya itself, with its impoverished, fly-blown streets of sand littered with refuse. We were only interested in water.

We stumbled out of the jeep, through a door in a mud-brick wall into a compound, and past a well – a hole in the ground with an old truck tyre over it. We sank onto a mat in the shade of an awning made of palm fronds. Two uniformed soldiers quickly brought water in deep enamel bowls that we impatiently passed around. I was limp and exhausted, consumed by self-pity. I kept sighing deeply and muttering, 'Stupid, stupid.'

Père Bessita spoke. 'I'm going to give thanks for our safe arrival,' he said.

'No thanks to *you*', I thought ungratefully.

We bowed our heads. Bessita muttered a prayer, then lifted a bowl in his bear's mitt and drank.

PART II

Tibesti

CHAPTER EIGHT

Vast Eternities

THE HUMAN BODY has not evolved to withstand sunlight. An active man in a shadeless desert will lose fifteen or twenty pints of water a day through evaporation. It has not been unknown in human history for men to dispose of enemies simply by staking them out in the sand. By nightfall, you are raging with delirious fever. Your body, unable to replenish fluids, takes water from its blood, fat and tissues. Your blood thickens, circulates more slowly and fails in its cooling effect. Your sweat glands, having excreted water all day in an effort to cool the body, fail from overwork. In a high fever, suffering from circulatory failure, you die. Even with water, you would eventually have died from sunburn. To survive in the desert human beings need to find shade, and eight to twelve pints of water a day.

The Mediterranean has been a desert, the Sahara a sea. Throughout geologic history, deserts have come and gone in cycles of hundreds of millions of years. The period of human evolution, the last two to three million years, has been a cool, moist phase of that history. But since the end of the last ice age 10,000 years ago, deserts have once more been on the move. The world is getting hotter, and it is going to take us a long time to adapt.

Though deserts do not teem with life, they are self-contained ecosystems with neatly adapted flora and fauna. Snakes have been around longer than humans – they have had 100 million years to get used to life under the sun. The horned viper of the Sahara has evolved a diagonal double-S-shaped shuffle which allows it to make only fleeting contact with the scalding sand. This does not mean that it can take it easy in the desert's oven. All life needs water, and wherever there is even a drop – in a well, from rare rainfall, even as mist – nature has found a way of harvesting it, with evolutions so fantastic they have tested the faith of devout Darwinists. There are plant seeds that can remain fertile for decades, waiting for rain. Certain birds have evolved protective plumage the exact shade of the shadows cast by rocks. But the most fully evolved desert creature is the camel. Unlike other warm-blooded creatures, it does not lose water through perspiration. It can survive for two weeks without food or water, and given food it can keep walking across the desert for many weeks; it can drink 200 pints of water in 10 minutes; its long legs help it lope across the great distances between water holes and keep its body above the hot layer of air at the desert's surface; its massive circular feet let it tread soft sand without sinking. I could go on – the camel is a miracle.

In humans, the most crucial survival mechanism is adaptability. In some parts of Australia Aboriginal peoples still live intimately with the desert, while in others whites live in underground caverns with washing machines and satellite TV. From the arid grasslands on desert fringes to the oases in their furnace hearts, hundreds of millions of people live in deserts, in tents and huts and air-conditioned condominiums: nomads, kibbutzniks, miners – even fishermen. Western Man has adapted the desert to *him*, irrigated and electrified it, made Las Vegas the fastest-growing city in America, and Palm Springs a desert St Tropez with swimming pools and neon-green grass.

The deserts are littered with the remains of earlier civilizations which often fell because they over-exploited water. Houses lie abandoned and sand-choked. The buried inhabitants have been mummified by the extreme aridity, their skin and hair and clothing preserved intact for thousands of years. There has always been human life in the desert, and lines of communication; today there are even more – railways and freeways as well as camel tracks. There is often no formal public transport, but there are usually lorries or herders. You can get through.

We are seduced by the desert in a way that turns upside-down the usual priorities of travel. We still think in terms that would have been recognizable to an eighteenth-century Grand Tourist, visiting cities and temples, recalling civilizations that rose and fell. But civilization exists in a tension with wilderness. Alone in wild places we instinctively pine for the security of numbers, but in cities our hearts long for the untamed. It is arguable that the city, with its institutions and neat streets, newspapers and cafés, represents the great achievement of humankind, but order brings uniformity and stagnation. While cities are centres of dissent and renewal, their individualism is metropolitan and social – the prod of the political pamphleteer, the strut of the style-victim. It is only in nature that one can come face to face with one's own deepest essence – an experience of recognition. Nature, the great parent, reminds you that you came out of the earth, and will return to it.

Nature at her most endearing provides glossy alpine slopes and glassy pools, but deserts are ill-tempered environments. They are primitive and dramatic in the manner of the tales of Gilgamesh or Jason, epic sites where you witness the impact of vast volumes of sun-excited sky on scarred landscapes. On some parts of this planet – the Sahara, the Arabian Empty Quarter or the Gobi's Taklamakan – millions of years of relentless erosion have created that received image of the desert, the sea of sand. But countless mountains and plains are also desert – bare hills resemble a carcass with its flesh seared away, and rocks of black and feruginous red stand out along ridges like scorched spinal chords, or are sand-smoothed into the oddly regular columns of a gods' battlefield littered with giant torsos and lopped-off limbs.

The desert is clean. City lights bouncing off clouds produce light pollution that blinds telescopes, but in the desert there is neither light nor air pollution; you inhale sweet, dry air and at night darkness drops like a trap-door. The sky is so devoid of atmospheric matter that there are few red sunsets, the stars do not twinkle but are steady, and the moon is bright enough to read by.

The desert is silent. An enormous silence simply *is*, and, until you have experienced it, you have not known noiselessness. I remember a dawn in Death Valley. The air was dense, still, scented with sage musk released in the night. I became aware of a sound like a sail beating the wind, but more rhythmic, a whooshing of air beaten and resisting, and saw with surprise that it came from the flight of a far distant bird. Its wings made the only noise for miles, impelling eddies of sound through the tranquil air. Short of a vacuum, true silence requires the absence of friction of air upon object – the emptiness and stillness found only in the desert. Hovering over the binaries of dust and sky, dun and blue, shade and sunlight, silence eventually becomes a sound itself: a sibilant blood-rush in your ears. It focuses the sense of self, it provides a relief from urban unquiet.

The desert magnifies solitude. It can seem uncomfortably inhuman, just as orbiting the earth has moved astronauts to awe and fear. So much of our human life is made up of the chatter with which we locate ourselves as social animals. We deny solitude, yet a preparation for it must be essential, because we have to die alone. Many contemplative traditions insist on solitariness. Certain Hindu and Buddhist tantric traditions prescribe meditation before corpses, the unavoidable messengers of the truth that everything decays and passes. Monks must fix their attentions on the atemporal and incorporeal, on union with the Divine. Jesus spent forty days and nights in the desert, beyond the limits of the town, beyond the goats and the lepers, without motorways rumbling in the distance or jets etching the skies. Alone with your spirit, your guardian angels and your demons, there are Blake's infinities glimpsed in a grain of sand, Marvell's vast eternities.

Aldous Huxley, celebrated for his dystopian *Brave New World*, lived for several years deep in the Mojave Desert of California. It was the time when he was under the profound influence of Indian mysticism, compiling his anthology of mystical writing, *The Perennial Philosophy*. I went to meet the American theologian Houston Smith, who remembered visiting Huxley in the Mojave. 'Huxley loved the desert,' he told me. 'He said he was happier there than anywhere he had ever lived.'

I asked if he felt Huxley had been attracted by the same things that had drawn mystics to deserts.

'Oh *yes*. He talked about how the sand and light cast a mantle of *sameness* over everything, bringing out the unity in all things.' Houston paused, his eyes glowing in the memory. 'It was simply a *golden* afternoon. I remember helping Aldous's wife, Maria, sweep out sand from under the beds – that's *glory* for you!'

Some years ago I climbed to a shrine on top of a hill in the Thar Desert of northern India. In India it often seems that no violation of the sky – realm of gods – is left unchastened by man, who caps every jutting hilltop with a cairn or shrine. The hill grew in size as I approached, was steeper than it looked, and made of a loose shale that fragmented underfoot. The climb was hotter and more difficult than I expected.

The shrine was a low, domed building with thick whitewashed walls. I ducked into the inky doorway, and stood while my eyes adjusted to the viscous gloom. The tight space was dominated by a relief of the elephant-headed god Ganesh, worn by reverent fingers into a primitive shapelessness. Ganesh was smeared with blackened butter and splashed with red and cinnamon powders; dessicated petals were stuck all over his body – a squat, powerful god presiding confidently from his dark abode. Through the narrow doorway the desert, a rectangle of bleached ochre, quivered. I felt the power of the place, and knelt and leaned my brow against the cool stone at the god's feet. My heart opened.

Our human response to the desert is complex – instinctive discomfort and fear alongside exhilaration, aesthetic ecstasy and awe. If you are alone you must be able to cope if anything goes wrong, you must have water and medicine and reliable transport. You keep elaborate maps with you and several compasses. It is not physically easy to feel at peace in the desert. The heat can be insupportable, and airborne dust and sand compound your discomfort. Solitude, too, brings suffering, your familiar domestic chit-chat replaced by an echoing silence which you fill with gibberish. It is only when all this nervous excitement settles down that the desert's numinous qualities emerge. You are *apart,* separate from your fellow people with their tiresome demands. You are somehow beyond all that.

CHAPTER NINE

L'Homme Moyen Sensuel

Those who have believed the messages of the prophets
will be summoned to enter paradise, where they will
live for ever in a garden with cool streams, luscious
fruit, silk couches and beautiful, unblemished women
– although all these will be as nothing compared with
the sight of God.

Willy Ardener, *The Pillars of Islam*

SOON AFTER OUR ARRIVAL a flood of parishioners entered the compound, wearing Sunday-best and bearing gifts. Officers of the Church bustled in to discuss matters arising since Bessita's last visitation, and the huge priest and these slighter servants of God inclined together in earnest conclave. Grave government servants arrived, neat and formal; old women greeted the Father like a messiah; respectful mothers saw him more superstitiously, a talisman to help ensure the success of their husbands and the health of their children, brought, scrubbed and barbered, to have their heads patted – though few received more than a forbidding scan from Bessita as they sat, awestruck and silent.

The community was also concerned to ensure that its pastor should not go hungry. Plates of lean meat under lacy bead-weighted doylies were handed over, and Bessita put them to one side with polite disinterest. He handed back bag-fulls of Christian paraphernalia – robes, Bibles, hymnals, tracts and gilt crosses on strings of turquoise plastic beads.

Among the visitors was a well-groomed and sober alcoholic who came to confess and plead for redemption (and a little financial assistance – for his children's education, you understand). Mistaking me for a missionary, he had grabbed my hand to kiss it before I could pass him over to Bessita. He received no handouts, and turned up again a couple of days later, unsober, slovenly and defiant.

The routine of Bessita's household did not match that of the town at large. By the time we had all risen, rolled up our mats, taken our turns to stand behind a low wall in the corner of the compound and wash from a galvanized bucket, and eaten our communal breakfast, it was 10 a.m. In Faya, the heat really has a grip by that hour; a sight-seeing stroll is impracticable, and in any case the town is shutting down for the long siesta. I found myself wading through streets of soft, deep sand at midday, half-faint from the heat, a source of amusement to the cheeky little boys who were the only other human beings abroad at that hour.

These days, academics tell us that the exotic, the other, is a form of colonial oppression – entire civilizations traduced by cheap cultural stage-scenery, the painted back-drop from *Aida* that masks the reality of modern Egypt; orientalism, the Desert Song, romantic fiction, Hollywood; Arab civilization travestied by advertisers to flog perfume and mediocre family cars. This new orthodoxy tells us that other cultures are not strange, but in their own terms normal, and that we must respect them as such.

But Faya *is* an exotic town.

Webster defines exotic as 'Foreign; strange', and Faya simply screams 'foreign': an oasis lying on the south-eastern edge of the jagged mountains that stretch for 2,000 miles through the heart of the Sahara, reaching their highest point at Emi Koussi, just 150 miles to the north in the Tibesti Massif. The only way to reach Faya is by camel, jeep or military plane. In its time it has been a wealthy centre of trade in camels, salt and slaves, but it is now best known for the ratty 'Libyan Market', supplying smuggled sardines and flip-flops.

Faya is also, by any Western standard, strange. Its streets of sand are littered with sardine cans, chicken claws and the inedible parts of goats; there are hoofs and scraps of furry suntanned hide everywhere, and starving dogs lick already clean-picked skulls. Buildings are the plainest mud-brick blocks imaginable, without decoration. (One afternoon, I took out some watercolours to make a few sketches of Faya and soon found myself surrounded by a ring of entranced children. None of them had seen anyone drawing before, and they

begged me to make drawings of the gigantic trans-Saharan *gros porteurs*
that fascinated them.) A few open doorways look grimly onto the
street; they are shops. Some have signs, scraps of plywood with crude
characters daubed on them in bright colours: *restaurant, pâtisserie,
alimentation*. In Faya, that is sophistication. One shop, a dressmaker,
showed peerless ambition: its plywood scrap was painted, in
inaccurate homage, *Galerie Lafayette*.

On my first morning in Faya it was – praise Allah – overcast, and
a soothing wind was blowing. I accompanied a nervous Estelle to the
schoolroom where she was to sit her entrance exam, then wandered
to the top of the town, where a shell-pocked and empty watertower
dominates. Sprawled around it is modern Faya – school, police
station, hospital – concrete and hideous.

In N'Djamena I had seen a tourist-office poster depicting 'the
Market at Faya', a whitewashed arcade with palm trees and
picturesque women selling dates. At last I recognized that the weary,
shabby central marketplace did bear a resemblance to the pretty
Mediterranean scene artfully conveyed by the poster. It was the only
place in Faya with even the remotest aesthetic pretence.

But what right had I to expect a remote and war-damaged Saharan
oasis to conform to European aesthetics? Arab houses are traditionally
externally blank and internally lavish, with public beauty reserved for
the mosque; Western notions of the public realm do not apply, and
never have, and the decadent Western usurpation of the religious
realm, with universities and museums as *ersatz* cathedrals, is rightly
repulsive to the stern Moslem soul.

Near the Libyan market was a square where you could join a *gros
porteur* northward to Libya or Algeria. One was waiting, and its driver
displayed none of the shiftiness of the men I had met in N'Djamena.
Yes, he would be happy to take me, but he would not be leaving for
about a week.

It was clear that if I wanted to return to N'Djamena with Bessita,
I had only two weeks to get to Tibesti and back. I was behind
schedule, but within budget. I decided I could afford to hire a jeep
and continue north to the rock paintings and epic craters of Tibesti,
close to the Libyan border. There was also a chance that when I
reached the far north, I would find a way to cross into

neighbouring Niger and continue my journey across the Sahara.

There is no Avis in Faya, no taxi rank, and the few jeeps I saw were government vehicles. But there were privately owned flat-bed pick-ups, I was told. I went to meet various local officials who I hoped would act as intermediaries, but made little progress; the sums under discussion were impossibly large. It was Bessita who explained to me that these helpful intermediaries would be exacting an arrangement fee of around 50 per cent – in the eyes of every official, I was a wealthy foreigner whom it was normal to 'tax' this way. The police chief told me to my face that I was a spy. A driver Bessita found announced aggressively that he was my 'only option', and muttered darkly when I told him to get lost. Days went by.

By mid-afternoon, the heat in Faya would reach a revolting intensity, evaporating water having humidified the air. You lay helpless and almost feverish, covered in flies, longing for the evening cool. Compared to the desert, I began to think, oases were unclean places. Returning from my midday exertions I would find everyone else sleeping as though they had been breaking rocks all morning: the Father on his back, his whale-like paunch rising and falling, a snore fluttering under the handkerchief that kept the flies off his face, Romain dozing, and Sylvestre and Thomas always in the sweat-smeared depths of Morphean stupor.

One day I was in the Libyan Market with Romain, who seemed to have attached himself to me. 'So where,' I asked him, 'are Thomas and Sylvestre? I never see them, and whenever I do they're asleep.'

Romain gave a lop-sided grin, and put a forefinger into a ring made from the thumb and finger of his other hand.

'What?'

'Fucking, Monsieur Martin. Every morning, every afternoon. I don't know why *you've* come here, but that Sylvestre had only one thing on his mind. *Il voulait goûter la chair arabe!*' And he went into his infectious, high-pitched laughter.

It took a moment for this to sink in. 'You mean, they're *at it* – all day long?' Having sex with locals is, of course, one of the most traditional forms of exoticism. 'But – who with?'

'Oh, there are plenty of prostitutes here, Monsieur Martin. Of course, it took a while to find a clean place, you know, with young girls.'

'You went with them?'

'I *found* them the place,' he said proudly. 'But I don't have any money. Thomas is with Monsieur Sylvestre all the time, but even he has to sit outside sometimes.' He added glumly, 'Monsieur Sylvestre would only pay for me once.'

'So you thought you'd try your luck with me, did you?'

'Oh, Monsieur Martin!'

'Don't "Oh" me. I know perfectly well you wouldn't be wandering around buying mangos with me if you could be *sauter les nanas* with those two.'

Romain let out a joyous whoop, and we kept on walking. A woman passed us, a woman in her mid-twenties with the characteristic cosmetic lip-black of Tibesti (I could not determine whether this was designed to attract, or conversely was a sort of labial yashmak). She was taller than average, well dressed and pretty.

Romain turned and said something. She shook her head, but coyly smiled.

'Do you know her?' I asked.

'No.'

'What did you say?'

'I said, "How about a quick one, since it would do us both good?"'

I laughed out loud. 'Was *she* a prostitute?'

'Er, no, I don't think so.'

In late afternoon, *Père* Bessita would raise his conical bulk, pout dopily and splash his face, and sometimes catch my eye with a wry twinkle in his own. My manic pursuit of a jeep and driver amused him. I had begun to see how this inscrutable bear subtly took a paternal interest in each of us. At night as we all gathered around a central dish of *boule*, the meat stew and semolina which Estelle had been preparing since morning (who would have cooked, I wondered, if she had not wandered into our lives and accepted her domestic fate?), Bessita would always notice my reluctance to tuck in, and push a meaty bone in my direction, saying, 'Eat, Martin.'

One day I picked up a sheet of paper on which Sylvestre had been scribbling an application for a visa renewal. It was an illiterate scrawl. I realized that Sylvestre was dyslexic, which put his dislike of my reading in a different light. I remembered that at twenty-two, I had tried to shed my Westernness, in my case to become absorbed into Indian culture. Here Sylvestre spent his afternoons penetrating African women, while I burrowed into *The Heart of the Matter* and Conrad's *Heart of Darkness*.

At night, we would roll out the nylon mats to sleep. Estelle ('the woman', as the others called her) would always sleep near Bessita – for protection, one presumed, from the unbound male carnality in the compound. Bessita's consolation came from the human voice. He was a radio addict, and sat every night by his transistor, listening as the studios in Paris diffused news of African wars.

At last I found a driver, a man who regularly drove goods and humans up to Libya in a prehistoric Toyota. He had a friendly smile, and seemed good-natured. We scheduled our departure for Sunday afternoon.

On Sunday morning, all were expected to be present and correct for Faya's triannual celebration of Mass. I arrived to find a gathering of Faya's Christian great and good, including the brawny, crew-cut *commandant* of the small French garrison.

'Not much of a parish, is it,' he said, pacing impatiently in the already-hot sun. *'Une paroisse maudite.'* He chipped cement off the whitewashed wall with a heavy duty aluminium key-ring.

Nearby stood the holed water-tower. 'Why don't they repair it, and give people some running water?' I asked him.

He shrugged. 'Some houses do get water – for a few hours a day. But they haven't got any money for repairs. Before the war, they even had electricity here.' His face brightened. 'But have you noticed that in the last three days they've started to clean the streets? Normality returns!'

I had not noticed.

Goats and donkeys wandered around outside the church, there were palm trees: it was a recognizably Biblical world. Bessita appeared, solemn and white-soutaned, a tall ship tacking in the midst of a flotilla of singing and clapping parishioners. I half-expected them to toss palm fronds in his path.

The church was long and dark, its narrow roof made from palm trunks, with a single glass roof panel that cast a lozenge of quivering white light on the altar. It flared on Bessita's crisp, bleached soutane, and occasionally flashed sparks across his face. Then sparks of music, and song flamed into life, a high-pitched whine from the swaying front rows of old and young women, with a fizzy pizzicato from the little girls' rattles (made from 'Rambo' insecticide cans). Three blasé-looking musicians banged drums fashioned from finned shell-cases. On the wall behind Bessita was a simple wooden crucifix, and the words *Que ma prière monte vers toi Seigneur comme la fumée de l'encens*. The altar was of whitewashed cement and mud, covered by an ironed cloth embroidered with 'PAX' and the image of a golden dove; on it, a metal chalice, a small wooden crucifix and two hurricane lamps.

'*La parole de Dieu*', Bessita informed us, and the congregation scrutinized its Bibles. As he spoke, his French was translated by an assistant into the lilting sing-song Sara of his audience. Brays and bleats drifted in from outside. Thomas fiercely chewed his nails. A faint smell of sweat hung in the air.

Bessita was uncharismatic, but I felt the power of the Church resonate from his bulky form, clad in white, embroidered with orange anchors – fisher of souls? His head leaned at a slight angle during the two-hour service, as though he were lost in thought, a man in a crowd oblivious of his surroundings, isolated inside the ritual.

There was less joy than I expected from an African congregation. At ordained moments it would start with the hesitancy of a hand-cranked engine. There would be some moments of unco-ordinated swaying and clapping, then it would suddenly combust into a rhythmic, uniform voice and motion. As abruptly, it would stop for the next bit of churchy business. It all felt rather Protestant.

The French *commandant* drove me back to Bessita's compound, and I told him about my imminent trip. 'Before you go, come over to the camp and I'll give you a couple of cases of water,' he said. 'We'll radio the French garrison in Bardai, let them know you're coming that way. Just in case . . .'

My driver did not turn up. I waited an hour, then I went to find him.

'There'll be a delay,' he said.

'Where's your jeep?'

'Ohh . . .' His jeep had been seized by the police. He and a friend had been to the *sous-préfet*'s garage to try and buy a spare part. 'My friend simply touched this part, and he was arrested for, er . . .'

58

'For what?'

He squirmed. 'Well, for sort of . . . looking as though he thought he might steal it.'

'He *didn't* steal it?'

'Certainly not, Monsieur!'

'But the police have impounded your car?'

'Yes, sir, it's so unfair, I—'

'Your friend wasn't trying to steal from the most powerful man in Faya, was he? They'll lock him up and throw away the key!'

'*No*, sir! He was only—'

I plodded through the throbbing afternoon heat towards the French camp, a mile outside town. Before I reached it, the *commandant* swept by in his jeep. As I told him the story his face darkened.

'But – really – I'm – I mean – *fuck*! Get in!'

Sitting beside him was a Fayan *haut fonctionnaire*, a fat, perspiring figure out of a 1940s Hollywood *film noir*. The *commandant* rounded on him furiously. 'I'm *fed up* with you lot! This man's been trying to get out of this shit-hole town for a week, and he's had nothing but lies and obstruction! Because he's not bribing anyone, I suppose!'

The *fonctionnaire* looked uncomfortable. The *commandant* fulminated on, jerking wheel and gear-stick angrily with his big arms as the jeep swept through Faya, spraying sand.

'I'm pissed off! This town gets everything from us, we drive you all over the Sahara in our planes like it was a fucking bus service and what do *we* get? A European can't even find anyone honest to rent him a jeep! Well, that's it! I'm putting M. Buckley under the protection of the French garrison! I'm suspending co-operation until this man is on a jeep and out of here!'

Within a couple of hours I was fixed up with a 'fully equipped' jeep and driver. We would leave the next morning.

That evening, the *commandant* drove me out to the garrison for a free gift of two cases of Evian. We were now accompanied by the local *sous-préfet*, the man whose jeep parts had been in danger of being stolen. They were the two most influential men in Faya.

'*Mon commandant* has a good life here,' the *sous-préfet* told me. 'Look at him: he lives like a prince – we have water for four hours a day, *he* has it for twenty-four!'

'That's your idea of being a prince?' asked the Frenchman. 'Someone who can wash any time he feels like it?'

'Ah, you're a big man here, *mon commandant.*' He turned to me. 'Wait till he's back in France – he'll be a nobody again. Back in the Parisian suburbs with his wife.'

'Look at him,' the *commandant* returned affectionately, 'he has four wives, the dirty bastard!'

'The Lord,' his friend said, 'has commanded us to be fruitful. We are His servants.'

'Yes . . . ' said the *commandant.* 'Well, I had a chance to serve one of your wives while you were away.'

The *sous-préfet* twitched. 'Oh yes?'

'Oh *yes.* Number two. I got to know her. *Charming* woman.'

'Well, that's very dangerous, *mon commandant.*'

'Dangerous?'

'It's forbidden in the Koran. You are condemned out of your own mouth.'

'What's forbidden? Condemned for what? All I did was give her a lift a couple of times.'

'Nothing to do with it.' The *sous-préfet* regained his composure. 'You're condemned.'

'And the penalty?'

'Seven camels – and you'll be declared *persona non grata.*'

The *commandant* laughed. '*Persona non grata,* what a disaster. My tour ends in two weeks and I'm out of this dump!'

'And what about my seven camels?'

'I'll give you an IOU.'

The *sous-préfet* turned to me again. 'He can't be trusted, you know.'

'I can't? Well, I'd like to meet a more honest man in Faya. Among you lot of thieves I stand out like a saint.'

'No,' the *sous-préfet* told me. 'I'm an expert physiognomist, you know.'

'And?' I asked, cast as the straight man. 'What does that tell you about *M. le commandant?*'

'Unfortunately, there's a difficulty here. You see, all white men look alike.'

I said a warm goodbye to Romain, and thanked Bessita for all his help.

There was a chance that from Zouar I would find a way of crossing into Niger, in which case I would not be returning to Faya or N'Djamena. I told Bessita that if I did leave for Niger, I would send back word with my driver, and asked him to let the Delacroix's know.

Bessita looked sceptical. 'From what I hear, it would be very unwise to go into Niger.'

'Well, a lot of people said that about coming this far, Father.'

CHAPTER TEN

The Waste Land

A S THE SKY LIGHTENED, faces began to appear in the blue shadows and several thumbs were stuck out, but my driver – his name was Mohammed – was in no mood to give lifts. Mohammed had never before taken this north-western route to Tibesti. And he had accepted a low price. I wondered how the *commandant* and the *sous-préfet* had brow-beaten him into taking me.

We soon passed the truck whose amiable driver I had spoken to in Faya. It had pulled over for its passengers to say morning prayers, and pious Mohammed stopped to join them. The driver recognized me, and suggested that we carry on in tandem with his truck. Mohammed shook his head. 'We don't have time,' he said.

'I know, you *should* get there first,' returned the driver cheerfully. '*Inshaallah*.'

'*Inshaallah*. You know, if you want, I could give you a guide. One of my passengers, Idries – that chap over there.' He pointed to a lean, moustached figure in a brilliant yellow djellaba and pants, smoking a cigarette. 'And that's his brother over there. You'd do all right to take them, you know. They know this desert *very* well.'

Idries and brother threw their bags on the back of the jeep, and climbed aboard.

We passed mammoth palm clumps bloated with dead fronds, and neatly ordered baby palms in fresh plantations. As the day advanced, light flooded down, cultivation ended and the desert began – a reg, one of the distinctive landforms of the Sahara, a gently undulating plain of hard rock and pebbles where we could just make out the tyre marks of previous vehicles. Next we reached a fawn sand plain with perfect orange crescent dunes arranged at odd but not quite random intervals, like an exhibit of conceptual art. There were animal tracks on the flank of a dune – life in this dryness! – perhaps a desert fox. Brick-red outcrops loomed like Victorian madhouses planted in the shimmering desert. Then further cliffs, pink like the body's inner flesh, and pink, conical hills strewn with black boulders. The sand was scattered, like a seashore, with powdered white sea-shells and smooth, fist-sized stones.

At nine, Mohammed halted to track down a worrying rattle in his jeep, and I went for a pee. As I trotted towards the decent seclusion of a rocky ridge, thinking it would also offer a view, my right foot slipped. I looked down stupidly to see the broken toe loop of my sandals, then blood. I felt a rush of pain. My toe had made contact with a fragment of glassy rock – in all this vastness of naked geology, this dainty shard had found something soft and organic to penetrate. The cut was deep, and under the bleaching sun I watched my blood coagulate into a fat worm on the sand. If I were alone out here in this yawning expanse, I thought, I'd be dead in a couple of days. No snakes, no vultures, just the sun to suck you dry. I felt a sudden compulsion to be back in the security of the jeep.

Half an hour later two men on camels materialized miraculously from behind some rocks. As they saluted us, smiling, I caught the white flashes of their teeth, and in that moment I *saw* the landscape for the first time, with its ochres, greens and mauves, the sandstone veins rich in minerals crushed to powder where the track crossed them, as though the torn sacks of a passing spice caravan had spilled saffron, turmeric, lavender.

Later we came across a legacy of war – an exploded Libyan tank, its carcass shiny and scorched, a surrounding aureole of ragged scrap. 'Irreparable', commented Mohammed dryly.

It was almost the first word he had volunteered since I had met him. Mohammed was thin and short, with a permanent, lugubrious frown. He was by nature very, very cautious.

We had our first puncture. To repair it the wheel had to be removed and its steel rings broken down to yield a sliver of shiny inner tube like fish guts, patched many times over. Mohammed's 'fully equipped' jeep carried no replacement inner tube. It was hard work, and I watched sinewy Mohammed sweat as he expertly drove in tyre levers and jumped on rims, until Idries pushed him out of the way and took over with equal expertness (the paying passenger was not permitted to help). The rubber around the hole was scraped with a razor blade for the tacky glue to take, and a black patch cut from an old tube was massaged into place.

I walked off across egg-yellow sand under a bold blue sky. I turned and looked back at the jeep, a distant white dot pressed between these blocks of colour. So this is it, I thought – I was actually making the journey that had started with a soggy map in a steaming bath, two years earlier.

The sun's heat was a weight, unpleasant and invasive. Even the silence was heavy, almost as tangible as the heat. I experienced the sensation I associate only with the desert. Your mind comes rolling up at you in big unruly waves, subconscious promptings take on the substance of thought-bubbles, swimming around you like jelly fish. You turn physically, to try and fasten your mind on something external, but there *is* nothing, and the Zen sound of no-sound is the blood rushing past your ear-drums. It is White Out.

It took exactly sixty-six minutes to mend the tyre. 'You left too late,' Idries told Mohammed. 'The midday halt is a long way ahead.'

This meant nothing to Mohammed, who had never been this way before. He was looking tired, and I suggested I drive. The passengers behind looked dubious. As soon as we reached a zone of soft grey sand, the jeep wallowed and stuck. The passengers behind banged on the roof in protest. Mohammed sucked his teeth and shook his head, '*Non, pas comme ça.*'

'I'm a good desert driver,' said Idries, pulling open the door and elbowing me out of the way. It was true. He drove briskly and precisely, shifting gear with sure, sharp motions.

At one-thirty we saw a sort of green gauze in the distance, a grove of acacia trees, the only vegetation in all that wilderness. You could tell that it was a traditional halt by the debris of sardine cans and oil filters. Nearby there were more wrecked tanks.

'We should have been here by eleven,' Idries said irritably, as he nosed the vehicle in among acacia branches. It seemed a bit rich that he was complaining, since if we had not picked him up he would still have been several hours south of here, sitting on a sack of dates in the back of a lumbering lorry.

Mohammed quietly opened the bonnet and pulled out the hot dipstick, tipped with steaming toffee-coloured oil. I watched the black, oily engine with its steady animal shudder. Mohammed wrapped his hand in a rag and took off the radiator cap and tipped in some cooling water. Satisfied, he crouched down, threw together some twigs and pulled out a little leather duffel bag with a small enamel teapot. Marsupial outer pockets held chipped glasses and tufts of fresh mint. I keyed open some tins of sardines, and after a hasty lunch we closed our eyes and slept.

In the desert, shade transmutes the base metal of the body into the gold of emotion. Shade delivers you from suffering, you become fond of it – you *love* it. At three-thirty we emerged reluctantly from the heat beneath the jeep into the greater heat outside it. Bars of white sunlight fell through the acacia branches onto our clothes, burning.

The infinity of dunes was relieved only by occasional metal way-posts, planted by the French Leclerc expedition in 1940. Metal does not rust in the Sahara. Left undisturbed, Gaddafi's tanks would still be there in 500 years.

We drove until dusk, when we pulled over for Mohammed's evening prayers. As we climbed down, a large truck lumbered towards us. 'My God,' said Mohammed, 'that's the third vehicle we've seen since morning. The desert's really busy today.'

We missed the roundabout.

The next morning, as we approached the Tibesti foothills, I had been looking forward to what my map showed as *le Rond Point Charles de Gaulle*. Where the track divided north and north-west someone had made a roundabout, named after the paternalist French president who had done so much in the second half of the twentieth century to keep the countries of North Africa – the former French possessions – firmly under the Parisian thumb. When the heads of independent nations rang him up for instructions, they called him '*Papa*'.

How surreally suburban, a roundabout in mid-desert. I wondered how it would look. A ring of old tyres? An actual concrete marker with the engraved name of the *général* half-obliterated by bullet holes? But as we reached the point where a tangle of tracks slewed north, Papa's roundabout mysteriously eluded us. 'The sand is loose,' Idries told Mohammed. 'Keep up your speed.'

We paused before the monumental spectacle of two tanks which, unlike many of the relics of war we had come across, were perfectly preserved, their guns neatly aligned in opposite directions. They looked as though they belonged at a barracks' gate – all that was missing was a ring of whitewashed stones. I insisted on taking a photograph, while the lads took a cigarette break. As I got close to the tanks I heard yelling behind me. All three men were jumping and waving frantically. 'Martin! Come back! Mines! Mines!'

Keeping my composure, I took the picture, then picked my way back, carefully stepping in my own footprints.

'Why did you go so close?' they scolded. 'The tanks are mined! Why do you think they're in such good condition?'

'I didn't know. Why didn't you tell me?'

'We're used to it! There are mined weapons all over the desert! We never go anywhere near those ones. We assumed you'd know, too.'

Steadily we crossed the wilderness, Mohammed, both fists gripping the shuddering wheel of his old Toyota, squinting without sunglasses into the primrose-hued haze. There was inexpressible beauty to be seen in the desert, but rarely during the day. By night there were dusky purples, delicate washes of pink, revolutionary reds, but by day there was mostly only hot, white light.

Much of my conscious energy was spent trying to keep the heat at bay. Barefoot, I kept my leather sandals as a cushion against the searing metal panels around the engine. I wore a constantly re-moistened bandanna round my neck, because evaporation chilled the jugular veins (a damp scarf suddenly applied to the neck produced a sort of giddy drug-rush). Still, the crown of my head became unpleasantly hot – I wondered what the temperature of my brain must be – and occasionally I would pour water over my head, relishing the iciness as it trickled down my back. Mohammed saw this as girlishness, but was mostly too polite to comment. For his part, he wore a tightly wrapped turban, the multiple layers of which, he insisted, kept his head cool (he was right, but it took me a long time to discover the efficacy of turbans).

I made a fetish of cold water. After the water shortages en route to Faya I had equipped myself with several twenty-litre jerrycans, in addition to the *commandant's* two cases of Evian. At dawn the water would be deliciously chill, and I would fill up my insulated canteen; it had to be kept away from any hot object such as the door or seat, dangled on a cord somewhere in the car's internal shade. Behind the seat I kept two other bottles wrapped in a damp shirt, and evaporation kept them relatively cool. Each time we stopped I would use this water to top up my canteen, then refill the bottles with Evian (the aqueous equivalent of champagne, you might think, but I later discovered that much Saharan well-water is as sweet and good as anything from the Alps).

While Mohammed saw these rituals as a weakness, I found his water habits – like Bessita's – a mystery. He would never accept my canteen, and when we stopped every couple of hours he would take chaste sips from his own supply, a water-sausage made from half an inner tube that was tied at each end with a few turns of cord and slung on the side of the jeep. One morning, tired, he fumbled when he was tying it up and it swung out of his hands, the puckered mouth disgorging pints into the sand before he managed to pinch it tight. He merely laughed.

Mohammed spoke little, and, after the manic babble in the back of Bessita's jeep, I was content with this. When he did open his mouth, it would be to say something interesting. 'I went three days and nights in the desert once, without food or water,' he announced one day, during the midday heat.

'When was that?'

'A few years ago. Near Libya. It was when I deserted from the army.'

It was news to me that he had ever been in an army. 'Why did you desert?'

'Because I didn't want to be in the army,' he said, simply.

I waited to see if he would say any more.

'They tried to conscript everyone. I wasn't going to be in their army, to get shot by Libyans. I live in Libya, but I wasn't going to fight for Gaddafi, either. A lot of Arabs and *toubous* fought on the other side, for Gaddafi, you know. They came from other countries too. Gaddafi promised them a separate state in the central Sahara if they helped him defeat Chad.' He paused again. '*I* wasn't going to be caught in their war.'

I was not sure whose war he meant – the Libyans'? The Chadians'? or everybody's? Mohammed was a typical Saharan, moving among three countries without a passport. Born in the Sudan, working in

Chad, raising his family in Libya, to whom did he owe his allegiance?

'So I walked away. I walked a long, long way. A black would have died in my place. Three days, three nights. Then I was safe.' He allowed himself an almost imperceptible grin. 'The desert isn't the enemy to us that it is to others.' After a minute or two of silence he added firmly, 'Gaddafi is a criminal.'

The white sand began to metamorphose once more into creamy-grey savannah, with coarse, low-growing grasses. '*Ici 'y a des gazelles*,' said Mohammed. '*Si on voit, on tue*,' – if we see 'em, we kill 'em.

Not on my watch, I thought.

We saw no gazelle. Slowly the desert reacquired landscape – I had crossed the southern sand wastes to the central Sahara. This was my first glimpse of the *hammada*, or rocky highlands, the desert's mountainous spine. Hills appeared, dark humps that swam around like a school of sleepy whales as we approached them. The sight of those black hills stirred the heart, for here was an end to monotony. Indistinct threads thickened into strings of palms, thorny shrubs appeared. Merely sensing that I was closer to water made something in me let go. A mountain peak soared hazy and grey in the north.

As the road scaled a rocky bluff we met a barbed-wire barrier. From a stone shelter the size of a sheep shed a handful of black soldiers appeared. One carried a Kalashnikov AK47 in his right hand and a fan of playing cards in his left. All wore the chilly expressions of working soldiers – but they could not keep it up. They were too pleased to have some company.

As I peed behind some rocks, I saw Mohammed talking to one of the soldiers, a tall fellow in fatigue trousers with a long, supple, naked torso. He was holding his arm out, long fingers gesturing into the distance.

'What was he telling you?' I asked as we drove away.

'He said from here on there are a lot of minefields, so don't leave the tracks. Also, *coupeurs de route*.' It took me a moment to work out the meaning of this: road-cutters – highwaymen. 'If we see anyone, we should keep moving.'

Mohammed took the advice seriously. A couple of hours later an old man emerged from a palm-frond hut set back from the track, and ran across the sand towards us flapping his arms.

'Come on, Mohammed,' I said, 'slow down.'

Reluctantly he let the jeep decelerate, and the panting old man caught us up. He was good-looking, with a fleshy face, strong features and a bushy white moustache – had he been dressed in a baggy khaki djellaba and pants, he might have been a Pakistani policeman. However urgent his request was, it was set aside while the formal round of greetings took place – the shaking of hands, the *Asalaam-o-aleikums*, the *how are you, how are your family, peace be on all of them*s, all rattled out in emotionless staccato.

'My baby grandson,' he told us as soon as decency allowed, 'is ill, very ill.' He fixed his strong hands with their thick fingers and broken nails on the window frame. 'My son is away, so my daughter-in-law set off last night by camel with my wife and youngest son.' It was 100 miles to the nearest medical facilities, at the next oasis, Zouar. 'If you see them on the way, could you pick them up?'

'Yes,' said Mohammed, taking his foot off the clutch.

'Peace be upon you,' the old man called out, letting go of the window frame and running alongside us.

An hour later we reached the home of our guides. Women and boys were labouring with ropes and canvas buckets to extract water from a well. I suggested we stop and replenish our supplies. Guarded-eyed women stood aside from their tasks, and, when they saw my camera, covered their faces. I looked at the water instead, so alive, so profuse.

The grey foothills of the Tibesti Massif undulated like a rucked carpet. Conical black hills were decomposing into custardy sand. The path steepened, and the jeep clawed its way up rust-red passes in first gear. This was T. S. Eliot's landscape, beating sun, dry stone, no sound of water, the shadow of this red rock.

First we saw the camel, then in the shade of a shawl spread on a thornbush the little family group – two women, a bundle at the younger woman's breast, and a boy of fifteen. The tiny child's eyes, tongue and nails were acid yellow, presumably with hepatitis. I put the women and baby in the cabin and climbed up behind. The boy stayed with the camel.

We mounted a black hogsback where the track was a bright red wound, dropping towards the gateway to Tibesti: a huge plain of golden sand ringed by grey cliffs. Directly ahead was a break in the cliffs where a single monolith rose like a sentinel, the entrance to Zouar.

CHAPTER ELEVEN

Gentian Violet

Massive stone boulders like sphinxes were arranged across the plain. Amidst this splendour was a scattering of palm-frond huts and whitewashed mud-block boxes. One of them was Zouar's town dispensary, deserted, its windows and doors sealed by metal grilles. Eventually we reached *le quartier*, where Zouar's small black population lives. 'I don't like it here,' said Mohammed. 'These blacks are all hooch drinkers.'

Our young nomad mother was terrified. She had never met a white man before, her baby was ill, possibly dying, and she was being conducted into the heart of the town's black quarter.

We pulled up at a mud-brick house and I banged on the corrugated iron door. Sleepy voices stirred, then a half-naked man appeared with a sheen of sweat on his face.

'*Docteur?*' I said.

His eyes slowly focussed. The door slammed shut.

'*Un moment,*' he said.

He reappeared, face splashed, wearing the uniform of an army captain and carrying a stethoscope. 'I'm sorry, I was sleeping.'

'I'm sorry—'

'Think nothing of it . . . Oh, yes!' Seeing the baby, he showed the enthusiasm doctors always manifest for advanced illnesses. 'Hepatitis has really got a grip on this little chap, hasn't it? Mmm . . .' He placed the stethoscope's chromium disc on the baby's tiny chest. Severe lung infection, too.'

'Will he live?'

'Oh yes – if we treat him. The only problem is drugs. I haven't got any. I never have enough drugs, but at the moment haven't even got a strip of aspirin.'

I was surprised. I had been told that thanks to various donor organizations, there was no drug shortage in Chad. And that, given that the northern tribes had a reputation for 'independence of spirit', the government made an effort to keep them sweet.

The *capitaine* shrugged. 'Once every four months a plane lands with my supplies. The next morning there's a queue of 300 people begging for drugs at the dispensary door – half of them with made-up symptoms. If I give them antibiotics, they take them for headaches. They'd be quite happy drinking saline solution and overdosing on morphine. Anyway, the plane's a month late. The cupboard is bare.'

And so we spent hours touring Zouar's unofficial pharmacies. The marketplace started unexpectedly from the bare sands, a ramshackle collection of poles like a scaffold for spotlights and scenery, actually designed to support woven-grass mats that deflected the sunlight.

A grimy stall sold bread, a couple of others tinned mackerel, soap, pencils. Across the way, women sold comestibles – flour, salt, dried dates, dried chillies. The women were half-asleep, basting in swathes of nylon printed with black and tertiary colours – twists and swirls of turquoise, acid yellow, aubergine, a depressing spectrum. I wanted some green vegetables, but could not make them understand.

'You won't find vegetables here,' said the *capitaine*. 'No one in Zouar grows vegetables.'

'Why not?'

'They're nomads. It's not part of the tradition. They despise the sedentary, they'd never dream of growing anything. It doesn't matter how many times I tell them they need vitamins. What the hell are vitamins? They only want meat, but they haven't got enough animals to provide them with the nutrition they need.'

'They prefer malnutrition?'

'*En effet!*'

'Let's buy some meat,' said Mohammed. 'Where's the butcher?'

'There isn't one,' the *capitaine* told him. 'You'll have to find someone to sell you a goat or a chicken.'

There was, however, a medicine stall, with its owner sleeping on the floor. He stirred, and opened a cabinet with a tiny key. It contained almost nothing.

We drove out to a prosperous home, a spacious hut fifteen feet long with a bright nylon mat on the floor. A man brought out a vinyl gladstone bag filled with medicines. Many were Libyan. He had all the drugs the baby needed – except one. I pulled out some cash.

'Where will we find the other drug?' I asked the *capitaine*.

'The only chance is the French garrison in Bardai. Are you going there?'

'Yes – but I don't know if I'm coming back this way.'

'Then the baby may not get better.' He shrugged. 'You decide. In case you do return, I'll give you a list of the drugs I need. Maybe the French can help. Now come this way, I want to show you something.'

He led me towards another hut. Inside, a young woman was lying on a mat. Recognizing the *capitaine*'s voice, she stared up at us with bright blue eyes.

'Blind,' he said. 'She wanted different-coloured eyes, so she decided to put gentian violet in them. It's an antiseptic, a beautiful violet colour – anyway, she thought it would turn her eyes from brown to blue. That's the depth of the ignorance you find here. I tell you, this is the most depressing thing I've ever come across. Look at her, she's beautiful, and at the beginning of her life.'

An old man was approaching the hut.

'That's her husband.'

'But he's *old*!'

'Yes, but rich. He can afford all the wives he wants. So he'd got himself this lovely creature – she's about fifteen, I suppose, sixteen. And now this. He's not very pleased. Ah well.' He squeezed her hand with a doctor's professional bedside way. 'Come on,' he sighed, 'let's see to this baby.'

A few minutes later the baby had been washed, injected and provided with rehydrating salts and vitamins. Its mother and grandmother, installed with relatives, were patiently instructed in its care.

'I'll be back tomorrow,' said the *capitaine*.

We returned along the valley, sand and rock, jagged wind-cast anvils. I asked if there were any rock paintings nearby. He looked at his watch, then at the sun. 'I know someone who knows,' he said.

Back in town we picked up a friend of the *capitaine*'s from near the marketplace, and drove east as the sun set behind us in a blaze of brilliant red, turning the pink sphinxes of the valley to the colour of blood. The young *toubou* tribesman led us up one of the gently curved rockfaces to a ledge, where he pointed at a frieze representing a hunt, with running human and animal figures.

Between five and seven thousand years ago this region was covered with grass and trees. There were giraffe and elephants in abundance, and lakes well stocked with fish. Some scientists believe that one of the earth's periodic changes in orientation – a lessening in its tilt from 24.14 degrees to the present 23.45 degrees – abruptly caused the Saharan region to grow arid. Others hold that the Saharan tribes cut down too many trees, and that their herds were too large – man desertified the savannahs.

It was the first time I had seen prehistoric rock art. The figures were not exceptionally pretty or graceful, but I am not sure that such terms are rightly applied to them. Several thousand years ago, humans used these jutting ledges to provide them with shade from the elements. Someone mixed pigments from blood and vegetable paste, and painted on the rock walls images of leaping humans and creatures. Were they decorating their homes? Or, given that their whole existence was based on hunting, did their paintings have a deeper significance? Perhaps these creatures leapt in abundance through the low grasses of that time, or perhaps they were already starting to disappear and the paintings were a religious invocation, to try and bring them back.

'Does anyone here paint these days?' I asked the *capitaine*'s friend. I was remembering Faya, with its total lack of decoration.

He shook his head vigorously. 'No, no, the practice has completely died out. But they knew a thing or two, eh? Our ancestors.'

CHAPTER TWELVE

The Crater of Salt

To continue north, we needed a guide to lead us through the minefields outside Zouar. That night we found ourselves on the edge of the plain bargaining with an old man who was demanding the equivalent of £100 sterling for the job. In local terms, it represented almost £1,500. I told him it seemed a lot of money for sitting in the car and doing nothing.

'Doing nothing!' He pretended to be offended. 'I might be saving your life – I'm risking my life to save yours.'

I pulled a sceptical face. 'I've heard the track's well marked.'

'No, no, it's *complicated*.'

'I think thirty's enough.'

He threw up his arms. 'OK, go ahead and kill yourselves!'

He had us over a barrel. In the end, we agreed on fifty.

'And I want the money *now*.'

I counted the notes into his wrinkled, clay-pink palm, and caught Mohammed's eye. I could see he thought the old villain had lost all dignity.

As the sun rose the next morning, birds began to flutter by, their grey-blue bodies glinting like opals. The valley steadily narrowed to a gully with sides 250 feet high, then ended abruptly, disgorging us onto a broad plain. There were various tracks before us, and though our guide waved in the direction we should go, it was obvious enough. As though reading my mind, he croaked, 'There are *plenty* of mines out there, uncleared. People lose animals all the time – goats, camels. They can't afford feed so they have to let them wander to find grass. Nobody knows where the mines are, so every now and then, *bang!*'

'Hmm,' I said.

Next he tried to start a conversation with Mohammed, who

showed his contempt with one-word replies. Abashed at last, the old man subsided into a sulk.

The path became confusing, and I had to concede (to myself) that without a guide it would have been hard to find our way. The climb through the next pass began, inching in first gear over jagged boulders the colour of bruises. It was terrible for the car, and as the tyres slipped off rocks and the jeep yawed, Mohammed worried about breaking his crude spring-leaf suspension. 'Very bad road,' he muttered like a mantra, '*very* bad road.' I ignored him: he knew the deal, and I was not turning back. Apart from anything, the *commandant* in Faya had warned the French garrison·in Bardai that I was coming. If I didn't turn up there or in Faya for days, they might send out a search party. I sat quietly reading. I had finished the Graham Greene and had started *Heart of Darkness*. At a hairpin bend we passed a thorn bush on which someone had brutally placed the head of a goat, so that it scrutinized everyone who passed.

The only human we glimpsed the whole day long was a scrap of a girl in rags, leading an albino camel. We passed along dry wadis of pocked orange stone like ancient henges and tomb-approaches. Towards eleven the rock settled into a white plain of perfect smoothness, but with grass-tufted fractures, so that we might have been travelling the Appian Way.

At midday we reached the *Trou au Natron*, or Natron Hole, the remains of a gigantic volcanic crater. As I trotted down the cinder slope towards the rim, all I could see was a mile-off underbite of jagged teeth. Abruptly the chasm yawned before me, 2,500 feet deep, 18 miles in circumference, and I saw that those far-off incisors were the far side of the crater. The floor was snow-white with salts – natron – but for a central laval night-black cone aligned with a dark-brown ellipse. In India, where numerous peaks, lakes and streams have been incorporated into Hindu mythology as lingams and yonis, this would be a pilgrimage site – Parvati's Womb, perhaps, a tantric yoni in a pool of milk or sperm . . .

Mohammed came alongside me and his face, usually so reluctant to give anything away, glowed like a little boy's. For a while we contemplated the great hollow. Then Mohammed asked in a puzzled voice, 'Martin, why did God make it?'

I thought of the scientific answers I had been taught – volcanoes, laval flows, supplementary cones; they all seemed hopelessly inadequate. I said, 'Who knows why God does things?'

He nodded, and turned back to the jeep.

In the east a mountain rose perfectly conical, an implacable Fuji sporting a natty halo of cotton wool. In the north-east a range of razor-backed mountains shaded from grey to purple to navy blue. Sheer and tall, two-dimensional, they seemed to be surveying all.

Is this, I wondered, what I came all this way for? Had I travelled to the desert for the space and silence, or was the journey itself the thing? I was discovering in the central Sahara a world far more remote than I had ever imagined. There were no jets in the sky, there were none of the conveniences of modernity. People lived self-sufficient lives. For all the obstacles of the journey, I felt more earthed, calmer and more focussed.

I began hiking around the rim of the salt crater. A fluffy cumulus cloud was immobile overhead, its shadow staining the pure white crater depths. At this altitude the air was thin and cool. Moisture nurtured a few grasses and some musty yellow and flame-red flowering shrubs that quivered in the breeze. I dislodged rocks which tumbled into the void, disturbing a rabbit. Two eagles with enormous wing-spans soared out from a ledge beneath my feet and flew across the crater, disappearing from sight long before they reached the other side. Eagles feeding on rabbits feeding on grass. I saw, or thought I saw, a gazelle. Behind me somewhere a camel groaned.

When I returned to the jeep Mohammed woke up, made tea, clucked about his suspension, and we finally drove on, descending now through naked hills with raw minerals bursting on their surfaces like sores on the flanks of a street dog. We began to pass a huge amount of rock art. Cattle and giraffe and gazelle were daubed or scratched on the vertical pink planes.

It was almost dark when we reached the military post outside Bardai. A soldier asked deferentially if he might have a lift into town, 'to try to buy some sugar'. 'We haven't been paid for four months,' he said, 'so we have to buy everything from the Arabs in the market on credit. Sometimes they sell to us, sometimes they don't.' I asked why he wanted sugar so much. 'We use it to make date wine,' he said. 'It's the only thing that makes life here bearable.'

'If you haven't been paid, why don't you leave?'

He laughed uproariously. '*Leave?* Don't you understand, those of us from the south are hostages here. If things with the government go a way the Arabs don't like, the Arabs can kill us. Some of us have been serving here unrelieved for eight years; we can't even remember what it's like to be down south, with water and fruit. And music, and women.

'If they find us on a truck, deserting, we'll be killed. If we tried to walk for it across the desert, we'd die. We're trapped here.'

At the iron gates of the French army compound, a tall Arab bade me wait while he summoned the officer in command. The captain who came out to meet me was about my own age and height, but rather chunkier, shirt-sleeves rolled up over knotty biceps. He greeted me courteously, but was confused by my conviction that the *commandant* in Faya had told him to expect my arrival. So much for the cosy feeling I had cultivated that the French army was watching over me.

'Anyway, I hope you'll be our guest for dinner? You'll want a shower, of course. And it would be our pleasure to put you up for the night. But—' He looked embarrassed. 'We don't have an air-conditioned room free, I'm afraid. Would one with a fan do?'

I assured him that it would.

The small fort was a single-storey building with a veranda running around it. Its mess, garishly muralled with regimental mascots, opened onto a view across palmy, cliff-locked Bardai. There was a sandy courtyard with a large and serious collection of bar-bells. The latrines were simple but exquisitely clean. Mohammed had opted to stay with his vehicle, setting to work on his bruised suspension. I enjoyed my first shower since N'Djamena. I had long ago stopped worrying about the dirt in my pores, stiff hair, the grainy sand on my scalp, but showering was a sensual delight.

My hair was still wet when I came into the mess, with its table laid for dinner. The men of the garrison were politely waiting for me to arrive, a kaleidoscope of races and types, a cross-section of the peoples once embraced in *la France outre-mer*: the WASP captain; a huge and hugely muscled adjutant of possibly Polynesian origin; a tall, shovel-cheeked North-African type; a stocky, crew-cut Celt and a slender, thoughtful-looking Asian, whose grandparents might have been born in Vietnam or Cambodia. They must have made quite a spectacle for the local coal-skinned, aquiline-featured tribespeople when they blasted through town in their open-top jeeps.

'Here's something you might just be able to remember,' said the captain, handing me a beer. I drained the glass with relish. Several of the men, I noticed, did not touch alcohol, and a little later I was the only person to accept a second drink.

The French were not allowed outside a twelve-mile radius of Bardai, which meant that they had not seen *Trou au Natron*, or the extraordinary display of rock art I had just passed, and they were keen to hear about them. I asked what exactly they were doing in Bardai, and was told, a little stiffly, that they were providing weapons-training to the Chadian military.

'Mind you, everyone here has been carrying a gun since he was old enough to walk,' said the stocky man with the red, crew-cut head.

'Do they use them?' I asked.

The Algerian-looking man sneered. 'Of course they do. Whenever people have guns, they get used.'

The crew-cut grinned. 'They do have a tendency to take a pot-shot, the old *toubous* – especially when there's a woman involved. It seems that a couple of years back they had a go at some Germans . . .'

'Ahem.' The captain intervened diplomatically. 'Frankly, they are very traditional. It's necessary to understand that this is a pretty wild place. It's always been cut off, but then there was the war, which a lot of the men fought in, so that some people have been slightly, erm, well I suppose one could say . . . brutalized. It also means that without too much government interference they have been left to their own devices so far as, er, customs are concerned.'

'What sort of customs?'

'Well, indeed, attitudes to women. Attitudes to the resolution of conflict. Feuding. That sort of thing.'

The conversation turned to political generalities. The captain and several of his men were remarkably well informed on world politics – rather a waste if their mission really was limited to a little light weapons-training.

'So you don't have any radar or listening equipment here, for keeping an eye on the Libyans?' I asked, impertinently.

The captain looked me straight in the eye. 'Absolutely not. Not part of our role, *Monsieur*.'

I had been unsure about whether to loop north and east back to Faya, exploring Tibesti more thoroughly; or to return to Zouar and look for a way of heading west into Niger. I decided that if the French could give me the medicine the child needed, I would return to Zouar.

Next morning, the beefy Polynesian presented me not only with what I needed for the child, but with a huge box of drugs, virtually everything on the list I had been given. It was extremely generous, and it brought a lump to my throat.

On the journey back we were all relaxed, calm and quiet – but we were always quiet. Night was falling as we crept out of the hills and onto the white, mine-strewn plain – Shelley's lone and level sands. We stopped for *salat al-'isha*, the kneeling men scarcely visible in the dusky desert, as though cloaked in a sea mist.

We reached the great wall of cliffs that ringed Zouar. The cliff-locked track seemed to go on for ever, apparently going from nowhere to nowhere. If the tracks ran out we would stay suspended here between the eternal planes of the canyon walls, in purgatorial, moonlit numbness. The moon cast toneless shadows into the vertical and horizontal fissures, and metamorphosing forms leaped off the rock – Easter Island statues, the severe geometry of Lincoln at Mount Rushmore, a gargoyle, a fist. As I watched the mostly grotesque shapes shiver in the rock, I reproached myself for conjuring the banal imagery of the popular imagination. Another man might see a Leonardo Madonna, or the face of Christ.

CHAPTER THIRTEEN

The Sky At Night

I SPENT A DAY IN ZOUAR trying to find a driver to take me into Niger.

Northern Niger was unsettled, and I knew that a Canadian had been kidnapped there by separatist guerillas earlier in the year; but with a good driver I thought it should be possible to slip across the erg, or sand-sea, in north-eastern Niger, and make our way down to the oasis of Bilma. There I would find a truck to take me on to Agadez, the halfway point between Cairo and the Atlantic. I would avoid the huge loop back to N'Djamena then north again to Agadez – and I would avoid several more ghastly days in the back of the Iron Maiden.

The people of this region once ran merciless protection rackets on the caravans that threaded across the Sahara – it was pay up or die. Today, any outsider stupid enough to find himself in need is likely to be mercilessly exploited. There was only one man in Zouar who could take me to Niger, and the price he demanded for the two-day journey, £800, represented in local terms a fortune – it was what I would have paid Mohammed for several weeks. I was not just being mean; I had budgeted according to local costs, and I still had a vast journey ahead of me.

I wasted a morning trying to negotiate: 'Take it or leave it,' he said.

As we drove away, I asked Mohammed why the man had been so intractable. 'Even if I'd paid half of what he was asking, it would have been a tidy sum.'

'For a start, a big chunk of that goes to the local *chef*,' said Mohammed.

'Of course; I didn't think of that.'

'And he thinks you're a rich man. A rich man should pay what he can afford.'

The conversation was beginning to have uncomfortable implications for my financial relationship with Mohammed; I dropped the subject.

Before leaving Zouar, we assured ourselves that the baby was on the mend. I handed the box of drugs over to the *capitaine*, and he grinned happily. 'You realize I've got here enough of the principal drugs I use to treat most of my serious cases for several months? It's a very generous gift.'

'Well, thank the French.'

As we prepared the vehicle to leave Zouar, our cantankerous old Bardai guide wandered out of the afternoon heat.

'Can you give me something for my stomach?' he asked.

'Like what?'

'Antibiotics.'

'Oh, for Christ's sake – if you're ill, go and see the *capitaine*.'

'You didn't give him all those drugs, did you?'

'It's none of your business, but yes, I did.'

'You shouldn't have. He'll go straight to the market and sell the lot.'

'Shut up. Push off.'

The old man obediently turned around and walked away.

I turned to Mohammed. 'He's absolutely *shameless*.'

Mohammed grinned. 'It pays off, you know. I bet he's got a fortune tucked away. He's a greedy old shit.'

We left Zouar with a new passenger seated between us: Mohammed's 'cousin' Fatima, a member of his enormous extended family who happened to be making her way towards Faya. Mohammed was in a good mood and spent most of the day chatting to her. I felt relaxed, too. *Inshaallah*, we would be back in Faya in time to meet up with Bessita.

On the passes south of Zouar I noticed many small roadside cairns, sometimes with an empty shell-casing for a spire. They must be way-markers, I thought.

'*Beaucoup de morts*,' said Mohammed. So the scores of cairns were marked graves. We passed one whose uppermost stone was the size of a man's head. Pulled tight over it, withstanding wind and sand, was a faded green military fatigue cap.

By full moon, the desert was as flat and white as a frozen lake. I asked Mohammed to stop and turn off the headlights, and I clambered up on the roof. He drove on across the luminous sand, with me clinging to the roll-bar and feeling the soothing slipstream on my skin. I seemed to be motionless at the midpoint of a moonlit disk running from horizon to horizon – the visible curved surface of the earth, a fragment of mind adrift in nothingness. Not nothingness, I thought, but space.

After an hour Mohammed stopped for the night, pulling off the track – a maze of ruts – and ploughing for twenty yards into virgin sand. It was warm underfoot, with a delicate crust like a beach when the sea has withdrawn. There were wafers of sandstone, but no scorpions, no vegetation. In the heart of the Sahara, no life save ourselves.

We tossed down our bedrolls, and Fatima laid her blankets on the opposite side of the jeep. Mohammed took an armful of the dried wood we were carrying and made two piles, one for Fatima to cook with and one to brew up himself.

Sipping sugary mint tea, we smoked and stared at the orange-pink embers and talked about nothing much – the trip over the rocky pass out of the oasis of Zouar, tomorrow's journey over sand to the oasis of Faya. Fatima brought the food, served herself directly from the cooking pot, then curled into her blanket and slept. And after eating, so did we.

In the early hours my eyes sprang open. I was weightless. Around me a lattice of constellations hung in the velvet darkness like ghostly vines.

Before dawn the sand grew cold. As we rolled our blankets, the distant sky greyed and drew closer. Fatima made a breakfast of bread, jam and tea, and I took her photograph. She chided me angrily, and I looked at Mohammed. 'You have to ask her husband first,' he said, with mild reproach.

The jam was sweet, the sky cool and subtle. On desert mornings, you so easily forgot the heat.

We set off once more across the uncanny ballroom flatness. Somewhere else there were trees and people, but here there was only this slab of desert, no blade of grass, not even an ant. The sky grew rosy in the east, then a molten sun tore itself dripping from the horizon.

Fatima dozed. Mohammed was humming.

I smiled at him. 'And why are you in such a good mood?'

He smiled back, showing his teeth. 'Because we're going back,' he said. 'Soon I will be with my family in Libya. I didn't like that trip to Bardai at all, you know.'

'I know you didn't.'

'I thought it was going to break the jeep. I don't like it here. Mines and highwaymen, and I don't know anyone. No, I don't like it at all.' He brightened again. 'But I'm pleased I came. To see a new place, to know this route to Libya. It's interesting.'

With a nod towards Fatima, he said, 'A man needs a young wife – it keeps you young! Next year I'm going to get a second wife. I've been saving up. Next year I'm going to get a fifteen-year-old.'

I looked at Fatima's young face. She cannot have been more than twenty-two. Already her skin had been lined and coarsened by exposure to the wind and sun. Her lips were black.

'Why do the women make their lips black?' I asked.

'If a woman's lips are red, it's good to make them black,' said Mohammed. 'Her hands too. *Il faut dessiner.*'

'What is the black stuff?'

He shrugged. 'I dunno. They buy it in the market.'

'And tell me, why do I have to ask her husband before I take her photograph?'

Mohammed's expression showed that now he thought I was being deliberately perverse. 'Because her husband isn't here to ask.'

'In my country you don't have to ask the husband.'

'Hmm.' He was unconvinced.

'And you know, with us, you can't take a second wife.'

'*What?*'

'No, my friend, no second wives.'

He unleashed a bellow of incredulous laughter that woke up Fatima. Luckily, she spoke no French. 'But what do you do when she gets old?'

'Well . . . you're meant to stay together.'

He let out another belch of laughter. 'When she's *old*?'

At Bessita's compound, Romain was the first out to greet us. 'Monsieur Mar*tin*!' he cried, throwing his arms about me. 'I was worried you might be in trouble!' I was pleased to see him, too.

I took Mohammed aside and paid him what we had agreed. He accepted the money with a nod. I gave him a tip, and he inclined his head again, and smiled slightly. Then he held out his hand to shake mine.

'You're a true man,' he said.

'Pardon?'

'A lot of whites don't know how to behave. They're always in a hurry, and they think everyone is trying to rob them. But you know who is trying to rob you and who is not; you treat the right people with respect. That is how a man behaves.'

I smiled. 'Thank you, Mohammed,' I said. 'I met a lot of crooks in this town before I met you. I couldn't have asked for a better companion.'

'Perhaps we will meet again one day.'

'*Inshaallah*,' we said together.

CHAPTER FOURTEEN

A Photograph

He who does not heed advice must expect ill fortune.
Arab proverb

TWO DAYS REMAINED IN FAYA. At dawn I went out into the fly-blown streets. '*Nasrani, Nasrani,*' chanted the urchins, '*perdu, perdu?*' – Foreigner, are you lost? An adolescent girl leaning in a doorway asked me for a coin, holding out a palm, pushing herself against the doorframe with a secret adolescent grin. The sun came up like a demon.

I complained in my diary, 'It is *bloody* hot, and it was *bloody* hot coming back from Zouar. I feel weak, I have a headache, I have a heat rash all over my body. Now I have to face the back of that goddam jeep again.' We must not repine.

A profound ennui seemed to have descended on the house. Everyone got on my nerves. Sex was all around; you could smell it. Thomas and Sylvestre were stupefied. In the heat, they picked themselves up and wandered around at a loss. What else was there to do? In the absence of any amusement you copulate, eat, sleep, and wake up to be pricked by sexual desire. The heat paralyzed us all, but their absorption in their bodies made me extraordinarily restless. Thomas dragged his muscular, fleshy body about the compound like some randy animal, and I began to feel morbid and nauseated at the sight of his flesh. He never stopped bragging, jeering, squabbling, whining. He and Sylvestre had made friends with a tall Arab wastrel with large hands and dirty clothes who hung around the compound. His eyes were dark and intense, the whites brown, with red and yellow depths. Sylvestre said to him,

'Shall we go *chez France*' – their brothel – 'or is it too late?' Sleepily, they would speculate whether they had the energy for another sexual fix.

Meanwhile, Romain sat to one side, sexually excluded, watchful. Sometimes he too would disappear for hours at a time. So even did Estelle. It was a place of secrecy, Faya, of murmurs and imagined or real conspiracies. It was said that when the men of the oasis were away – as they often were – their wives committed riotous adultery. By night, figures stole up alleyways, slipped over walls. It was also claimed that when the husbands found out, throats were slit.

I had some small objects stolen from my bags – a mouth organ, a chain, a Swiss Army knife. The thief was clearly one of us, someone with access to the inside of the house, but Bessita showed no inclination to investigate. He moved around like a slow, inscrutable bear; and then that tizzing radio that he never turned off, his remedy for silence.

Only the Woman – no one but me called Estelle by her name – was exempt from the general lassitude. She never stopped working, her burnished mahogany skin covered in a permanent gauze of sweat. In a loose black T-shirt and a red batik skirt tied tight around her waist she flip-flopped through the afternoon, lugging pails, her prominent bustle of a backside swinging as she chopped meat, crouched to stir a pan, or bent over straight-backed to pound washing, so that we could see her swaying breasts.

Sylvestre leered at me. 'The Chadian government tells people they should try to have four wives, if you have the means. That's the way it is here – because of the war. Too many women.'

In the evening I took Romain to a bar to buy him a beer. He tried to chat up the waitress and was doing quite well until her husband arrived.

I bought him too much beer, and he gave me Part II of his life story: sucked into the war at fifteen; drugs, whores, killing. He had ended up the man the officers would tell to take prisoners behind a rock to shoot them or slit their throats – an executioner. 'When you're fifteen, Martin, you know nothing. They can turn you into anything. I was so deeply sunk into killing, I no longer knew what was real and what was a nightmare. I still dream about those days. I just dream about blood. Just – blood.' An uncle had found him. 'He cried that day. "What have they turned you into," that's what he said.

He paid to get me out of the army. He was a born-again Christian. He introduced me to the Lord. So now I'm born again, I witness the Lord Jesus Christ!'

He had spent more than a decade in the Valley of Death, and emerged with a sort of innocence.

The sun was rising when we reached the Erg du Djourab, the dunescape south of Faya. The jeep was bucking and I felt sick. At nine we stopped for breakfast – hunks of plain bread, but good. The sun was white, the sky a faint blue gridded with clouds like crocodile hide. A beautiful morning.

The locals had no notion of the desert's beauty. I had shown Mohammed a Western postcard of sand dunes. 'Beautiful, isn't it?' I said. He frowned at it, seeing only sand.

Two army trucks closed from the rear with unsettling urgency, overtook us too fast across the unstable surface, their drivers exhilarated, the turbanned soldiers in the back gripping their rifles like crucifixes at Mass. Eros and Thanatos. I thought of Romain's words: 'It's good to kill. You learn a lot.' My pulse was racing, my face running with sweat. I was becoming feverish.

A wind began to blow, and soon visibility dropped to thirty feet. I watched the indentations – camel tracks, tyre tracks – fill with sand. In moments, all traces of our passing vanished.

In mid-afternoon we left the Erg behind us, looped around the minefields of Koro-Toro and arrived in the jumble of block-houses that had so repelled me on our way north. We dropped our guide, pulled up at a well and wearily piled out into a wind that sprayed sand in our teeth. My temperature was high, but I had swallowed some tablets. I sat in the shade of the jeep and savoured the immobility and the silence.

Some nomads, inclining their heads against the stinging grains, were herding a few dozen camels to water them. Another camel was harnessed to a pulley, drawing up a bulbous black goatskin that leaked mercurial arcs. Driving the camel was a young girl of twelve or thirteen, with quick, bright eyes and a brilliant smile. A big truck stood nearby, one of the trans-Saharan monsters that bristled with every conceivable water container, and some that were inconceivable. Its passengers were coming to from their siesta.

As the herders brought two large, troublesome camels to the well, I watched a small boy of seven or eight use a twig switch to keep two dozen thirsty animals, the rest of the herd, at bay, his big eyes flicking nervously between the camels and his elders. I took his photograph, and one of the girl at the well.

Now, the *gros porteur*. I had not so far been able to snap one of these trucks. In Faya, I had twice been prevented by angry drivers and passengers. Stealthily I lifted my camera and at once a number of white-clad men sitting in the lee of the truck began to yell and wave their arms. I felt ill and impatient. I flapped a hand dismissively in their direction, composed my picture and released the shutter.

Minutes later I was standing at the rear of the jeep changing the film when I heard raised voices behind me. I carried on with a show of unconcern, then felt an odd tingling in my lower back. I sensed that I should turn round at once.

Two men, Arabs in turbans and full-length robes, were yelling in Romain's face, flashing wild looks at me. Romain stood with an arm outstretched in my direction as a barrier. One of the men had his hand resting on something under his robes, which I knew was the hilt of a dagger.

It was Romain's face that alerted me to how bad things were. His mouth had hardened into a single, narrow line. His eyes had in them an expression I had not previously seen, an utter, ruthless coldness. He looked extremely dangerous.

'What is it?' I asked, trying to keep my voice steady.

Without taking his eyes off the men, Romain told me, 'They say you shouldn't have taken their picture. They say they are going to kill you. I said first they'll have to kill me, and it's sure that I'll kill at least one of them.'

His words landed on me like a punch in the chest. The two raved on, gesticulating wildly, eyes bulging, spots of white spit on their lips. Their violence contrasted with Romain's sprung, almost feline stillness. Suddenly Bessita appeared around the corner of the jeep, a six-foot-something giant in the uniform of an army lieutenant, asking in his calm, grave voice what was going on. The hail of words continued, with Bessita reasoning. The exchange lasted perhaps five minutes. At last, still flinging murderous words and looks at me, the men withdrew.

88

'You should be more careful,' said Bessita tersely. 'They would have killed you.' He glanced towards the retreating men. 'Let's go.' He called out to Thomas and Sylvestre, who had been watching from a distance. 'You two, we're leaving NOW!' The tone of his voice made them both start running.

Romain was grinning, adrenalin and relief rushing through him. He hugged me. Bessita said, 'They would have killed you and thrown your body in a well.'

'That's true,' said Estelle, who had been sitting in the front seat throughout the exchange. She too was grinning ecstatically.

'Who were they?' I faltered, unable to take in what had happened.

'Ex-army,' said Bessita. 'They were demobbed yesterday in Moussoro. These are dangerous men, they have been at war for years.'

'What did you say to them, *mon Père*?' My voice was shaking.

'I told them if they killed *un blanc* the army would track them down and kill them. They didn't know I was a priest, I just look like an officer to them. They would assume I was armed. I think that's what swung it.' He started the engine.

'What did they say to you?' I asked Romain, as we pulled away.

He grinned. 'They said the white men killed their grandfathers, and now the white men should die. They say your life is worth less than camel shit, your blood wasn't good enough to stain the earth of Chad.' He laughed. 'You're lucky one of their women wasn't present – then they would have had to go through with their threat. Not to do so would have meant too much loss of face. Even if they'd seen Estelle it would have made things worse.'

'He's right,' said Bessita over his shoulder. 'These men are like that. You should thank Romain. He saved your life.'

After twenty minutes we pulled up at a hut that sold tea. As everyone climbed out I remained in the vehicle, in shock. A boy came to the back door and begged for alms. I screamed at him furiously, and his eyes widened in fear. Romain came over to me and spoke to me quietly. 'Don't yell at him, Monsieur Martin,' he said, 'he only wants a few coins. Come out here with us and sit in the shade.'

PART III

To the Atlantic

CHAPTER FIFTEEN

Days and Nights in N'Djamena

IT WAS TIME to move on. I had already spent too long in Chad, and I was determined to go to Niger overland. It turned out that Yves Delacroix was planning a government inspection north of N'Djamena, close to Lake Chad and the border with Niger. If I waited a few more days, I could travel with him.

I rang Penny and found myself unable to communicate to her the things I had seen, or to express how I felt about them. London seemed very distant. Soon, the first of my letters – intense, slightly deranged documents – would arrive there. Heaven knew what Penny would make of them.

The afternoon of my return from Tibesti I had taken to my bed at the kindly Delacroix's, and been self-indulgently ill for several days. Puking, I became intimate with the bathroom in the peaceful early hours, the lineaments of the toilet's chaste white porcelein, the hair in the bath, the squeezed toothpaste tube. Lying in bed, I slept and read. I had moved on to *War and Peace*, a loan from the Peace Corps library.

A frequent visitor to the Delacroix's was the Military Attaché at the US embassy. At thirty-five, Dick was the head of US military operations throughout central Africa, a disarmingly youthful career diplomat with a freckled Tom-Sawyer face.

'Dick, everyone here thinks America is ready to move in, there's a lot of talk about US hegemony. Is it true?'

'No, Martin, the US can't afford to step in here and do what the French did here, i.e. be *everything*. But we do have an interest – free trade, human rights. The last ambassador put a stop on military assistance because of the human rights situation here. Now that has improved, and we're helping again.'

'You mean selling them weapons?'

'My, you are a cynic. We're sending people here to train the Chadians in mine removal. We call it Hearts and Mines.'

I snorted. 'And *I'm* the cynic?'

'Martin, we have a democratic president who wants to do good works. Mine removal doesn't cost us much, and it does a great deal to improve people's lives in the world. And it's training; it helps make people self-sufficient.' He lowered his voice. 'There is a perception that the French don't want their supposedly "ex" colonies to do anything for themselves, right?'

'You hear it everywhere.'

'I do not, of course, have an opinion. But you're as aware as I am that the Brits get praised for supposedly encouraging a higher degree of self-sufficiency in their colonies.'

'So the Americans want an image of Brit co-operation rather than French high-handed arrogance?'

'All I'm saying is, don't call what *we're* doing cynical and selfishly motivated. We are helping people to help themselves, promoting human rights and encouraging free trade.'

'Free trade.'

'Sure, free trade, is that some sort of dirty word in your book?'

'Dick, it's imperialism. People in some of the world's poorest countries get invited to a gambling club where they play against a bunch of rich people who have also fixed the roulette wheel.'

'Heh, how about some facts here? This place is land-locked, a tiny economy, no infrastructure. If Chad gets to sell its oil, say, on the world market, it has a chance of improving the lot of its people. And make no mistake, Chad *wants* to be in a global marketplace.'

'And it gets there on your terms. It becomes a client state of Uncle Sam. It's exploitation however you look at it.'

'Yves, we have a commie in our midst.'

Two things Chadians and Americans have in common are a tendency to bear arms and a corresponding interest in guns. One evening the conversation turned to firearms, and Yves acknowledged that he owned several. Dick's ears pricked up at once. Yes, said Yves, one for self-defence on my travels, another to hunt, another to protect the house.

Let them be brought forth! cried Dick. We had all had a few drinks.

Hermana looked nervous. 'Please don't get those things out, Yves,' she said.

'It's all right!' he replied, a little impatiently. Women meddling with men's affairs.

Yves and Dick began to caress the shiny metal-necked, wood-sheathed objects, instruments whose music was death. 'Oh, beautiful,' sighed Dick, sighting down the blue barrels and caressing the breasty stocks.

Suddenly there was a crack and a whistle as the rifle in his hands went off. He was holding it up, away from himself, like they tell you to. There was a rustle as the bullet tore through the tree at the end of the garden and vanished over the Chari Hotel.

Dick froze, then turned to Yves in horror. 'You keep it *loaded*?'

Yves shrugged sheepishly, aware of Hermana's furious gaze. 'In case of an intruder,' he said, attempting defiance, 'it has to be ready.'

'With children in the house?' yelled his enraged wife. '*Put them away!*'

Such were the pleasures of evenings at the Delacroix's. Days I spent writing and navigating the world of the Rostovs and the Bezukhovs. Sometimes I took the children to the pool at the Novotel with another frequent Delacroix houseguest, Georges. The youngest Delacroix, intrepid, water-winged Chuti, took invariable delight in hurling himself into the pool at the deep end, frightening the life out of us and necessitating splashy rescues during which he screamed with pleasure.

Georges took me – much to Hermana's disapproval – to N'Djamena's nightclubs, where French soldiers slow-danced with their African girlfriends, hands sunk like claws into pillowy buttocks. Georges had uncanny knowledge of this twilight world, and would whisper to me, 'See the man with the moustache in the Lacoste shirt with the girl in the red? He's an attaché at the French Embassy. She's a little whore called Lucie . . . See that fat man with the girl over there? He has the M— tractor concession. His wife's in France. The girl is nineteen – lovely isn't she? Just got here from Cameroon.'

Georges did not seem that interested in the girls himself. His was a Proustian pleasure in charting the nocturnal life of N'Djamena, listing conquests, cataloguing careers, tipping bartenders who whispered gossip about diplomats and NGOs while they wiped glasses. At the

Novotel pool, Georges would exchange friendly greetings with the swim-suited couples around us, then regale me with outrageous stories of their domestic arrangements.

One Sunday at the pool he said, 'I'm expecting four Romanian whores to show up today, it should be amusing.' Four women appeared on the far side of the pool and began to display themselves, to the sun and the assembled men. 'They've been brought over by a Lebanese disco owner,' said Georges. 'They're here today to make sure everyone gets the message.'

The most attractive of the four dived into the pool, then stood at the shallow end, her hands behind her neck, holding back her long black hair, firming her breasts. Every male head turned.

Another day, a Sunday, we drove with the whole family out of town to a restaurant with a swimming pool overlooking the Chari river. As we returned, Georges and I were in the back of the pick-up, a child each in our laps. Georges drew my attention to the western horizon. The setting sun was sliding out of a bank of soft grey cloud, a pale yolk seeming to tear itself free from membranes of vapour. 'It looks as though it's being born,' he said.

Georges looked around him and smiled at a memory. 'This road hasn't been tarmacked that long, Martin. Not so many years ago I remember driving along here flat out, as fast as we could go. The driver had been wounded in the leg, we had to get him to hospital. Ambush. Yves was wounded, too. I was OK, so I was driving.' He smiled again, dreamily. 'Things are different now.'

Several times I took the moped out to meet Romain. As we had expected, he had failed his driving audition with Bessita. Romain lived with his wife and daughter in a house – a mud-brick box with a tin roof and a low-walled outside privy – in N'Djamena's slum suburbia. He saw me as his friend and potential saviour, petitioning me endlessly to find him work, and I tried with everyone I had met in N'Djamena. But they were jaded by requests of work on behalf of a thousand other honest, reliable men – there were too many men in N'Djamena, and too few jobs.

On my last day, Romain climbed on the back of the Yamaha and we puttered to the central market, where I was replenishing stores for the journey ahead. I bought him gifts – a watch, some clothing, and for his wife a bolt of cloth and some housewares. We drove back to his house with him on the back of the Yamaha clasping brightly coloured washing-up bowls and sieves. I gave him some money, too, and thanked him, and bade him goodbye. I felt sad, but I knew he felt sadder. He was losing touch with someone from the world of the rich and powerful, someone who might have helped him out of his impoverished existence on the edge of N'Djamena.

CHAPTER SIXTEEN

The Government Inspector

IT HAD TAKEN US two days to reach Noukou. Now Yves was ensconced with the *sous-préfet* and the sub-prefectural books. I was sitting on a wall in early morning sunshine. I was reluctant to lean back not because of the blood-red pigment in the brick, but because of scorpions. All around me were vistas onto hills of orange sand. Crows croaked, a camel grunted, childish voices drifted from a nearby playground.

There was a wind today – thank God for this wind, we'd need it by lunchtime.

A lovely, turquoise, swift-like bird, its ogee tail extended by a single, very long, very fine feather, executed an astonishing mid-swoop turn in which it collapsed and briefly fluttered like a scrap of paper on a wire barb, a dead thing, before hardening into a firm, aeronautic shape and soaring in a new direction.

An old man approaches me and speaks, he wants something. He is toothless, his skin the caramel parchment of a dried fig. I cannot understand him, and we laugh.

I call the *sous-préfet*'s boys. 'He wants to know if you have any plastic bags,' they say. 'Do you have a spare one?' Noukou is remote enough for carrier bags, ten a penny in the city, to be precious, hoarded.

The old man and the boys move away. Near my feet, incredibly, growing out of the sand, are mushrooms. I try to imagine a passing shower that has watered them. Or a passing mule? Or man? The boys return unbidden, bringing me an armchair, which they plant in the shade of a thorn tree. Blue sky, crumbling colonial ruin, orange sand.

The two-day drive here had been hard. Yves sat in the driver's seat with his rifle wedged behind him and his colleagues wedged beside him. I was told to sit behind, in the back of the pick-up. Really, it's more comfortable up there than crammed in here, they lied, sitting in the air-conditioning and roaring with laughter. A soldier perched on the roof, bolt upright, a red chequered Arab scarf wrapped round his face, copper-mirrored lenses over his eyes, his fist gripping the barrel of his rifle. As we roared over the desert I bounced on the bedrolls under a cloudless sky. We stopped for lunch in a village where we were brought spit-fried lamb, and the soldier laid his rifle beside his prayer mat to incline to Mecca. The July sun beat down and I felt ill. A huge bird swung out of a tree and Yves braked and grabbed his rifle, pop, pop, pop, pop, but the big bird glided serenely on.

It was the night of the second day when we reached this town, our destination, Noukou. In the centre of a large courtyard sat a fat man, a Sidney Greenstreet in *Casablanca*, a man-mountain. He almost stood to attention when Yves arrived, and clapped his podgy hands. Servants scampered into the night and hasty preparations were made for food, while we sat around the table and drank water and watched waves of enormous insects immolate themselves on the pyre of the paraffin lamp or paddle in our water bowls. An hour and a half later, still no food, and I realized nothing had been prepared in advance; there were no phones here, and the date of our arrival had been vague.

Late in the night it arrived, *boule deluxe*, a Vesuvius of semolina with lavas of orange fat forming pools around boulders of mutton. And tea, and more tea, and talk, until I had to excuse myself and go and lie on the prepared mattress. When I woke up it was after dawn and Yves, Government Inspector, scourge of bureaucratic impropriety, was auditing hard, the sweat streaming down him, in a small hot room crammed with men and marbled, ribboned ledgers.

It took ten minutes to explore the town. The market was a deserted fretwork of poles, and the handful of shops were dark doorways where strings of razor blades and shampoo sachets dangled. The streets were of deep sand, and walking was a dreamlike trudge.

At the edge of the wadi I looked down to the far-off palm trees that covered its flat floor like a carpet of moss. In their midst was a lake-bed, glassy – there had been rain. As I walked down the track, a string of mules climbed past me carrying biscuity new-baked mud-bricks, the muleteer greeting me with a grin. I approached the ring of palms, and what had from a distance been a beckoning Islamic paradise harshened into something no easier to penetrate than a barbed-wired Cotswold or Californian farm. It was all private date gardens, delineated by razor fences of dried, tangled thorn bush. Children raced between these zones through secret gates, but I had to follow common paths that stayed far from date palms or any well. The lake-bed was shiny brown like an old unsilvering mirror; moist, tacky, swelling around my feet into a chain of tiny atolls. Two small boys approached and shook my hand. Tinier children ran up. Ullulation throbbed unceasingly through the still air. Attracted by the children's cries, black-robed old women emerged from the gardens and beckoned me, offering not water (which I badly wanted) but enamel plates piled with yellow dates. And at last, a chipped enamel beaker of sweet well-water. Then I had to go back into the sun and climb that hill.

I sat in the shade of the arched doorway, staring past yellow splashes of drying dates, to the ridge of sand on the far side of the wadi.

'There will be a truck to Nguigmi any day now. Yes, probably today.'

So the *sous-préfet*. But a truck never came. Yves Delacroix finished his work and prepared to depart. 'I'm sorry to leave you waiting here,' he said, 'but a truck will soon be along.' He was not one for sentimental farewells. 'Come to Chad again soon. And write kind things about us!'

The days passed, and no truck came to carry me on to Niger. If I felt uneasy at outstaying my welcome, no one seemed to mind my being there; perhaps they welcomed the presence of a new face. I was a friend of a powerful government servant, and the guest of the most powerful man in the district, who, an impeccable host, seemed content to have me stay.

Afternoons in the sun-sedated oasis: imperceptibly, the temperature becomes impossible. I take a pillow and flatten myself on the floor. When I awaken, the *sous-préfet*, the *commandant* and a colleague lie sprawled around me on their cushions, playing cards.

'We're going to keep you here,' the *commandant* says. 'I'm going to make you my prisoner. You'll never see England again.'

'I think you'd get bored first,' I reply. But something inside me tightens. I am among friends, but the *commandant*'s tease touches a real fear. A handful of dust.

The truck came – then left.

'It's loading with salt,' the *sous-préfet* told me. At a salt mine a few kilometres away. Don't worry. Pack your bags.'

I did worry. All day long, I imagined the truck leaving without me. But any request I made for information was met with a reproachful look, as though it was somehow impolite. Afternoon came and I read *War and Peace* while the friends, Noukou's élite, played cards, cupping their hands, rocking gently, humming little phrases over and over, flicking the still-new cards and taking pleasure in their laminated springiness. I dozed, and when I woke it was cooler, and two men had left. The *sous-préfet* was sitting cross-legged in his white djellaba, his toes curled, testing his radio's batteries. He had a pair of tiny crocodile clips from which red and black barley-sugar wires ran to a little light bulb. Each time the bulb lit up, he grinned gleefully to *le commandant*.

'Finished playing?' I asked.

Le commandant reached for his leather belt and holster and stood up. 'Time to go and get some rest,' he said. It had been a taxing afternoon's cards.

After he went, I accompanied the *sous-préfet* into the dusky courtyard, where his boys were spreading woven plastic mats on the ground. As the purple dusk thickened we talked of games. Monopoly, where you land on Gare d'Austerlitz and buy hotels on the Champs-Elysées; Scrabble, where Noukou had recently trounced the bureaucrats of neighbouring Moussoro; and chess.

'I can't play chess,' said the *sous-préfet*, 'but I love to watch it. You know, it's always *guerriers* (men of war) and *politiciens acharnés* (obsessive politicians) who love chess. I remember during the war, a bunch of them would get together at the Novotel, take a room and spend the whole day playing it.' He picked up the plastic kettle and began to wash his feet, toes, ears, nose and mouth, slooshing and spitting several times. 'It's all about strategy, you see, that game.' He slooshed and spat again. 'You know, without the war, we might have been able to do something for Chad. Achieved something for the next generation. But it didn't turn out like that.' With a mild groan he picked himself up and crossed to a nearby mat. 'Somebody should write a little book about it, there's enough to tell. The deaths of friends. All the wasted years . . .'

He faced Mecca and his two boys joined him, standing behind him on the mat so that together they made an isosceles triangle. Meticulously, in a strong, confident voice, he began to lead them in prayer. 'Allah ho Akhbar, Allah ho Akhbar, Allah ho Akhbar . . .' God is most great. They bowed, straightened, knelt and prostrated themselves, sat back on their haunches, made a second prostration. By word and mouth they expressed the meaning of the word Islam: 'submission'.

Darkness fell, an absolute, inky darkness. One day, I thought, the city fathers of Noukou will be wealthier than they are now, and will install street lights here, believing they are doing good.

Prayers concluded, a single oil lamp was placed in the doorway and food was brought to where we sat. The ochre glow of the burning paraffin threw long shadows, so that our movements seemed to be those of giant cloaked figures. We ate the boule in silence. From the direction of the village a bright light raked the sky and, seconds later, an engine roared. Out of the darkness a figure appeared and spoke to the sous-préfet, who turned to me. 'The truck is about to go,' he said.

Unbidden, the sous-préfet's sons shouldered my bags, and we all climbed to the top of the village. Out of the gloom we saw a giant truck, its black bulk only distinguished in the darkness by busy torchlight beams around it. A curved bonnet was open like the maw of some monkey-god and men were thrusting their arms inside it as though in tests of faith or fidelity. Others manipulated hoses like the slack tentacles of a giant squid, directing trickles of water and diesel from brass taps into the multiple bellies of the god. The truck was surrounded by mechanics, apprentices, businessmen, brokers, spectators of all ages – no women were present – and last and least, the passengers who would travel on top of its load of freshly-mined rock-salt destined for the tongues of Nigerian cattle. The truck's master came forward and shook hands, the sous-préfet's, the commandant's, lastly my own.

Hausseman was tall and immensely broad, his body a rounded V, one slab of buttery muscle. He had, he told me, despite his name, no French blood. He had a huge jaw and an ear-to-ear grin, and I liked him at once. His cab was a small country, a glazed amphitheatre like the observation car of an American loco. I looked around, wondering with how many people I would be sharing the space. I had seen trucks passing with six or more faces looking down on me from the cab. But the *sous-préfet* had, I suspected, spoken, and the driver, who would be coming this way again, needed to please him. He and I would be alone.

At last he switched on the headlights and gunned the engine, and we moved off into the night as I waved farewell to my Noukou friends, who were quickly lost in the dark.

CHAPTER SEVENTEEN

Hausseman

'I LIKE TO DRIVE at night,' Hausseman said, 'it's cooler now.'
Driving, he told me his life story, a turbulent tale recounted
without bitterness, laced with aphorisms that must have been turned
in the mouths of his grandfathers and great-grandfathers and now
emerged time-smoothed like pebbles.

'My father is an important man in our town, but now he's in
opposition – his enemies are in control. But he still has influence. A
few years ago he was turned against me. An uncle, a man my father
trusted, turned my own father against me. I had been to my father and
told him, This man you trust is cheating you, telling you lies. My
father was furious; an old man's pride. We argued – who'd lied, which
lies. Is there a man alive who never told a lie, by God? Even to his
mother! It's like they say, "Between your truth and my truth, the
truth is in the middle"! It was complicated, my uncle had manipulated
everything. In the end my father threw me out. I walked out of his
house with one pair of trousers, by God! I had nothing. I went away,
to Libya. There's always work in Libya. A rich country? For a poor
man, yes. But a poor man will put his hand in a fire to pick up coal!
I worked three years in Libya, but it's no way to get rich. They keep
the women locked up in Libya, you know. Of course, there are
brothels. Anyway, I made some money in Libya, made some contacts,
learned how to do this – truck-driving. I worked for the oil men in
the desert, the French and English. I'm glad you're English, by the
way, I hate the French. When truck parts are worn out so that the
Westerners don't want them, we can take them away, overhaul them,
sell them. The English would say to us, Have them – but never the
French! My God, what harm would it do them to distribute some
charity? What you throw behind you, you'll find on the road ahead!

Not the French. And they treat you like idiots. I'm an intellectual, you know. I've got something up here, I think, I reflect! But now I'm a truck-driver. Still, not everyone can do this – driving these big ones across the desert. It's more complicated than it looks. A beauty, no, *la Merc? Elle est puissante, la vingt-six vingt-quatre!* But by God, driving her takes some skills. To know how to drive the truck, you have to know how she works. These Mercs are worth a fortune. If I lose the owner's investment, I'll never work again. And then there's the desert, sand dunes, storms, land mines ... And that's the least of it. Every policeman is a crook, and every town you come to – however insignificant – they want bribes, as though the *préfets* hadn't already taken everything! The owner gives me so much to pay bribes, but the cops always want more, more than's reasonable. God knows, they don't get paid, but what am I supposed to do about it? And there's always war. Soldiers, rebels. I've been threatened, hijacked. This is a dangerous job. When I'm away, my wife says, "Hausseman, I'm never happy, I can never sleep, until I see that you are back. Then I think, how long until he goes away again?" But I've always returned home safely, God be praised. My father sees me working like a labourer, he's rich, and he sees me work. But I have never asked him for anything. I've gone home and bought a house near his, not rich but it's a big compound, I support about ten people, by God. First I bought the land, then slowly I found the money to build. It's like they say, "And you'll eat a meal without the sauce"! But I've done it all with my own hands, I've asked him for nothing! People respect me, they know I'm honest. Now he has to respect me too.'

I grew drowsy. I started awake in the grey dawn and found myself alone in the cabin. Quietly, I climbed down. The surrounding dunes were tufted here and there with shrubs, and the silence was broken by a gentle whirring of flies – this was not true desert, but the southern edge of the Sahara, just north of Lake Chad. The vast olive-green *gros porteur* had simply stopped in its tracks, confident that it would obstruct no traffic. Its load of salt-rocks were covered with a blue tarpaulin. Arranged on hooks along its gunwhales were sand-ladders of great length and width. A diaspora of men lay sprawled around the truck like the victims of an explosion. For the first time I got a clear look at my fellow passengers, whom Hausseman had told me were half a dozen students from Burkina Faso. They wore colourful djellabas, flashes of turquoise and purple. God knew they must be poor to be taking this way home, the long, dangerous northern route around Lake Chad. Hausseman's two mechanics, one tall, one short, were sleeping in their olive overalls on nylon mats. And big Hausseman lay in front of his truck, like a one-man protest. The ridges of the rutted road supported his head like a pillow, and his feet and the small of his back. He was breathing deeply.

'I lost time loading the salt yesterday,' Hausseman said when he woke. 'I have to press on. I want to get to Nguigmi before dusk; there are bandits outside the town.'

The huge Mercedes was twenty years old, and if it had ever had power steering, it didn't now. I watched the triceps on the underside of Hausseman's arms – muscles I didn't seem to possess – standing out like hogs' backs from the shallower slopes of his arms. He wore a black singlet, and around his neck and under an arm a leather cord with a collection of what looked like leather pouches. These were *giri-giris*, he told me – charms of immense power (and in some cases expense), which had thus far kept him out of harm.

We pressed on through midday, our roof-top passengers becoming slowly heat-stunned. In the afternoon we reached Lake Chad.

The track took us not around but through the lake's northern shore, slicing off fingers of wadi that thrust north into the desert. The lake does not have a permanent shore; a seasonal body of shallow water – its average depth is three or four feet – it expands and contracts according to rainfall. Even at its most swollen, much of the lake's edge is a series of swampy shallows, and maps show only half of the lake's area as permanent. The maps are never accurate, for Africa's fourth-largest lake is shrinking. In 1870, the lake covered something like 10,800 square miles; recently this has dropped by 90 per cent to 1,000 square miles, or 2,600 square kilometres. The frontiers of four countries meet at this diminishing water resource; hundreds of thousands of people depend on it for their livelihood, and its disappearance is causing distress. The diminution is blamed on reduced rainfall and the desertification of the Sahel, which in turn are blamed on tree-cutting and overgrazing. No one has the answer.

It is extraordinary to think that this is all that remains of Mega-Chad, the ancient sea that covered much of the central Sahara, Egypt and the Sudan. Its legacy is a grey crust of dried clay and rock artworks that ironically depict fish. Of course, there is no permanence in nature, of mountain, ocean or even continent; everything is churning, forming and reforming, a human eternity but a geological blink. Yet something fundamental has changed. Americans divert vast rivers into suburban swimming pools; Russians drain a sea; Libyans suck dry the prehistoric water deposits under the desert; Egyptians drive canals into the desert.

Crossing the fringes of the lake was not easy. The old Merc would slither to a halt in soft sand on the slopes that led down to the lake-bed, and the giant sand-ladders had to be broken out. No question of pushing the massively laden vehicle out of the sand: Hausseman's mechanics had to dig around the wheels to get the ladders in position, while he used the engine in its various forward and reverse gears to rock the truck until it gripped the ladders and hauled itself onto firmer sand.

It was early afternoon when we stopped to eat our first food that day. Hausseman filled a bowl with something resembling cereal, in fact a sort of dried and grated radish, and crumbled on top of it brown egg-sized balls made from dried groundnut oil. He added raw onions, oil, salt and a tin of Canadian herrings. It was excellent.

As evening approached we were driving through no man's land, having left Chad behind us. We knew that a fortnight earlier this border town had been attacked by rebels – or bandits – when two policemen and two customs officers died. As we drew close to the town, Hausseman grew wary.

'This is where any bandits will be. Not in the desert, miles from anywhere – *here*, outside the town.'

'Are we really in danger?'

He looked at me, his expression excited and sober at the same time. 'It's the trickiest part of the journey. We haven't got anything to steal, but they'd kidnap you and make you a hostage. So this is where I put my foot down.' As the sky grew pink ahead of us, we thundered towards Nguigmi. We reached the town without incident.

I had to make my official entry into the Republic of Niger, the mostly desert country due north of Nigeria. So few were the travellers coming this way, and the customs on the Chadian side of the border had been so cursory, that I expected the customs post to be a dusty backwater. But there were rebels in these parts, a sore thorn in the government's flesh. A quick-witted policeman took great interest in my presence. What was I doing here? What did I hope to gain from a trip to Niger? He kept me waiting ninety minutes for my stamp.

Hausseman waited too. At last we drove his truck to a lock-up, then walked – Africa's ubiquitous small boys having arrived to bear our bags – to the compound where he habitually stayed in Nguigmi. The streets were leafy, the air almost cool. Gratefully we went into the corner of the courtyard and soaped our filthy bodies.

I sent a boy to bring us all cold drinks, and, as the night settled, a brazier was lit and dinner prepared. News of our arrival had circulated quickly and people began to arrive to give greetings.

The news in Nguigmi was bad. A few days earlier, men in unmarked trucks had torn into the town, firing wildly. They had murdered twelve people and burned down several buildings. These were clearly reprisals for the earlier killings of the four men, and it was generally believed that the latest killers were members of some government military force, exacting a general vengeance on a town believed to be too sympathetic to the separatist rebels.

Niger is, like many African countries, a colonial carve-up of earlier tribal boundaries. In 1885, the Conference of Berlin denied Portugal's claims on the Congo basin and gave France and Britain untrammelled access to vast territories (the British were the major beneficiaries). These stark Cartesian lines are an enduring source of inegality. Of course, there were empires and wars among West Africans before the Europeans arrived – many powerful and wealthy kingdoms had risen and fallen – but the modern frontiers were crude colonial conveniences, which it suits modern governments, made in the image of the (almost) departed colonialists, to uphold.

Modern Niger is dominated politically by an ethnic group from the country's south-west, the Songhai-Zerma; but almost half of the population are Hausa, who are also the most numerous people in neighbouring Nigeria. The desert north of the country belongs to the Tuareg – with a clutch of *toubous* in the north-east where Niger clips the corner of Tibesti – while the people of the south-eastern border with Chad are the Kanuri.

The bloodshed in Nguigmi was a result of the same phenomenon

I had met in Chad – the perception among Arab northerners that the country was run for the benefit of black southerners. Nguigmi, a frontier town between the sedentary, agricultural south and the nomadic, pastoral north, had thus become a target for northern rebels.

We left soon after dawn the next morning, stopping at the *poste* on the western edge of town. 'This is where they murdered the cops,' Hausseman whispered as he climbed down with his papers. We drove on, Hausseman peering anxiously at the road. 'They're around here, in this area, we could be stopped anywhere. It's a matter of chance.'

After twenty miles or so a large silver spanner appeared in the dust ahead of us.

'Aren't you going to take it?' I asked.

'No! It's *very* dangerous here! It's a common ambush technique – put something in the road that drivers will stop to pick up.'

I remembered Romain telling me that the Libyans had wired pistols and new hammers to mines. Glistening traps.

Hausseman was sweating with tension. Only after an hour did he start to relax.

'I never get through that town without thanking God. My wife hates me to come through Nguigmi. "I pray for you to get through safely," she always says. Thank God we're past the border. Soon be home!'

Thirty miles or so from Nguigmi, the landscape began to darken. The dry, pale orange earth became a deeper shade of rust. Spikes of lime-green grass appeared, then trees, then woodland. I felt a sense of relief and release. We had left the desert.

CHAPTER EIGHTEEN

An Interlude

IT TOOK TWELVE STRONG MEN seven hours to unload the truck, working in teams of three or four in the humid heat on the riverbank, tossing and shovelling the sharp, chalky salt-rocks, then toting them down to the river. Another team reloaded the salt on a raft made from oil drums, and, guided by a wire slung from bank to bank, ferried it to Nigeria. On the far bank it was unloaded, then loaded on another truck. Hausseman was an aristocrat here. One truck provided work for two dozen men. Five other trucks were being discharged alongside his, and more than a hundred men and boys laboured on either bank. Please God, I thought, let them not build a bridge here.

Other travellers came and went, tradesmen, tinkers, straddling the unsteady oil-drum raft. Stalls turned out hot food, wisps of blue smoke from meat and corn grills drifted in the air. Beside the raft, a dozen small boys played, their naked bodies sleek and shiny in the mud-brown, sun-starred water.

The palm-fringed river was perhaps twenty-five yards wide. 'Come on,' said Hausseman, 'let's swim to Nigeria.'

Niger to Nigeria. Yet this river dividing the two countries was not the Niger, the black river, once believed by Europeans to be a route to fabulous cities in the African interior. That lay ahead of me, further west. This was the humbler Komadugu, pouring towards Lake Chad.

Hausseman undressed and waded into the clinging mud, his brown body round and strong. I, an albino-bodied spider with brown head, arms and feet, neurotic about possessions – passport, money, all that nonsense – carefully piled my clothes before plunging in after him. The water was warm and velvet, silt-thick. The current was strong, and we had to swim hard against it. I stepped out on the Nigerian shore, an illegal entrant.

Hausseman scraped up some fine sand with a forefinger and brushed his teeth. He did not want to linger. Maybe my white body was too conspicuous, for there were customs of a sort on either bank. I swam back behind him, but refused to leave the water. After the desert, to be immersed in liquid silk! He gestured to me with his arm once or twice.

'No,' I grinned back, 'leave me alone!'

He returned to the water's edge, stern-faced. 'There are spirits in the water, you *have to* get out.' Perhaps I seemed scornful, for he yelled, 'You *must* get out! You can only stay in so long! The spirits will be waiting, then they'll drown you. Men have drowned here! You must get out *now*!'

'OK, keep your hair on.'

We walked up the bank and found a patch of soft ground. Skeletal doum-palms swayed overhead, and a man in the crown of one dropped fruit to us. We picked at the gingerbread fruit with our pen-knives and ate the meaty brown flesh. The day bent and unbent, we slept, woke, ate a mango, drank tea. Commerce across the Komadugu continued.

Hausseman led me to a grass-roofed hut where a trio of dopey customs officials sat up and looked at me with interest. I was as respectful as they were officious, and Hausseman laughed and told them not to meddle in matters that did not concern them. He must have known them well, because they took it from him, subsiding sleepily on their wire chairs and string beds.

'Come, Martin, I want to introduce you to another friend of mine.'

A few yards away a queue of men sat in the mild afternoon sunlight outside a golden grass hut. Hausseman crouched past them and looked inside.

'Hausseman!' Female voices.

He gestured for me to follow, and we crawled on grass matting, a handful of men making way for us. Two women were greeting Hausseman with gold-capped grins. For me there were wide-eyed stares.

A woven-grass partition screening off a third of the hut rustled, and a man emerged on all fours, smiling cheerfully. A moment later another head appeared from behind the curtain.

'Hausseman!'

'Mama!'

She was a big woman, with a broad leonine face. She came out, pulling an indigo batik robe tighter around her and lifting her elegant hands to smooth back her braided, beaded hair.

I was introduced, and we sat cross-legged while Hausseman told them his latest news. It felt rather like a polite tea party. But after a few minutes, Hausseman said that he had to leave.

'We're holding up the queue,' he told me.

The women protested noisily, 'Been here all day and haven't visited!' Mama made Hausseman the sort of indulgent reproach you might give a mischievous child, telling him to come back soon. And we were back out in daylight. Hausseman looked embarrassed.

'They were obviously very pleased to see you,' I said.

'Yes, well, I can't be wasting my time with these old roadside whores.'

'But a few minutes ago you were so keen for me to meet this special "friend" of yours.'

'Ah, well, *fin bref*. That woman, you know, she's really something.' I kept on looking at him, and he let out a laugh. 'Martin, who is the most beautiful woman in the world?'

'I don't know.'

'The woman you are in love with.' A pause. 'Ask an old man who is the most beautiful woman in the world, what will he answer?'

'Tell me.'

'He will say it's a woman at the beginning of her pregnancy; or a woman breastfeeding.' He laughed again. 'If you hear me say bizarre things, it's because I spend a lot of time with the old men, Martin. They teach me a lot. Life is full of paradoxes. For example, it's a funny thing, you know, some men are allergic to women. You see them trembling outside brothels. It's a strange world.

'When you love women,' he continued, 'they despise you, to the benefit of seducers who despise *them*. Hah! What a joke on us all. Their parents lodge them in a hut and feed them rotten cereals, but when they get married they demand a villa, Western food and new clothing. I tell you, women are feeble. No, not physically feeble. Up *here*. No brains. Unreflective. And no . . . no strength of character. How many times has one seen a man cuckolded in his own bed?'

Nominally Hausseman had two wives, but the first he had caught 'en flagrant délit'.

'I came home from a journey unexpectedly and caught them at it. I was going to kill him. So he was gibbering and trying to swear on the Koran, and my wife was screaming, "We're all going to die!" and I *was* going to kill him. Then I thought, Why should I go to prison because she has an itch between her legs? I said to him, "If you ever talk about this to anyone in this town, you're a dead man." Well, he's kept quiet so far. I see him sometimes and he sort of grins at me the way a whipped dog wags its tail. Her, I sent packing.'

His second wife, he said, was quiet, reliable – and had doubtless learned a salutary lesson from his treatment of the first.

I thought of my fiancée, Penny, so far away, across the desert of which I was on the southern shore. The last time I had spoken to her, in N'Djamena, she had sounded emotionally far away, and imaginatively remote from the world I was moving through, the strange tales I was barking down the phone during our infrequent, expensive, unsatisfactory conversations.

Hausseman's seven-year-old son flew into his arms. By stark contrast, his remaining wife was undemonstrative. Knowing her place, she smiled shyly to see her husband back from the fearful void.

At the rear end of a walled compound Hausseman had built a row of three houses, each house comprising a single room. He lived with his family in one, and provided the others rent-free, the first to a friend, the second to his personal shaman and his wife. He had learned from his powerful father that a man of substance should have people dependent upon him, distribute largesse. Since Hausseman was estranged from his father, this was a replacement extended family.

In front of Hausseman's small house a grass canopy provided shade. Under a lychee tree there was a brass standpipe, the only tap for the three houses. The walled courtyard bustled with visitors, the children and their friends, chickens, dogs, tethered sheep, birds and lizards. Hausseman's wife and sisters and the shaman's wife, with her thousand-lined face and shrivelled breasts, ceaselessly pounded mortars with wooden pestles.

Hausseman's seven-year-old son loved to stalk the lizards with a stick. But the next morning, as we sipped our coffee, the boy was solemn.

Hausseman grinned at me. 'He's upset. I had to box his ears.'

'Why?'

'Because the little fool has been chasing the lizards. He shouldn't, they're dangerous.'

'How are they dangerous?'

'They have genies in them. I woke up in the middle of the night and he was running around all over the room. I said, "What's got into you?" and he said, "The lizards are chasing me." I said, "You little fool! You spend all day hunting them, does it surprise you that they come after you at night?" The little idiot. So I gave him one round the ears, to teach him a lesson. Now he's upset, he's afraid of me – it's natural.' Hausseman smiled affectionately. 'A father has to be severe. If I educate him right he'll grow up like me – he'll be polite to everyone, he'll know how to behave. Otherwise, God knows, he'll turn into a little trouble-maker!'

CHAPTER NINETEEN

Sahel Taxi

T HE PEUGEOT STATION WAGON made furtive sideways lurches
and lunged at slopes, the engine making a din like a hammer
going down a drainpipe. Then it shuddered to a halt.

The driver threw open the bonnet and sanded-down the spark
plugs, and we limped on. At the next small town we stopped at the
bus stand while a mechanic went to work.

I dipped sweet buns in sweet tea in the shade of a spiny succulent.
People passed in sunhats with broad brims and pointed crowns that
bulged like swollen parsnips. A man in a blue djellaba and white hat
called out from the door of a building shaped like a cross. It was a
mosque, he an *immam*, and men drifted from the teashops to pray.

I was struck by the beauty of these people in whom Arab and African
blood was mingled. It was shared by those with and without regularity
of features, a spectacle of bright smiling eyes and wide smiling mouths.
How unlike London.

At last the share-taxi was ready to leave. We crammed back onto
its ragged nylon seats, noisily speculating about whether it would
actually start. The driver arrogantly jerked his column gear-stick and
the 504 leapt forward like a gazelle.

Melted ridges of sand erupted into the orderly Sahelian landscape
of young grass and tall trees; it was like a giant's golf course. We passed
an unwalled roadside sun-shelter where eight slender men stood
immobile, silent, an assembly of Giacomettis. At immense height the
sky was stencilled with delicate white traceries; nearer the earth there
were armadas of swollen cumulus clouds. I watched the sun stain
them pink and saffron as it sank.

The car juddered to a halt. We laughed tolerantly as the sulky driver
picked up a plastic bottle holding four litres of petrol and poured it in

the tank. A man behind me muttered, 'Look at this car, the back door is held down by string, and it probably has been for ten years. As long as it moves, that's good enough for them. They spend nothing on maintenance, they carry no spares, they chuck their money about on other things – drink and whores.'

After dark we pulled into a town. I bought water, sold in chilled plastic bags like those people take home from fairgrounds with goldfish in them. Chilled water, electricity – civilization was drawing closer. I felt quickened at the prospect of showers and cold beers, yet repelled by the prospect of ordered urban anonymity.

At nine the car ran out of petrol again. This time we were marooned in darkness in the middle of the bush. 'I couldn't afford any more petrol,' the driver insisted.

I expected a riot, but only the man behind me was emboldened to utter, 'What did you spend all the money we paid you on?' He was ignored.

At length, an old pick-up wheezed up behind us and three male occupants peered cautiously out. Yes, they said, they could sell us some petrol.

'I told you,' the driver said, 'I haven't got any money left.'

'Fill her up,' I snapped. 'I'll pay.'

We reached Maon at 1 a.m.

Standing in a sweltering plywood cabin in the post office, I got through to Penny again. It was wonderful to hear her voice, but it was filtered through something like iron wool, and we had to bark emotionlessly to hear each other.

Why did we want to get married in the *desert*? Most of our friends thought us mad. The desert summoned an image of remoteness, insufferable heat, an utter absence of all those things that make a wedding what it is meant to be – most notably, perhaps, family. Ah, well. After the honeymoon we planned to invite our families to a blessing, to be held in the church in the garden of the London square where we lived. The vows repeated, friends doing readings, a cake, a marquee on the lawn. And that is what we did. But that was then, and this was now.

Penny and I had set our sights on Mauritania. Nouakchott, the capital, looks out at the Atlantic from the edge of the Sahara, on the same latitude as Faya. The Land of the Moors was not an overwhelmingly obvious place to celebrate a Christian rite, but that was, to some extent, our point. The British Consul in Nouakchott, a friendly and competent woman by the name of Nancy Abeiderrahmane, had warned us not to expect too much.

'Nouakchott is not exactly *picturesque*; but people are very friendly here. You could have a party on the beach, hire a nomadic tent and some musicians – I'm sure lots of people would come.' Red tape would be minimal. 'You get a local *immam* to bless you in the eyes of the community; they haven't got round to marriage certificates here yet.'

But the priest in Nouakchott had gone AWOL, Penny now yelled from London down the crackling line 'He's vanished without a trace, darling. Nobody knows where he is.' Maybe we should consider Dakar, the capital of Senegal, south of Mauritania. Or perhaps Mali, or Morocco.

I had a heavy heart when I put the phone down; I would not see Penny for another two months. I was leaving her with a vast amount of organizing. Meanwhile I was turning north, back into the central Sahara.

The road north to Agadez was threatened by rebels (or demobbed exrebel bandits), so vehicles left only once a week, accompanied by a military convoy. Early on my fourth morning in Maon, a convoy of buses escorted by gun-mounted pick-ups pulled away from the bus station.

I was next to an army parachute instructor. He sat bolt upright, his rifle barrel gripped tightly. 'There've been several hold-ups lately,' he told me, 'about twenty kilometres from town.'

'Did they hurt anyone?'

'There've been a few murders, and they tend to rape the women. I can't quite see what they're after. Maybe I'm missing something.'

I looked at him hard. 'What do you mean?'

'I can't see that they're rebels. If they're rebels they oppose the state, and they should be attacking the army. But if they did that, they'd probably get killed. They prefer soft civilian targets instead. So why do people call them rebels? It's making them into something they're not, and giving the impression that there's an armed rebellion going on when actually it's nothing of the kind. Frankly it's beyond me. People who attack civilians, murder and rape are just bandits. They deserve to die.'

CHAPTER TWENTY

Agadez

Just get me to the church, get me to the church,
for Gawd's sake get me to the church . . . on . . . time.
Alan Jay Lerner, *My Fair Lady*

I HUMMED HAPPILY as the bus rattled through the red semi-desert south of Agadez. The Sahelian fringes were less barren than they looked. We passed many thatched grain stores, crude buildings built on stone rings to keep out pests, and trees heavy with doves like plump white fruit.

Time was running out before my appointment at the altar, and I was only halfway across the Sahara. Reaching Agadez would be a milestone, albeit one that told me I was still far from my journey's end. Meanwhile, I was headed towards the Aïr Mountains, back into the Saharan heat.

Agadez was once one of the wealthiest cities of this desert, back when Saharan wealth was built on slavery. During the mid-1800s, Europeans began to take an interest in the network of caravan routes that webbed the Sahara. Perhaps they would provide access to the central African cities that the dreamy reports of peasants and the rapacity of European traders had inflated into El Dorados. At the same time, terrible stories were emerging of the Saharan slave trade.

As the British Empire had turned economically towards *laisser-faire* and India, it lost interest in the mostly American-oriented phenomenon of slavery. During the first half of the nineteenth century slavery was being phased out, and with the enhanced zeal of the converted, people were disposed to be horrified by the number of slaves reportedly perishing on the long treks across the Sahara. With the sanction of the British government, the Richardson expedition set out from Tripoli in

1850, aiming to explore the Saharan trade routes and gain information about the slave trade. Reaching Agadez with great excitement, the explorers were crushed to discover a city of merely 7,000 inhabitants, surrounded by decaying buildings and traces of vanished splendour. Perhaps succumbing again to imaginative gigantism, they speculated that the population may once have reached 50,000.

Human appetite for myth often exceeds our stomach for reality, and the cities of the Sahara have always disappointed European explorers. Most notorious was Timbuctoo, a city of legendary wealth described by René Caillié, the first infidel to reach it and return in one piece, as 'the most monotonous and barren scene I ever beheld'. It is easy to imagine that the Richardson expedition might similarly, during months in the remotest desert, have allowed Agadez to swell in their imaginations into the kind of gold-rich civilization the Spanish had discovered and destroyed in the Americas three centuries earlier.

On what monuments did Westerners base their pictures of the fabled oases? Crusader castles, the Pyramids, ancient evocations of the Seven Wonders of the World? With hindsight, it seems clear that given the physical conditions in the Sahara, the enormous distances, the difficulties of transport and the small populations, nothing made possible or even necessary the development of great fortified cities in the desert. Nevertheless, in an environment where most people lived in tents, where the most basic commodities needed to be transported hundreds of miles by camel caravan, it would have been awesome for the nomad to come to Agadez, gaze up at a seven-storey mosque, and visit a teeming marketplace stocked with cloths, jewels and fruits that shone like jewels. It would have been to encounter civilization.

As the bus moved through the haze south of Agadez, the earth was turning from copper to gold. I did not expect fabled wealth, but I was arriving with great expectations. I knew that shabby Agadez had experienced an odd sort of renaissance in the late twentieth century as the place where Westerners doing the principal trans-Saharan crossing – via Algeria – would fall gratefully on cold beer. It had been one of the focal points of the Paris-Dakar car rally, and, by the late 1980s, was the centre of a burgeoning tourist industry. But all that had ground to a halt with the Tuareg rebellion.

The coach pulled up at the *autogare*, and baggage was released from the nets on the roof-rack. Guides, porters and taxi-drivers mobbed the bus, shrieking at the tops of their voices. The Agadez I found myself in that afternoon was a place of shuttered travel agencies and deserted souvenir stalls with peeling signs. When I reached the Hotel de l'Aïr, a two-storey building opposite the mosque, half a dozen slumbering touts woke up and ran towards me. I pushed past them into the hotel and spotted the entrance to the bar.

I walked in to find chairs on tables and the bar-top patinated with dust. The receptionist shuffled in after me. 'I'm sorry, Monsieur, but we haven't bothered to open. There are no tourists, you see.'

'But you do have beer?'

'*Mais bien sûr, Monsieur.*'

I followed him into the dining room, where a refrigerator in a corner was secured by a padlock. The tall ceiling was supported by four massive pillars.

'Magnificent,' I said. 'Such high ceilings! In the days before air-conditioning they knew all about how to make buildings cool!'

He pointed enthusiastically at the rafters. '*Monsieur*, when this was a palace, that's where they used to hang people.'

I went to the police to register my arrival, and was warned about the political instability. A tentative cease-fire was in place, but earlier in the year an Italian tourist travelling north – where I wanted to go – had been shot by rebels. Agadez was where the government writ ended, and I went further north at my own risk. This was not what I wanted to hear, but the hotel manager and even a sad underemployed travel agent agreed. It was unthinkable to go into the Aïr Mountains, the rebels' lair, and what was more, no driver would be willing to take me.

On my second night in Agadez, the bar opened. The couple now lounging there had apparently had it opened on their behalf. He was a Frenchman of considerable physical beauty, she a comparably perfect Nordic-Californian, with radiantly white skin. The man, Jean-Yves Brizot, turned out to be something of a local celebrity. Agadez has a reputation for silverwork, and he had built up a jewellery workshop selling expensive pieces directly to the French house of Hermès.

I had found my key to the Aïr. Jean-Yves grinned. He did that a lot, his big square jaw-muscles drawing back his lips to show multiple rows of brilliantly perfect teeth. 'The man you need to meet is Michel Ange,' he said, and turned to Alison for agreement. It was her first visit to Agadez, and she had met Michel for the first time the previous night. She looked troubled. 'I don't know,' she said, 'I didn't like being around a lot of young guys with guns. I didn't like it a bit.'

Unperturbed, Jean-Yves turned back to me. 'He's a guide and driver, and he's just got back from a trip to the north. If anyone can take you up there safely, he can.' He beamed. 'Michel is an excellent fellow. How much can you afford?'

I told him, and he winced. 'It's too little. It's the *absolute* minimum. I doubt you'd be able to negotiate it yourself – it's the sort of rock-bottom price you might get after a couple of days' hard bargaining. But if it's all you've got, *ça va* – Michel needs the money. I'm not prepared to bargain for you, but I'll tell him what I think is reasonable.'

Michel was out of town and I spent the day going around Agadez on the pillion of Jean-Yves's Soviet bloc 500cc motorbike. He blasted through the narrow alleyways of the old town, scattering chickens and small children, and took me into some of the old buildings, fantastic labyrinths of mud-brick boxes, where Bertolucci had shot the scenes of Kit's imprisonment for his film of Paul Bowles's *The Sheltering Sky*.

We were drinking coffee in the courtyard of a small restaurant in the old town. Though we were in the shade, our metal seats were hot. In summer, everything in Agadez is hot.

At last Michel came to join us, a slender man with dark curly hair and a wispy Mongolian beard. He seemed to be my only hope for getting into rebel country; I hoped I would like him. I laid out in great detail the journey I wanted to do, and we discussed the number of miles involved and the amount of petrol needed. Finally we broached the subject of money. I was getting used to these bargaining sessions for vehicle hire, and I reckoned I could get the measure of a man from the way he handled his side of the negotiation. 'How much would you want for such a journey?' I asked.

'How much can you afford?'

I told him, and I saw his eyes flash. He looked at Jean-Yves, who nodded almost imperceptibly. Clearly Michel trusted Jean-Yves. He sighed. 'It's a fraction of what I'd normally ask, but it's do-able. I have to feed my family somehow, and this is better than sitting at home.'

'How safe is it?' I asked.

'Safe?' He grimaced as though the question were absurd. 'It's *perfectly* safe. Don't worry, nobody mucks me around.'

'It's just that . . . Everyone says it's dangerous. And there was that Italian who got shot—'

'Look, I was a rebel myself until the cease-fire! You'll be quite safe. As for that Italian, they didn't kill him, you know. They just shot him in the arse!'

I moved into a room in Michel's house. Each of Michel's two wives – he had recently taken the second – occupied a separate set of rooms on opposite sides of a courtyard. (Co-existence was not proving popular, he told me, with his first wife.) Goats skipped under the acacia trees and naked children played with toys salvaged from engine parts, as the courtyard was also a workshop. Under an awning to one side a number of young men were crawling over a new-looking Land Rover. Michel grinned. 'They took this off some British tourists. They had four vehicles; the rebels only took two of them.'

'Very generous,' I said sarcastically.

One of the men working on the jeep turned to me. 'Some people would have murdered them and taken all four,' he said darkly.

Early the next morning, Michel introduced me to Reesa, who would be accompanying us as a guard. He was a tall, good-looking man in his mid-twenties, with a magnificent smile and a strange, shuffling walk. I was not sure why we needed a guard if everything was as safe as Michel claimed, but we were in too much of a hurry to debate anything.

Michel and I had worked out on a sheet of paper what stores we needed to buy. He went off to collect a barrel of diesel and Reesa and I went to the market for the usual desert rations – tea, sugar, salt, oil, tinned sardines, macaroni, onions, tomatoes, tomato purée, chillies, beans. We bought extra tea and sugar as gifts, and great bundles of tobacco leaves. 'The old people love chewing these,' Reesa said, 'but we need lots of ordinary cigarettes, too.' It would turn out that he had an ulterior motive for this suggestion. I bought three cartons.

Back at the jeep, Michel surveyed the pile of tinned fish and vegetables without enthusiasm. 'No meat?'

'Oh.'

'Never mind – we'll find a shepherdess and buy a goat.'

As I climbed onto the blanket-covered bench-seat, I felt something hard beneath me. 'What's this?'

'Oh, sit forward a minute.' Michel tugged at the blankets and pulled out a Kalashnikov, which he handed to Reesa.

'What's that for?' I asked, shrilly.

'Security.'

I stared at the steel-blue weapon, with its stock not of wood but of marbled flesh-pink plastic, like a child's toy. 'I thought you said where we were going was *safe*?'

'Yes, but *le Kalash* guarantees it.' He swallowed sheepishly. 'Anyway, I always carry this. Didn't you see it yesterday?' His tone of voice gave him away.

'It wasn't there yesterday.'

'Yes it was.'

With Reesa squeezed in between us, we set off. There was another passenger, Wali, Michel's fifteen-year-old apprentice mechanic, who was expected to travel on the roof.

'Is it your weapon?' I asked Michel.

'No, it belongs to the rebellion.'

Reesa pulled out the clip. 'Anyway,' he said, 'we shouldn't be needing this. Anyone who comes after us is a fucker. I mean, they wouldn't be a true rebel. Some fucker with a gun, or police or a bandit, but not a rebel.' He paused. 'I'm ready to die for the rebellion, you know.'

How reassuring, I thought.

He removed thirteen shells, one of them tracer – 'You see the red ring on the cartridge base?' – tested the springs, reloaded, checked the firing mechanism, showed me the safety catch – single shot, auto, safety – and reinserted the clip.

'You'll see a lot on this trip,' Michel was telling me. '*Real* people, the Tuareg. No policemen asking for permits, none of that nonsense.' He pointed at the gun. '*That's* our permit. That's the police where we're going.'

The story of the Tuareg rebellion is an unhappy one. Niger is one of the world's poorest countries, and the Tuareg, the inhabitants of the desert north, are one of its poorest communities. Yet Niger's principal industry is in Tuareg territory – the uranium mine at Arlit. And tourism, which had been growing into a major sector, is entirely based on the desert around Agadez. During the 1980s the Tuareg saw few of the profits from these industries, which were funnelled into the pockets of the élites of the capital, Niamey. Many rural Tuareg tribespeople did not officially exist in any government data, education and health facilities in the north were appalling, and the numbers of Tuareg teachers, doctors, policemen and soldiers were negligible. Southern blacks dominated the civil service, and were growing in numbers in Agadez. They tended to treat the Tuareg with contempt.

By the late eighties, resentment had reached boiling point. The rebellion began in 1990, an ill-led guerilla movement, fighting for control of the Agadez region, the Aïr Mountains. It lasted six years. Perhaps one or two thousand Tuareg died, and larger numbers of black government troops. But the government dominated in terms of troop numbers and equipment – including French assault helicopters – though they were unable to flush the Tuareg out of their mountain strongholds. Finally, the Tuareg were compelled to sue for peace, with nothing gained. The government's only real concession was a promise to absorb Tuareg men into the armed forces, but in the twelve months of the cease-fire, they had done little. As I reached Agadez, an uneasy truce was in place, with occasional flare-ups and a number of 'freelance' former rebels operating as bandits. Many Tuareg – like Reesa – grumbled that the peace was a farce, and that at the drop of a hat they would take up arms once more.

The occasional figures we passed were invariably dressed in the traditional Tuareg colours, blue for the men, black for the women. They would be tending their fields or following small herds of goats. The rains were late this year, and the previous year had been dry. It meant there was a shortage of grass for grazing, and the animals were scrawny.

Twenty miles out of Agadez we saw a truck pulled over at the roadside. Reesa loaded a bullet into the chamber of the Kalash with a metallic double-click, then sat up straight, holding his gun forward, so that he would be seen by everyone in the truck. 'Pull over on the left,' he told Michel. We stopped, and there was an unmilitary moment as I fumbled with the broken door catch and Reesa clambered over me.

He walked towards the truck, gun at the ready, with his strange shuffling walk, dragging one foot (I would later learn that the injury dated from his having been captured and tortured by government troops). When he reached the truck there was some chat, then everyone relaxed and hands were shaken. Michel let out the clutch and pulled level, Reesa climbed in and we drove on. This routine was repeated perhaps five or six times a day, and we would all be tense until we knew whom we were dealing with. I wondered how our intimidatory behaviour affected the other people, villagers or travellers trying to earn a living out in the wilds. But it had been this way for years. At least we turned out to be good guys.

No one but me seemed disturbed by the inherent absurdity of our armed posture – anyone who wanted could pick us off from one of the buttes we passed under, and with thirteen bullets Reesa was not going to be much use in an exchange of fire. When he dozed, as he often did, his AK47 would slowly tilt to the right until it was pointing in my ear, and I would push it vertical again. I also got into the habit of twisting the weapon around to check that he had not accidentally switched off the safety catch. When I had a chance, I raised with Michel my doubts about our ability to defend ourselves. 'Look,' he said patiently, 'no one's going to shoot us without knowing who we are first. And when they see our gun, they have no way of knowing how many bullets we have, or guns – or friends behind us. It's like I told you, the gun is a symbol, we don't *have to* fire it. It's a passport.' He laughed and clapped me on the shoulder. 'Don't worry so much, *Monsieur* Martin!'

He continued to find my discomfiture over weapons amusing. He also saw the funny side of the shooting of the Italian tourist. The incident had occurred because the Italian's Tuareg driver, a member of one rebel gang, had driven into the territory of a rival rebel gang and refused to stop when challenged. 'When they caught the man who shot the Italian,' Michel told me, 'they said, "Why did you try and kill the white man?" And in his defence he replied, "But I wasn't trying to kill the white man – I was trying to kill his *driver!*"' He spluttered with laughter. 'You see, it's OK to kill Africans, it's only wrong to kill whites – so you're perfectly safe here!'

We came into a camp where Michel had several relatives. Over two acres of flat sand there were six pairs of Tuareg huts, made from interwoven branches like the skeleton of a coracle and covered with mats made from woven palm fronds. Each was surrounded by a low, thorn-bush corral, which prevented the goats and chickens from escaping.

Michel introduced me to his relatives, two good-looking men in their early thirties in ankle-length blue shirts and indigo turbans. The women wore traditional black blouses, with two white circular panels embroidered on the chest. They sat in the shade of one of the huts, weaving baskets from palm fronds, and greeted me in a friendly way, looking me in the eyes and asking Michel questions about me. They were much more self-confident than the black wives I had met in the south. In another hut, an old woman was scraping a goatskin with a solution of tamarind pods, which cured the hide so that it could be made into a water-carrier, or *guerba*. Nearby an old man lay alone, coughing. Michel had known he had bronchitis, and had instructed me to buy some cough mixture from Agadez. The old man spoke no French, so I pompously issued instructions to one of his sons on the correct dosage.

'It's OK,' he said to me sharply, 'we know how to use cough mixture. We're not stupid, you know – just poor.'

Michel's cousins taught me to abandon my useless sun hat in favour of a four-yard turban wound around my head and covering my mouth and nose. At first it seemed uncomfortable, even counterintuitive, to wear all that weight on my head during intense heat and to gnash on a piece of dry cotton. Within a couple of days I had grasped that the cotton gauze prevents dry air from entering the mouth. Dust is also excluded, but, most importantly, the lips do not crack and the mouth does not dry out, so the total loss of humidity from the body is diminished. If the turban is soaked, then even at midday when the sun is doing its worst, evaporation keeps it cool for a couple of hours – and so the head, and so the entire body.

In Chad I had made a fetish of keeping cool with water; I now formed a deep bond with my new love, four yards of indigo-dyed cotton.

That night, Michel announced that he would take me to meet the most beautiful woman in the world. 'Extraordinarily beautiful,' he said, 'and extraordinarily fat.'

The woman in question turned out to be the village matriarch, holding court from a wood-and-string chaise longue by the light of a single oil lamp. In the circle around her, men sat on chairs while women reclined on beds, and the sound of their energetic chatter and laughter cut into the deep silence of the desert night. The great beauty would receive questions from the floor, as it were, and fling back a witticism that had the company in hysterics. Unfortunately, I could understand nothing.

In Tuareg society the family unit is matrilinear; daughters live with their mothers, and men leave their own homes to enter this womanish *foyer*.

'What do you think of her?' Michel hissed after a while.

'Well, I . . .' I certainly did not find her a beauty, but I was uncertain whether mercurial Michel had meant that literally.

The large lady had affected not to notice our existence. Suddenly she deigned to acknowledge us. 'So what does your white visitor think of the Tuareg?' she asked Michel.

The question was translated and I fumbled for something to say. 'Well, I've noticed that Tuareg women are much more independent.'

'*What?*' she asked.

'Than in the south, I mean.'

Michel translated, and the matriarch's eyes hooded as her mouth pursed in utter disdain. 'Hah! Hausa women are slaves. Silence and servitude is what's expected of them, and if they behave any differently, they're beaten. I don't know who to blame, though, the men for their crudity or the women for their passivity.'

There were mumbles of agreement from the men and women around her.

'Mind you,' she added, 'don't make the mistake of thinking Tuareg men would be any different if we gave them the chance!'

There were grunts of laughter.

The independent status of Tuareg women is rare in the Islamic world, but, as even nomads like the Tuareg are brought into the ambit of urban culture, Islamic observation is increasing and the traditional independence of the Tuareg women is diminishing.

Michel made our excuses and led me away from the tiny pool of light. Soon the silhouettes of date palms loomed against the moonlit sky.

'Don't believe that balls about Hausa women being slaves,' said Michel. 'They know they're *meant* to be slaves – it just makes them slyer. They know how to get their way!'

We walked through vegetable gardens until we reached a steep bank, and clambered a few feet down onto a bed of virgin white sand shining brightly under the rising moon. The broad wadi, perhaps fifty feet in width, had come to life a few days earlier, as it did every year, when the rains finally came.

'At this time of year you have to be alert when crossing a wadi,' said Michel. 'There's not a cloud to be seen over your head, and without warning a wall of water suddenly comes surging through it, and if you're standing in the middle, forget it. The water has come down from storms in the mountains miles away, gathering speed and picking up rocks, bushes, entire trees – even dead animals, and sometimes some that are half-alive. You even find fish here, that have been brought from God knows what lakes in the mountains. And when it's passed, the sand looks like this – white and clean. Look, no thorns.'

Free for once of the threat from the desert's ubiquitous thorns, we threw off our sandals and trotted on the moist, spongy sand. Here and there small pools of water remained, and by the light of the torch we made out some tiny, tadpole-like fish.

'You see,' said Michel, 'there is life in the desert.'

CHAPTER TWENTY-ONE

Desert Storm

WE SET OFF AT DAWN, and by eight-thirty the engine was backfiring badly. Wali jumped down and began to clean the plugs. The engine coughed and off we went again. We broke down nine or ten times that day, half the time with flat tyres, the other half electrics. I did not complain: I could not have afforded a jeep that worked. And it gave Michel, one of the most renowned mechanics in Agadez, much time to wax on his vehicular preferences, to wit the classical perfection and eternal verity of the Land Rover. 'Ah, *les Landrovairs*,' he said, in the tone of a man recalling a former lover. 'Land Rovers are good, but Range Rovers – ah! *La suspension*! There's nothing better than *un Range* in the desert, you know. And before this *putain* of a war I had three Range Rovers – three. Hah! Then the war came, the tourists went, and one by one I had to sell the vehicles to support my family. Now I have this old shit-heap. Can you imagine a tourist wanting to go anywhere in this shit-heap?

'But you know the advantage of a shit-heap? No one wants to steal it! Drive around these mountains in a decent vehicle and it's just a matter of time before someone takes it off you – army, rebels, bandits, whoever. Or they'll whip out the spare parts – if they need a fuel pump they'll take yours, even if they're generous enough to leave you the rest of the vehicle. But a shit-heap like this – who'd touch it? I tell you, there've been times I've been stopped by some bunch of villains and I've thought, *This is it, then* – but they took one look at this *vieille pourriture* and waved me on. Look at the doors.' The doors were stamped metal forms devoid of linings, armrests or handles, and opened by a tug on a loop of wire. 'How many times have I been stopped and had the army strip the vehicle to the bones searching for weapons or drugs. They always do the doors first – but if the doors already look like this, they don't need to. It saves hours!'

The car stopped again. We climbed down and looked under the bonnet. The engine was encrusted with filth, its every organ and artery dented, bent, pitted, flayed or decomposing.

Michel pulled off the fuel pump and held it between thumb and forefinger, like a fishmonger sniffing a rotten fish. 'Fucking old Toyota shit-heap,' he said, matter-of-factly. He tore the husk of the pump in two and shrivelled gaskets oozed out like intestines. 'Rotten seals. But I haven't got any spares. These'll have to do.' He sighed, and set to work. 'The British make the best cars, you know. Unfortunately, the Japanese make the cheapest spare parts, and since nobody in the desert can afford a new car, people get Toyotas. Tell Land Rover that when you go to England. Tell them that if they made their spare parts cheaper, everybody would buy *les Lands*.'

I promised that I would.

On the occasions when we were actually moving, our eagle-eyed guard Reesa spent most of his time asleep, his head lolling on my shoulder. Whenever we stopped he would shuffle away from the jeep and crouch alone in the bushes, enjoying a solitary cigarette.

'What's he up to?' I asked Michel.

'He has to get some distance off to watch out for anyone approaching.'

In late afternoon the sky darkened dramatically. The view ahead of us was deceptively reminiscent of the English Lake District, a vista of rain-swollen clouds lowering between twin hills about 400 feet high. The clouds slowly turned chocolate-brown, their bases gall-yellow, while the clouds in the south-east went a melodramatic purple. An acidic light suddenly burned through them, flooding the valley, turning the hills an unnatural orange and the shrubs a bright, chemical green. A powerful wind gusted, then rain swept across the surface of the sand like a sinister fog, engulfing us with a drumming that increased into a furious hail.

Michel nosed the jeep into the foliage of a tree. Mauve-veined branches and emerald leaves writhed and thrashed the windscreen. The lurid sky had turned a dull grey, though garish yellow flashes sometimes tore through. The runnels in the sand filled with water,

which deepened and rose until the once-level desert resembled a
stormy sea. Still the rain intensified, beating on the windscreen until
nothing could be seen through the glass.

The wind veered around to the east, driving thick cords of rain into
the jeep. In the absence of a window-winder, I undid my turban and
tried holding it against the rain, while Reesa took a pair of pliers to the
brass stump that had once held a handle and tried to raise the window.
He failed. The wind repeatedly tore the cotton from my hands, and
soon we were soaked. Then we had to get out to help push back the
car, which was now threatening to sink in the rain-sodden sand. The
engine, of course, had failed.

Fifty yards to the east, the normal drainage system, a wadi a
hundred feet wide and three feet deep, had become a torrent of
astonishing violence, entire trees shooting past like cars on a highway.

Slowly the storm abated, and the grey curtain passed over us. Wali
stood shivering in his soaked T-shirt; he had no other clothes. Michel
found himself another shirt, and I opened my bag and handed dry
shirts to Wali and Reesa. The veil of cloud in the north lifted to reveal
a radiant, clean sky of freshly washed teal blue and hills flecked with
white where the light struck dozens of running streams.

Michel dried out the electrics and managed to start the car. The
plain was a serpent's nest of rivulets in full spate. After clawing our
way for a few miles through this inland delta, we came to a river of
boiling water tearing through a grove of tall, leathery shrubs. Michel
appraised the situation, keeping well clear of the edge. 'We won't get
across that today,' he said. 'Maybe in the morning, definitely by
midday. We camp here tonight.'

Reesa, our professional soldier, assessed our position. 'Can anyone
approach us here?'

Michel raised an eyebrow. '*Attack us*? We're in the middle of two
thousand streams, and no one knows we're here.'

We spread our turbans and shirts on thorn bushes to dry, and
searched among the flood detritus for firewood. Not all desert trees
burn well, some being fast-growing shrubs with hollow trunks.
Michel, reclining like Napoleon on a folding military bed, would tell
me, 'That one's OK,' or, 'That one's crap.' We hacked the branches
open to expose the dry cores and built a fire which Michel used petrol
to ignite.

'In your country,' he asked me, 'do thieves rob anyone equally?'

'Er, yes,' I said.

'Well we don't. We just rob the rich. I mean, if some big American comes here with jewellery and a new four-by-four, they deserve to have it robbed. They're asking for it, they're not being . . . sensitive to the social situation here. Honestly, if someone like that turned up here, I'd rob them myself.'

Reesa had propped his AK47 against a bush, and I watched him contentedly make tea: bring the water to the boil in the teapot once, twice; take it off the fire; add sugar to a glass, pour in some tea; mix it up and pour it into the teapot; then repeat the process several times with minor variations, sweetening and cooling the tea.

'That's not the way you made it last night,' I said.

'Ah,' said Michel, 'there isn't a fixed, mathematical technique, no precise time or even personal method! It's a function of the heat, of how much tea you have and so on. It's different every time.'

'Yes,' Reesa agreed.

The fire burned fiercely and the wind whipped around us. We kept changing our positions to avoid the smoke being blown into our eyes. Reesa left the circle, and as dusk settled a question occurred to me.

'Michel, how Moslem are the Tuareg, exactly?'

'What do you mean? We *are* Moslem! Didn't you know a Tuareg was one of the Prophet's – peace be upon him – right-hand men!'

'But I've noticed you don't observe as much as some Moslems I've met. I mean, I've travelled with people who stop and pray, er, religiously, I mean five times a day.'

'Ah. You know, the Tuareg are *croyant* (believing), *très croyant*, but not very *pratiquant* (practising). Islam recognizes that it's more difficult for us in the desert. Conditions are tough. It gives you every help, though – if there's no water to purify yourself, you can use a stone, or even earth.'

I nodded, but was not convinced. Could the desert dwellers' neglect of religious austerities really be explained by a sort of spiritual sick-note that let them off? I was more persuaded by the theory that religions exercise a tighter ritual grip in towns, where houses of worship can be built and preachers can keep a close eye on congregations. Nomadic societies are by definition freer from such static forms of control. The nomad's soul is freer.

While we had wiseacred the sun had vanished, and Wali was still hard at work changing inner tubes. Reesa had disappeared. I took a stroll towards the puffs of grey smoke and found him crouched on his hams, so that at first I thought he was at his toilet. Then I saw that he was smoking hasheesh, in a fat joint made from a shiny pink envelope. So this was why he took a walk every time we stopped. It also explained his constant good humour – and fatigue.

'You want some?' he said. I was reluctant to accept because I knew Michel disapproved, but I crouched beside him and took it.

The blue smoke curled into the dusky sky. Reesa was grinning. 'This stuff gets you through a lot of shit, you know.'

'You smoked it when you were fighting?'

'Sometimes.'

'It can't have been the best thing for staying alert.'

He grinned again, a hasheesh-slackened grin. His eyes were shining. 'Martin, take it from me, when you've had hasheesh it improves your aim enormously, you can't miss, the bullets just fly to the target like, like . . . like . . . '

'You sure it doesn't just *feel* as though they do?'

Back at the camp Michel scrutinized us both. Obviously he knew perfectly well why Reesa kept disappearing. We turned in. Stars began to fill the sky, which turned from black to grey as a sliver of white appeared in the V between two groves of trees. A symmetrical and stately moon began to rise. By the time it was overhead, the others were asleep.

CHAPTER TWENTY-TWO

Hasheesheen

A
T FIRST LIGHT the river was down to only a foot. We loaded the jeep and edged towards the brown stream.

Wali jumped down to lock the wheels in four-wheel-drive, then walked ahead of us using a branch to plumb the depths. Michel inched the jeep forward. When we reached the far side, there was a mud bank ten or twelve feet high to mount. Three of us got behind the jeep and helped heave it forward, while it slithered around, threatening to crush us under its big rear wheels. But they only splattered us with black, shiny mud.

The mud dried in seconds. The tall sky was evenly blue and utterly cloudless. The rocky teeth and mesas of the Aïr posed sharp-edged in the clear air, sitters waiting for the camera's click, their colours those of industrial chic – rust-orange, glossy cinder blacks, greys as soft and indefinite as the cashmeres of Milan. Between the mountains, barren plains stretched for miles, but we spotted the occasional, primitive hut, surrounded by a few skinny goats.

'I fancy some cheese,' said Michel, 'and I know the people here.' As we descended a sandy track he grinned at me. 'Wait till you see the woman at this place. She's a beauty. And,' he nudged Reesa, 'unmarried.'

We were greeted by a woman with the near-freakish beauty of a fashion model: enormous eyes, a tiny pointed nose and a smile that cracked her face in two. Her hand rested on the rim of his window, the coarse, grey-palmed hand of a desert-dweller, the nails cracked and the dried skin ingrained with dirt. Her face was mapped with fine sun-lines, but somehow they only accentuated her remarkable beauty. I gawped.

'So, do you have any cheese?' asked Michel.

'We have some we made today,' she said, turning towards the hut.

'She's absolutely amazing,' I said.

'Not her,' he said scornfully, '*she's* the *sister*.'

A nondescript, oval-faced woman walked towards us, pulling her veil halfway across her face, careful not to conceal a flirtatious smile. 'Now that,' whispered Michel, 'is a real beauty.'

The first sister brought over several square pats of white goats' cheese, pressed between grids woven from hollow grasses. 'Do you have any old ones?' asked Michel. 'I want to show *le blanc*.'

She went back into the hut and fetched several shrunken versions, brownish, resinous and hard. 'They're about six or eight months old. They're a delicacy, you know. They take them into the towns and sell them for a good price.'

We contented ourselves with the day-old cheese. It was better than most of what passes for goats' cheese where I come from – fresh, fluffy and sweetly sour. A slight grittiness came from a small percentage of sand – excellent for the digestion. On chunks of crisp, slightly charred unleavened bread it was very good indeed. In return, we made gifts of tobacco leaves, tea and sugar – in rather generous quantities, I thought; a beaming Michel was trying to impress. He still had only two wives out of a possible four.

We made slow progress; the Aïr are nothing like as rugged as the mountains of Chad, but Michel's despised *pourriture* could manage only an hour or two without a breakdown. Towards midday we rumbled into a small oasis with stone cottages reminiscent of a Cornish fishing village. Boys were driving a camel back and forth to operate a pulley, which hoisted a bulging goatskin of water from a well. The few acres of the oasis were well irrigated by silvery streams running along cupped mud channels, watering maize and wheat and grapefruit as well as mango and citrus trees. At the centre of the grove was a house, shady and cool, and so tightly embraced by citrus trees that you could reach through a window to pluck an orange or lemon.

Farm boys lounged in the moist shade of spreading mango trees. In the crowns of the date palms the yellow-khaki fruit were bursting, and one of the boys filled my hat with dates. Michel told him I had a fondness for mangos, and he clambered into a tree and dropped me a dozen of the ripe yellow fruit.

When we left, I paused on the narrow lane where camels passed and irrigation ditches disgorged unneeded water. Water and camel urine mingled underfoot. Acacia trees, twenty feet high, met overhead, transforming the lane into a green tunnel. Flies buzzed busily, and there rose a smell of delicious decay, the ripe urinous summer rot of an English woodland. We all knew that this was paradise. At the end of the tunnel, we emerged into blaring sunshine.

The road decayed quickly into a track that tipped through savage hills of shattered rock. After an hour we descended a steep cinder slope towards 'la cascade'. This was the plunge-pool of a waterfall, a tiny dark lake under a towering cliff. On a beach of coarse pink sand we undressed – the others shyer of exposing their bodies than I – and waded in. The green water felt icy at first, then warm. Knots of caramel-coloured camel dung floated on the surface.

'This is where we came in the war, to rest,' Reesa told me. The water filling the pool flowed over a lip of smooth, black rock. I climbed up and followed the waters twenty feet back to a second pool. Within a deep cleft, plunging water had carved a perfectly round bowl like a vast black calabash. I dived and swam in its dark and soothing waters. To one side was a niche just wide enough for me to sit in, gaze up at the circle of blue sky above and feel the water cool on my feet.

Michel erected his camp bed in the shade beside his jeep and slept. The rest of us climbed the cliffs over the pool. We could see the source of the cascade, a winding river descending through a series of small valleys. To the south-east the river carved a green groove, where camels fed as they had done for millennia.

Reesa stared into the distance. 'The army attacked us one day at Timia,' he said, jerking his head towards the north-east, 'just over those hills. It was an ambush, someone had told them where we were. Anyway, we had twenty-one dead.' He lit a cigarette, as did Wali, a chain-smoking fifteen-year-old. 'And this is where we brought the wounded. Some died, but some got better. It was like a holiday here, away from the fighting. Swimming, joking . . .'

I imagined the scene. A bunch of young Tuareg guerillas armed with AK47s, fighting off a force of superior numbers, armed with superior weapons – including, finally, the French assault helicopters.

Reesa pulled off a boot and shook out a stone.

'Good boots,' I observed. They were orange Timberlands from America.

'Yes, these are new. Somebody got me them from Niamey. I wish I'd had boots like this when I was fighting. You wouldn't believe how fucking depressing it is to fight without good boots.'

'So what did you used to fight in?'

'Shoes that rapidly fell apart. Then flip-flops. It wasn't easy.' As often happened when we discussed the war, he became heated. 'And now we've stopped fighting, and for what? What have we gained? If it was up to me we'd go back to war tomorrow.'

By mid-afternoon he had fallen into his habitual doped stupor, lolling between Michel and me as we drove away.

'He's a lot of use as a guard,' I said.

'Hasheesh sharpens the eye,' said Michel defensively.

'You must be joking. Look at him – you could pour him into a bottle.'

'No,' Michel insisted. 'In a fight, it works differently. It does something to you, it makes you a better fighter.'

'Really.'

'He's had a hard time. When the army caught him they tortured him. I mean, they almost destroyed him. Only recently he was a young man, full of life. Now he's the way you see him. It happened to a lot of our young men. Or else, they died. We killed a lot of the fuckers, though. Some of our boys vowed that for every one of us they killed we'd kill ten of them. A lot of killing. My God, the things I've seen. Murder, torture. Seen *and* done.'

'You tortured them, Michel?'

'Yes. You try finding a friend of yours cut into a thousand pieces like a melon. All you can think about is revenge.' He became angry. 'And I don't regret it, I don't regret it at all.' He paused. 'I don't want to talk about it. Are you giving him money?'

'Pardon?'

'Have you been paying for Reesa's hasheesh?'

'No,' I said.

'Well, don't. I didn't think he had this much. He must have nearly run out by now. Don't give him any money.'

It was evening when we reached the oasis of Timia, driving along a broad white riverbed overlooked from a hilltop by the picturesque ruins of a French fort. The slopes that crowded around the town were heaped with boulders, while along the riverbanks stone defences protected gardens pregnant with date and banana trees from spating waters. Timia itself was a pretty jumble of square, mud-brick houses. Alleys darted among the houses in whimsical zig-zags, and I was reminded of sun-baked villages on a Greek island. Black-clad women stood at wells drawing water, and children played in open squares. It was the prettiest oasis I had seen.

Michel led us to a compound where he said we could sleep. Small boys crowded around us, soon chased away and replaced by adult visitors. There was the man who wanted to sell me silver necklaces, the man who wanted me to cure his toothache, and the tiny, dwarfed and crippled man, carried on the back of a friend, who wanted me to arrange for a Western charity to provide him with an electric wheelchair. 'I've seen a picture of one,' he said.

'It wouldn't work here,' I told him. 'The tyres would sink into the sand. How many people in Timia have bicycles?'

'None,' he replied, crestfallen.

Another man sat beside me, with bulging, blind, blue eyes that matched the blue of his robes. '*Globes terrestes*,' I heard a child chant.

'That's what the kids call me,' he grinned, 'because apparently my eyes look like the earth seen from outer space.' He had been sitting silently beside me, listening with a quiet smile on his lips. After a while he said, 'Can I ask you a favour?'

'You can try.'

'Will you let me touch your face?'

He ran his fingers over my face, softly, softly, then carefully felt my right arm and hand. And sat back, smiling with satisfaction.

More silver-sellers arrived, gaggling round me, spreading necklaces and pendants and yelling at the tops of their voices. I turned to Michel. 'I hardly have any money with me,' I said. This was true. I had just enough currency left to pay Michel off and buy a bus ticket to Niamey.

He shrugged. 'I know. But the tourists don't come here any more.'

I went painstakingly through every man's wares, and chose something from each: a dagger, some silver pendants, an old Tuareg tobacco pouch, some carved stone charms. They tried to bargain, but in vain; it was a buyer's market.

As yet another silversmith arrived, I looked at Michel ruefully. I had no more cash, and I would have to trade cigarettes. I went out to the jeep to fetch the cartons of cigarettes I had bought, and Reesa followed me out. Two cartons – forty packs – were missing. The truth emerged: he had been swapping them for hasheesh.

I told him he was a thief. His eyes flashed, and for an instant I felt fear, remembering the incident at the well in Chad. Reesa threw a look at the wall beyond which Michel was sitting and hissed, 'M. Martin, I'm sorry, I *had* to sell them. I *have* to have hasheesh. You mustn't say anything to Michel. He doesn't understand.'

'Why didn't you ask me, instead of just taking?'

'That's what I am doing – Martin, please, you must give me some of those cigarettes. I need hasheesh.'

I looked at him sternly and his face broke into a broad grin. Reesa was so likeable and irrepressible that it was impossible to be angry. With a sigh, I handed over the third carton. 'But for God's sake don't tell Michel I'm paying for your hasheesh,' I said.

'I won't. You're a good friend, Martin.'

Partners in crime.

That evening Reesa and Wali went off to a wedding party, and Michel and I sat by the brazier. A question suddenly formed in my mind.

'Michel, will *Globes* ever marry?'

'Difficult. I don't think so. He hasn't got any money, you see, and being blind he has no way of earning money. But you remember Akoli, the little cripple? Did you notice how well he was dressed, that he had a good haircut and he smelt good? Well he's wily, you know, a good businessman, he buys and sells, and he's worth money. They'll find him a wife OK. But the blind man – I doubt it.'

The next morning we paid a visit to the head man. It was soon after dawn, and the old fellow was sitting stiffly inside his small square room, reading the pages of a Koran. He barely looked up as we took off our shoes and ducked inside, but went on reading aloud, rocking slightly. Then he laid the text down and turned myopically towards us. 'I'm praying to help my people,' he said, wrapping the loose pages in a green cloth. 'I rarely go out these days. I pray, but there's no rain.'

'We have come to pay our respects,' said Michel.

The old man gave a slight nod. 'Does your friend have any money to help us?'

'No,' said Michel.

He nodded again, and turned to me. 'At least go home and tell your people how we suffer,' he said. The rains are late this year, and maybe they won't come at all now.'

'But we went through a great storm on our way here,' I said.

He waved his hand dismissively. 'There's been rain further south, but here in the hills it's patchy. By now there should be grass coming up, the well-water should be rising. We've had droughts for several years now, so we need a deluge to replenish the wells. Most of them are unuseable, and now we are only finding water in the deepest ones. Food is expensive, and our goats are thin. We used to be able to buy a sack of maize for two goats, now we need five or six.'

The staple foodstuff in the village was maize flour, grown in the south of the country and traditionally purchased by selling cattle and sheep. But the lack of rain meant poor pasturage, and the animals of the Aïr were suffering.

He told me that until a couple of years earlier they had sent their camels across the desert to Bilma to buy salt — the legendary salt caravans, a forty-day round-trip. After they returned, they would head south to Zinder, where the salt would be traded or sold for maize. 'But now the camels are too weak to make the journey. We have very little maize. It means we're hungry too. But what we need is rain.'

The road north of Timia leads out of the hills onto a huge sandy plain with volcanic cones hundreds of feet high. The land is covered with burnt grasses that resemble fine blond hair. There is little water, and habitations are rare.

We stopped at Assoudé, the ruins of the town that preceded Agadez in eminence in the Aïr Mountains. In the centre of acres of fallen stones and jutting rafters was a large, ruined mosque. A bluff overlooks the town in the west, and I climbed it to take a photograph. I saw a single carved pictogram of a man, a camel and an unloaded bale. Once upon a time this town was a nexus of camel caravans, with water supplies, a huge market, a Sultan, a standing army. I visualized the straggling camel trains, the bazaar, the brothels, the long lines of black slaves bound for work in North Africa, Egypt, Persia . . . .

The northernmost point I reached in Niger was the desert town of Iferouâne, close to the Algerian border, where golden sand-coloured buildings stood in sand streets planted not with palms but with acacia. It was a day of paralysing heat. We hiked out to a rocky bluff where there were tall human figures engraved, then, exhausted, slept in the shade of a huge tree.

North-east of here was Temet, one of the places I had most wanted to visit, where some of the most fabulous dunes in the world are found. Temet could only be reached via a dangerous, unpopulated, waterless road that led to one of the key rebel fortresses. I agreed reluctantly with Michel that it would be unwise to make the attempt. Maybe next time . . . .

CHAPTER TWENTY-THREE

Of Mere Being

Full many a flower is born to blush unseen,
And waste its sweetness on the desert air.
Thomas Gray

TWO DAYS LATER, heading south, as we crossed a river by a grove of palm trees, a group of people came rushing towards us.

Michel let out a peeved 'tut'. Already we were carrying five passengers, mostly rebel soldiers returning from leave, whom it was difficult to refuse.

'There's a sick woman!' people with frightened faces shouted at us, 'a friend – our cousin – my aunt.' A tall, imposing deaf man took charge. '*Une femme est gravement malade,*' he told us, in a high-pitched nasal voice. 'Three kilometres away. We need you to help.'

With a sigh, Michel gestured for the deaf man to clamber in and guide us. He pointed west, and we drove away from the little throng, up a track that led towards some low hills.

It was a terracotta-coloured village with a pretty little domed mosque. As commander of the vehicle, foreigner, presumptive man of knowledge or at least wealth, I was led into a dark room where a small, frightened-looking woman was crouching. Half the village crowded in behind me.

'What happened?' I asked.

It seemed that she had been reaching under a mat and something – a snake or scorpion – had bitten her.

I led the woman outside. There was a wall of spectators. Children forced their way through adult legs to gawp and chatter from the front row – it was street theatre. I called sharply for quiet, and a respectful silence fell.

The woman was in obvious pain. I unwound a filthy rag and saw her hand, swollen to the size of a boxing glove. The bite was between her index and middle fingers. I held her bulbous mitt lightly between my own hands and squeezed with my thumbs. A thick cord of puss spurted out like toothpaste, and she cried out in pain and surprise.

Her temperature was high, her pulse dangerously slow. The wound was infected and certainly spreading to her blood. I cleaned and bandaged her hand, and turned to Michel. 'She needs a hospital. We'll have to take her to Agadez.'

The enthusiastic babble began again. We were brought tea, and I could see Reesa and Wali basking in a glow of importance. After a few minutes the message came back that the woman could only travel with an entourage of three – two women to attend her in hospital and a male chaperone. Michel had a fit.

'I'm not a fucking taxi! This is the Tuareg for you. If you've got a vehicle they just expect you to be a voluntary bus service. Everyone expects a lift, and no one ever offers to pay a *sou*. The extra weight slows us down and uses much more petrol. I'm not doing it.'

We argued, and I won. We sent back word that we would be willing to take *three* people – the patient, one female attendant and a man.

It grew dark, and word spread quickly that a healer was in town. Families brought out their halt and lame, and I handed out antiseptics and antibiotics for infected gums and open sores. I was no doctor, and I know I ran the risk of actually doing harm. I always travel with spare drugs, but what if, for example, I gave penicillin to the rare person who is severely allergic to it? On the other hand, how could I deny people the drugs that would bring relief from suppurating infections and malaria attacks?

A woman approached me with a heavy bundle wrapped in the folds of her black headscarf. She showed me the long and wasted body of a child of nine or ten. His head was swollen to twice its normal size and covered with livid pink sores on which someone had plastered a green herbal medicine like boiled grass. He was crying piteously. Appalled, I called Michel over. A look of revulsion twisted his face. It appeared that the mother had already made one journey to the hospital with her child. The doctors had sent them home, saying there was nothing to be done. Now I was her last hope, and I was helpless. She pulled the cloth back over her child's head and carried her burden away.

Now a grinning Tuareg woman was brought before me. She was holding a weak, skull-headed baby obviously on the verge of death. As I questioned the mother she grinned idiotically. She had an attractive mouth and brilliant, stupid eyes. She was twenty-nine, and this was her fourth child. The first three had died. She was, people whispered, mad.

'They say she doesn't know how to look after her babies,' said Michel. 'She forgets to breast-feed them. She's been married several times, but the men always leave her. She just wanders around.'

The woman gave a burbling laugh.

'Look at her,' said Michel with disgust, 'She's loving this – being the centre of attention. She's fucking crazy.'

'Maybe, but we have to take this baby to Agadez.'

It was only as we were leaving that I discovered the woman with the infected hand was not coming with us. 'Her husband says it's four of them or no one at all,' Michel told me.

I looked at him in horror. 'He'd rather his wife died? There might not be another vehicle for days.'

'So now you know what these people are like. Come on, we've got enough passengers. Let's get out of here.'

I found the woman and gave her some antiseptic and a bundle of sterile bandages. I provided two courses of powerful antibiotics and gave precise instructions on when to take them. And we abandoned her.

A couple of hours later, Michel pulled up in the middle of the desert. It was time to eat. The Tuareg woman grinned foolishly around her, the mewling baby forgotten in the crook of her arm. Reesa was chopping up a leg of lamb. It would be the first meat anyone had tasted since we left Agadez.

'Who bought that?' I asked.

'It's a present to you from the husband of the woman with snakebite,' Michel said, 'to show his gratitude to you for treating her.'

'The same maniac who won't let her travel to Agadez to save her life?'

Michel shrugged. 'Once he'd said she had to travel with three companions he couldn't change his mind. He would have lost face.'

To everyone's relief, the next day was cloudy. We juddered towards Agadez with our cargo of rebel fighters, the madwoman and child, and somebody's sack of dates. Towards dusk a coppery sun flooded the desert hills with clean, cool light. The overladen jeep wallowed through a wide riverbed of powdery white sand and suddenly I felt at sea in the desert, as though I were passing through and not over it – the jeep was a ship yawing and pitching in a white sea. We passed a young palm tree with a blue and golden bird in its crown and scraps came to me of a Wallace Stevens poem I first read twenty years ago. The last poem Stevens ever wrote, it is called 'Of Mere Being':

> The palm at the end of the mind,
> Beyond the last thought, rises
> In the bronze distance,
>
> A gold-feathered bird
> Sings in the palm, without human meaning,
> Without human feeling, a foreign song.
>
> You know then that it is not the reason
> That makes us happy or unhappy.
> The bird sings. Its feathers shine.
>
> The palm stands on the edge of space.
> The wind moves slowly in the branches.
> The bird's fire-fangled feathers dangle down.

In Agadez we took the madwoman to the hospital. 'This child is nearly dead,' the doctor told me, 'but don't worry – we'll look after it now.' I saw the woman admitted to the female ward, paid for some drugs and left. When I returned the following afternoon, the woman was crouching on the floor in a corner of the ward, the child having received no treatment.

I read the riot act, threatened everyone with the sack, and the sulky orderlies – all southerners who obviously despised the Tuareg –

reluctantly began treatment. Over the next few days the child grew stronger. But God knew what its chances would be when the mother was discharged.

Reading in Michel's courtyard, I was struck by a few lines from the Koran, some words from the Prophet whose inheritance at birth was six camels and a slave girl, who was an orphan at six, went on to be a camel driver, later saw the light, and founded a great and moral religion.

Know that the life of this world is but a sport, and a play, an adornment, and something to boast of amongst yourselves. And the multiplying of children is like a rain-growth, its vegetation pleases the misbelievers. Then they wither away, and thou mayest see them become yellow; then they become but grit.

CHAPTER TWENTY-FOUR

Sand

Philosophy is the stray camel of the Faithful,
take hold of it wherever you come across it.
The Prophet Mohammed

Only he who understands is sad.
Arab proverb.

A COUPLE OF DAYS AFTER our return from the Aïr Mountains, Abdal, one of Michel's cousins, arrived at the compound. I had already met him once, on the night we had visited Michel's family north of Agadez. Now, travelling with two camels, Abdal was on his way back to his family camp, a hundred miles to the south.

He was a tall man, with a quick mind and a wolfish grin. He had been educated in Agadez, and, in the years before the rebellion, had worked in the uranium mines at Arlit. Abdal had fought in the rebellion and seen it fail. He was utterly disillusioned.

As we talked, the idea emerged that I should spend some time travelling with him in the desert. I wondered if we could turn it into an expedition to explore the east–west network of cliffs and dunes in the desert to the south of Agadez. At walking pace, we could take our time to look for signs of prehistoric settlements. I knew Abdal was not a rich man, and I offered to hire him and his two camels for a week. He readily agreed.

It took another day to prepare, and we set out together in mid-afternoon heat. Abdal walked – and walked. 'He's a camel,' Michel had told me. 'If I ever need to cross the desert I just climb on his back.'

Abdal was used to walking solidly for two days to reach his encampments. The women or children would ride, but he would lope. He found it difficult to slow down just for my benefit, and I, reluctant to lose face, did not want to ask him to. So he strode across the desert, and I struggled to keep up, sweat pouring down my back. He did not notice my difficulties.

After two hours, Abdal stopped.

'Would you like some water, Martin?'

'What about you?'

'Yes, I'll have some water.'

'Me too,' I said, attempting nonchalance. My hands were shaking as I lifted the bottle, and I had a strong desire to empty it.

We walked on. It had been a day of pulverizing summer heat, and now the sky was clouding over. We had been walking for three hours when it started to rain. At first it was delicious, but then it became cold. The rain stopped after only half an hour, but we were both soaked through and the low sun was too weak to dry us. By eight, when we stopped for the day, I knew it for sure: in the Sahara, for God's sake, in *midsummer* – I had managed to catch a chill.

Now, still only four hours out of Agadez, we hit a twin crisis. One of our two twenty-litre jerrycans had sprung a small leak as we loaded it at Michel's house. Under the layers of moistened coir that kept it cool, a crack was spreading. This was the 'brand new' jerrycan I had bought at Agadez market. 'When they get old,' Abdal said, 'the plastic gets brittle from the sun and starts to crack. So the men in the markets wrap them in insulation to cover up the repairs and tell you they're brand new.'

'Bastards,' I muttered, pointlessly.

We still had one twenty-litre jerrycan and two *guerbas*, goatskins. To return to Agadez to replace the jerrycan would mean losing at least a day.

We also discovered that we had packed almost no sugar. The seriousness of such a discovery for a Tuareg cannot be over-emphasized. We had tinned fish, pasta, bread, fresh vegetables, fresh fruit. Abdal did not give a damn. 'I can't go on without sugar,' he said.

Tuareg life is one of deprivation, and addictions help to make it tolerable. Alcohol is banned by Islam, so for most, the available addictions are tea, sugar and tobacco. 'I get headaches,' Abdal went on. 'It's worse than running out of water. We Tuareg are used to going without water. *But we must have tea.*'

'I have evaporated milk!' I said triumphantly. 'That's sweet!'

Even as I pulled the can out of a bag, Abdal was saying dismissively, 'That brand has no sugar in it.' I stared at the label. 'Made in UK', it said, 'No added sugar'.

There were no shops – or any permanent settlements – for many days' walk in any direction but back the way we came. But we might, Abdal thought, find a Tuareg to sell us some sugar. 'People *sometimes* have a little extra,' he said doubtfully. If all else failed, we would make a beeline for his camp; they had sugar there. We decided to carry on.

That night Abdal talked again about the rebellion.

'It was inevitable we'd lose when the French gave the government assault helicopters. Don't ask me why they did. There were meant to be people high up in French public life who supported the Tuareg – who understood our predicament. After all, they were the cunts who started all this with their criminal frontiers that cut the Tuareg lands into fragments and gave contol over us to the blacks. There are French people who feel bad about that. Anyway, what chance did we stand? There were too many rebel factions, we were never united, never trained properly, never clear about our strategic aims. We'd make some hugely significant advance like the capture of Iferouâne, then give it away because our troops didn't *understand* its strategic significance. We made the government's life a misery, but in the end we couldn't defeat a national army.

'As far as I can see, we gained nothing. They killed us, and we killed them, mostly soldiers from the south who didn't have the slightest idea why they were fighting. Plenty of murder, torture, bestial behaviour – that's what you get in a war. And the stupid thing is, we haven't gained anything but the death of a lot of young men – no, not only young ones. They killed old people and civilians. There was an old man who was shot because he happened to approach a scene where the rebels had ambushed the army. The government soldiers found their men dead, and then this poor old fellow appears in the distance, coming over a hill, bringing vegetables to market in Agadez on his donkey. There's a war going on, but he thinks, "I need to sell these vegetables, and I'll be safe, who's going to shoot an old man?" They started firing before they could even see who he was. Shot him to pieces.

'What this war has brought about is the utter impoverishment of the Tuareg. Everyone had to give his money to support the rebellion. When they came and asked for money, you gave it. You gave your sheep, even your sons – they joined the rebellion as soon as they were fifteen or so. Commerce collapsed, and people sold what livestock they had left to keep from starving. You had the army killing animals and poisoning wells. As a community, the Tuareg have been comprehensibly impoverished. *Our* rebellion has crushed *us*.'

I woke up the next morning with my teeth chattering. A chill: unbelievable. I plodded after Abdal for two hours, then he ordered me onto the camel. It was the most uncomfortable ride I had ever known. When a camel stands, its back and neck come together, throwing the rider forward so that he expects to be catapulted into space. At the same time the camel's neck meets a spike on the front of the saddle, which stops it tipping the rider any further forward. The saddle itself is made of wood padded only with layers of blankets. There are no stirrups with which to stress your body, since you rest your feet on the camel's neck, which you rub with your toes to make it go faster. The camel's lope tosses the rider backwards and forwards through an arc, so that you pivot from your hip like a metronome.

After an hour in the saddle, I felt worse than ever.

At lunch we reached a well, and Abdal unloaded the broken jerrycan. It had now lost half of the water in it, and it seemed that one of our goatskins was leaking, too.

We had made a late start that morning. Now the heat was fierce, and I felt faint. I tottered to the well and stared at the silver coin of water in its depths – sixty feet down.

Abdal was contemplating the horizon. 'Someone will come along soon,' he said. 'Sometime today, I should think. Look at the animal tracks – someone brought goats and donkeys yesterday, by the look of it. So they'll probably be back.'

I didn't understand. 'Why do you want to meet someone?'

'Because we need somebody with a rope.'

'What for?'

'We haven't got a rope – to get water. I meant to tell you to buy a rope and a bucket in Agadez, but it slipped my mind . . .'

I bit my tongue.

Eventually, somebody did come. We refilled everything, but the broken jerrycan quickly spilled its guts and the punctured *guerba* – despite attempts to repair it – kept leaking.

We dozed under a tree, waiting for the noon heat to pass. It was unpleasant to move before about three. Abdal's shoe had fallen apart, and he used an odd piece of nylon and an old nail to try and repair it. Here was a man who could walk thirty miles a day across the desert in broken flip-flops. I could not lend him my boots – his feet were twice as big as mine.

We loaded the camels and set off, arriving at a place where three men were digging a new well. They were not Tuareg, but hired black Hausas, men with strong sinewy bodies and ready grins, celebrating because today, after fifteen days' digging, they had struck water. It had been a lucky dig, they said, not too much rock, and the water not too far down. I climbed over the perilous mounds of grey earth and peered into the darkness. At a certain angle, you could make out a faint glimmer of light.

Abdal told me that anyone would have the right to use the well, though the man who had paid for its sinking would have priority. In times of drought, when the water-table dropped, it would be a different story: the rich well-owner's investment would guarantee him sole access to the limited supplies.

We walked on, south-west. I was feeling better after sleeping, but Abdal was still obsessed by the prospect of sugar-deprivation. He was determined to make for a camp the people at the well had told him about. Perhaps we could buy or trade for some sugar.

We reached the camp at dusk. Dogs and children in scraps of Tuareg blue came running out to meet us. Then a man emerged slowly from a domed tent, pulling on a shirt. 'He's a cousin of mine,' said Abdal with satisfaction.

After the pleasantries, Abdal mentioned what we wanted, which clearly did not delight his cousin. He went back to his tent and returned with a bag of sugar the size of a woman's fist.

'He had to help me,' Abdal said as we walked away. 'Firstly he's a relative. Secondly, a few years ago, when he was in trouble, I gave him a goat. That's worth a lot of money, a goat.'

'You were very generous,' I said.

'We have to stick together. That's why it's so mean of the bastard just to give us this lousy little bag.'

I decided to go without tea, so doubling Abdal's sugar supply. I was more preoccupied by water. With our reserves almost halved, we needed to find a well every day, and the journey became a time-wasting zig-zag from one well to another. Sometimes one would be shallow enough for us to reach it. We would knot together a series of camel halters and dangle Abdal's metal kettle into the hole.

My main source of frustration was the loading and reloading of the camels. Abdal was used to making journeys of a few days at most, to and from his family encampments. He did not possess the big, multicoloured leather saddlebags that can sometimes be seen on the camels of wealthy Tuareg. Like most desert-dwellers, Abdal loaded a camel simply by piling on blankets to stop its skin from being chafed, balancing his baggage on either side of the camel's hump, then binding it with scraps of cord, knotting and tightening them by hand and trying to make sure that there was as little movement as possible. The skilful balancing of the variety of different shapes and weights took about forty minutes. When it was not going right, the painstakingly knotted ropes had to be unknotted. In addition, it would be necessary to stop several times to make adjustments once the camels' motion had shifted the baggage around. Occasionally it would be necessary to start the process all over again. We loaded at dawn, unloaded at midday, reloaded in the afternoon and unloaded again at night. It took at least two-and-a-half hours a day.

What was more, our second camel, the female, kept giving us trouble. During loading she would suddenly stand, toppling luggage in every direction. Once, unhobbled, she made a break for the desert and took half an hour to recapture. Walking, she would rear her head, snarling, threatening to kick or bite. We could not work out what was upsetting her, and wondered if it was the slow leaking of water from the jerrycan and *guerba*. 'Camels hate being wet,' said Abdal; 'they hate rain. Camels get along just fine in any amount of heat, but get them wet and it upsets their thermostats. Eh, baby?'

But the female was in no mood to be placated. Partly it was the constant presence of a stranger that disturbed her. Partly it was the fact that she could smell the grass on the wind from the west, where she knew the familiar encampment and her babies were. And we were heading south, into the sand.

By contrast, the male camel was a delight. He was mild-mannered and curious. This camel (he was called something approximating to Dapple, for the sprinkling of sandy pigmentation on his nose) also had a weakness for chewing tobacco – ready-chewed. When he caught the scent of tobacco he would sniff at Abdal, and Abdal would lift the felt flap of his nostril and spit some tobacco in. Wide-eyed, Dapple would raise his head and strut happily forward. I got into the habit of chewing the odd wad myself, just for the pleasure of having Dapple's soft muzzle nuzzling me in search of his next tobacco high.

One morning we visited some shallow wells. 'The water here is superb,' said Abdal, ' – if there is any.' The wells were far from any grass, and in this season he expected them to be dry or sand-choked. They were neither, and we filled the jerrycan and *guerbas*. Desert water can be undrinkably saline, but is often sweet-tasting and utterly pure, sometimes from ancient subterranean lakes. It can be a cloudy orange in colour, or silver-blue. This, we both agreed, was exceptional water – clear, clean-tasting and sweet.

We set off to explore a range of sand dunes. Abdal knew the area well, though he had not been there for perhaps twenty years. He looked around in horror as we walked, 'People have cut down all the trees!'

I looked down and saw a large axe head in carved granite. In the course of our journey I would find a number of prehistoric artefacts, including an arrowhead and part of a bronze necklace. We visited sites where there had obviously once been settlements, and saw traces of human activity scattered along what seemed once to have been a major route. Occasionally we saw indications that once there had been rainfall and trees. At one rock we saw signs of domestic work, where pestles had ground hollows into sandstone. As sand dunes move, they leave a flotsam on the harder ground beneath, and I discovered a shattered pot with a handle. It was extremely old, and unlike anything presently being made in the desert.

We walked under a tall ridge, the skin of the rock seemingly charred black and splintered. The camels were skittish. They did not like being so close to the cliffs. The presence of water in the hills meant that there was life there, too, a food chain from spiders and snakes to small rodents, gazelle and hyenas, and sometimes we saw their traces in the sand. Abdal thought it was the smell of hyenas that made the camels nervously sniff the air.

Towards midday we found ourselves crossing a plain of soft, pinkish sand. This was not absolute desert. There had not been enough rain recently for grass to grow, but there had been grazing here some years back. Herders had cut down all the desert trees for firewood, and the land had had no chance to recover. The only trees we saw were midgets. Abdal shook his head sadly.

'This can't go on. When I was a boy, there were lots of trees here – big, mature trees of all the desert varieties.' The waxen, thorny trees of the desert are valued for their respective properties. They provide shade, camel fodder, arrow shafts, products for leather-tanning, dyes, indications of subterranean water. 'People are destroying the old trees, and cutting the new ones before they have any chance to mature. Look at it here. The desert is spreading.'

It was a cloudless day and we had been walking, as we did every day, since dawn. Now the sun was approaching its zenith, and we had no shade.

'We'll go up this ridge,' said Abdal.

It was twelve-thirty. I had been on the male camel for an hour, and we now plodded up the soft sand of a ridge. The glare from the desert's surface was blinding, the heat simply surreal. My turbanned head was scorched and my body limp. I could not understand how Abdal kept going, his sandalled feet striding in the fiery sand. A strong wind began to blow. It was as though, on the hottest summer day, someone had stuck a fan heater inches from your face.

At around one-thirty we found a tree some seven feet tall – tall enough to offer some shade. I almost fell off Dapple, and helped Abdal unload. The Tuareg always carry swords, an apparently archaic affectation which, I now discovered, had great practical use. With a few swings of the yard-long steel blade, Abdal cut away enough of the lower branches to make a space for us to lie under. But the tree did not provide enough shade, and we needed to erect an awning – an old blanket. We searched for rocks to weigh down its corners and the fierce wind tried to tear it out of our hands.

I was the first to climb into the shade. I reached for the jerrycan.

Up among the branches Dapple was eating. I saw his head bobbing against the blue sky, and brittle half-chewed leaves drifted around me. I drank, closed my eyes and fell instantly asleep.

Perhaps it was the extraordinarily unpleasant sensation of heat in my feet that woke me. I groped for a pair of socks, moistened them and pulled them on my feet. The effect was agonizing, as though someone were forcing knitting needles into my veins. Then came a relaxation, as my blood cooled.

Abdal was still sleeping beside me, breathing profoundly. After half an hour he came to, sat up and reached for the jerrycan.

We had both had enough. We decided to turn east, towards Abdal's family encampment.

We walked in moonlight until late that night, when we reached an area with enough grass to feed the camels. I was usually chef, but I had no energy that night for preparing food. We ate what was usually breakfast – boule, the Tuareg staple, made as usual from maize, but with dried dates, dried goats' cheese and water.

I woke at four, aware of a strong wind blowing from the east. The next time I opened my eyes it was six-thirty and a sandstorm was raging. My sleeping bag was full of sand blown in through the seams. Squinting around, I saw that Abdal had erected the blanket-awning against a tree and was crouching in its lee. He had already fetched the camels.

I joined him, accepting a plate of boule and a cup of tea.

'It took me an hour to find the camels. The wind had covered their tracks.'

'Could we have lost them altogether?'

'It sometimes happens. But eventually they find their way back to the camp.'

'And what about us?'

'We'd survive.'

It was no morning for chit-chat. Conversation was for evenings, after a meal. When we were exhausted, we hardly spoke at all. And when walking, it was only in the cool of the first hours that we talked. This was the rhythm the desert imposed. 'L'homme propose, Dieu dispose,' as Abdal pointed out.

'Do you really believe that?'

'It's obvious, isn't it?'

We plodded on through the sandstorm. I wore sunglasses with protective leather patches around the lenses and I had bound up my head in my turban, but sand still forced its way into my eyes and mouth. I was ahead of Abdal, who was leading the camels, when the angles we were walking at diverged. Eyes front, narrowed against the insistent sand, we became unaware of each other. I looked around and Abdal was nowhere in sight. Then I saw him, disappearing into a haze of sand off to my left. Panicking, heart racing, I ran towards him. With the noise of the wind, he had noticed nothing. If I had lost him, I could never have found him. The wind covered our traces within seconds. All my water, maps, compasses, my GPS satellite navigator – everything was on the camels. I would have been alone in a sandstorm, 100 miles from Agadez.

Slowly my nerves settled, and I considered what my prospects would have been had we become separated. I would have stayed where I was, in the shade of some thorn bush. Sooner rather than later, Abdal would have sensed my absence and turned around, and his intuitive sense of the landscape and wind direction would have brought him back close to my position. The wind would have dropped and we would have found one another.

But what would have happened otherwise? How long would I have waited before deciding to set off alone for Agadez? I would have had to walk by night, using the moon as my compass. What chance of finding a well, or some herdsmen? How would I protect myself by day? Could I have made it 100 miles to Agadez? Was there a chance that I would have walked right past the oasis?

The desert was hazy, formless, difficult to walk on. I thought of all I had learned in Chad and Niger about how to survive in the desert. It was not enough.

We spent two days at Abdal's camp. Abdal, an intelligent, educated and dignified man, lived in circumstances any Westerner would consider abhorrent; the water had to be fetched by camel and donkey from a well a two- or three- or four-hour walk across burning sand, then stored in inner tubes swung on a leafless tree stump in the sun, and transferred, scalding (by small children in scraps of off-cut blue, barefoot among the thorns and goat turds), into goatskins inside the hut to cool; the water whose rarity means that washing as the West understands it is not practised. I could go on about the poverty, poor diet, lack of privacy, the stoicism and misery that comes out of the lack of basic medicines, and I suppose I might, if I believed the Tuareg had much to gain from being admitted into the happy consumer wonderland of Western culture; but they do not.

It took two days to walk back to Agadez. On our second day we carried on through the midday heat, to reach Michel's house, utterly exhausted, at 4 p.m.

The previous night we had sat up late, discussing the future of the Tuareg. Their herds were decimated by the droughts of the 1980s and 1990s. Aid agencies have not helped replace their animals, feeling that the husbandry of animals for human consumption should not be encouraged in Third World countries, being a less efficient means of generating nutrition than the growing, say, of maize. The crisp-shirted geniuses of the UN and the EC fail to recognize that goat-rearing is the only agriculture possible in the desert; and that even if irrigation made settled cultivation feasible, it would be asking the Tuareg to change their cultural identity overnight.

Abdal's gloom about his tribe's future extended to his sense of a loss of cultural identity. 'They are forgetting Tamachec, and learning to speak in French – and even Hausa. The boys are wearing T-shirts and American caps, and see the turban as something to do with the *old*. And what have we, the old, proven to them with our rebellion? Nothing. It's not surprising that they turn away from us. But the government wants to kill our culture. It sees us as trouble-makers – which we are, because we insist on our rights.'

I suggested to Abdal that some people saw the Tuareg as former slave-owners who deserved little sympathy. And perhaps their best economic interests would be served by their integration into the economy of the modern world.

He gave his lupine grin. 'Oh, yes, the modern economy. How long do you think it will be before it reaches us? How long do we have to live on in abject poverty?

'Look,' he said, 'we have a history, a culture, a language – though the government doesn't want our children to learn it – we are what we are. Why should all that be lost? So that we can drive cars and eat food in plastic wrappers? No. That *is* imperialism! I do not accept that argument.

'We simply want an equal chance to exist in the world. We want the government to let the aid organizations come to the north. How many NGOs do you see in Agadez? They are not allowed to come here. We want assistance sinking wells so that we do not have to walk eight hours a day to water our animals. We want some dispensaries, so that our people do not die because they have some pathetic ailment the doctors laugh at in big cities. We want the government to do a census – do you know that most of us do not exist on any electoral register? How can we have needs if we don't officially exist? We want our children to be taught in their own language. That means we want some teachers to be Tuaregs. There are hardly any Tuareg teachers, and just a handful of Tuareg civil servants. The bank managers in Agadez are all Hausa – the policemen, too. How many doctors are Tuaregs? *None*! How can we ever have any autonomy if the government keeps us uneducated, powerless? We say, let us build up our herds to where they were before the droughts. Let us educate our children and enjoy a little higher level of health. We will find our own level in society. Then we'll see about the "modern economy".'

He paused. 'Go on denying us our basic human rights, though, and there will be another war. And that, to be frank, would finish us off for good.'

In the few days that remained in Agadez I lazed around in Michel's house, or spun around the streets of the old town on Jean-Yves's motorbike. On my last day I was invited to the wedding of a rebel chief. It was held on a tract of flat scrub a few miles from town, where some trees along a riverbed offered patchy shade. A score of tents were pitched in a ring half a mile wide.

I wore a shirt and tie, earning the derision of Jean-Yves, but I felt I should be formal. The other guests almost invariably wore new robes of indigo, their prestigious newness proven by the crystalline dye that glittered in the sunlight and rubbed off on their hands and faces.

The day started dully, with people wandering along the riverbed staking out the shadiest spots. I joined a group of Jean-Yves's friends, and we discovered that no general catering had been organized and that none of us had brought any water. We dug in the orange sand of the wadi, and, two feet down, found water, orange-stained but sweet and good.

As the midday heat faded, the festivities began with camel-racing. The boys wore indigo or turquoise robes and their camels were draped with ceremonial saddles of sliced and braided leather dyed many colours. The girls wore traditional wedding finery, glittering white tunics with panels of orange embroidery on the breast. The boys moved around in self-conscious knots, nervously adjusting their turbans and stroking flat their robes, the bolder ones forming expeditionary pairs to proposition girls who were wandering in threes and fours, their arms securely interlinked. There were also counter-insurgencies: I saw a group of girls run forward and pull a young man off his camel, to his inexpressible horror and humiliation.

Goats were slaughtered, their bright red blood arcing and splattering on the dry sand, the corpses spitted and roasted. Dancing was performed by girls in a circle of drummers and pressing boys, an unsophisticated jog to the accelerating beat of the drums, the girls' eyes losing focus as they sweated and became mildly entranced, the boys grinning and nudging, hungrily watching their jigging flesh.

As it was the wedding of a rebel chief, his troops had to make a grand entrance. They came on pick-ups mounted with big guns, roaring around the camel race-track holding their Kalashnikovs aloft and firing bursts into the air. At each crackle an accompanying ullulation burst from the throats of the women, as though the firepower was some sort of phallic thrust which must be greeted and celebrated by ecstatic tonguey gurgling. For minutes the air echoed with rattling gunfire. Men emptied Kalashnikovs and pistols into the sky. 'Well, there goes the rebel arsenal,' I said sarcastically. 'I hope for your sake you lot don't need to go back to war this year.'

'It doesn't matter,' a big man with a scarred face told me complacently. 'We have seized enough arms from the army to fight them for a decade.'

The crackling went on. The excitement was undeniable – the thrill of all that potential death pouring into the sky.

'Stupid,' I muttered, still the Western know-it-all. 'The rounds could fall back to earth and kill someone – it happens at weddings in Pakistan.'

'That cannot be,' the big man told me confidently, 'for when the bullets reach the sky they dissolve.'

As I gave him a withering look, another man said, 'It's all right, they are only firing blanks.'

A bullet whistled past us.

'That wasn't a blank,' I said.

'That wasn't a blank,' they both repeated meditatively.

Then someone thrust a Kalash into my hands, and I fired it into the sky as joyfully as the next man.

CHAPTER TWENTY-FIVE

To Timbuctoo

A s I stood on the quay at Mopti, a man leaning on the rail of a boat called down. His face contrasted with the black ones all around us by being pink-white, topped off with a goatee and a Tin-Tin tuft. The language was the inevitable lingua franca – English – but I caught a hint of an accent, maybe Dutch. What he yelled was, 'You'd better get your ass aboard if you're heading to Timbuctoo. This is the only boat going out today, and it's filling up fast!'

Ever since Cairo I had been pushed southward, and the Niger river is the meandering southern frontier of the Sahara. Politics had kept me out of the vast desert tracts of Libya and Algeria. I had made a sort of ragged progress through the central Sahara in Chad and Niger, but now I doubted I would get much further north.

Every leg of my journey had been attended by delays. I had been denied access to northern Niger, crossed from Chad between two bouts of murderous mayhem, and wasted days in Maon waiting for an armed convoy to Agadez. Now I was running out of time.

After Agadez I travelled by road to the Nigerien (as opposed to Nigerian) capital Niamey. I wanted to pick up a boat at Gao, inside the Malian border, and follow the River Niger west, the entire breadth of Mali, to Bamako. But amidst talk of 'rebel' attacks and armed convoys, the land frontier with Mali was closed. I would have to fly to Bamako, then backtrack as far east along the river as possible. I should at least reach Timbuctoo.

I arrived in Bamako under a purple sky, monsoon rain swelling the river to a bloated oily orange. Vehicles formed static queues as undrained streets became canals, and in the street markets people soaked to the skin bought and sold mangos and cabbages. Before leaving town, I wanted to arrange my Mauritanian visa. The Mauritanian Consulate was in a northern suburb, a building under conversion by a squad of builders into an altogether different building, while the staff continued to work there. I walked a series of planks across pastures of soggy concrete to a villa populated by men and women dressed in flowing robes and apparently unaccustomed to the quotidian tiresomeness of visa requests. (But it is every traveller's guilty secret that it is nice to be a novelty. As my mother once said to me, 'Of course you like travel – when you're abroad in some remote place, you're special. When you're home you're just like everyone else.')

As another day of monsoon rain dawned I set off for the bus station to travel north to the port of Mopti. In the bus station the ticket seller had been bribed to misdirect ignorant travellers to a particular bus. I found myself sitting in a foetid minibus, reading *War and Peace* and watching men wander in and out of the gents' lavatory, where a brothel was conveniently located. The hours went by. It took me and three or four other naïve passengers all day to work out that we had been conned, and at dusk we had a violent argument with the now drunk and abusive ticket seller, who refused to have our bags taken off the roof-rack. It was almost dark when we dragged our luggage towards the official Mopti bus-stand. The bus quickly filled, and those of us who had waited all day looked at each other in astonishment, realizing we were real country cousins.

It was a decrepit machine, a mobile skeleton of engine, chassis and wheels with a series of movable objects – seats, window frames and so on – semi-attached. Filling up proved to be the only thing it could do quickly. The top speed was twenty-five miles an hour, and we were frequently overtaken by farmers on tractors.

The man next to me began to converse in excellent English. 'I am a police inspector,' he told me. He lifted his shirt and patted a gun butt. 'Don't worry, with me on board there won't be any problems.' The only problem we had was *time*. The journey lasted all night and most of the next day. By the time we reached Mopti, at 3 p.m. the next afternoon, it had taken me thirty-two hours to complete what should have been a nine-hour journey.

Mopti was a pretty little port, with a colonial-period old town, a quaint fishing quarter and a well-stocked market. The harbour was brisk with slender, twin-pointed *pirogue* canoes poled by straw-hatted fishermen. A few larger, double-decked *pirogues* bore freight north and east, up the River Niger to Timbuctoo and beyond.

I could tell from the noisy throng surrounding it which boat was about to leave harbour. The long, stone-flagged quay assumed a yellowish cast from the late afternoon light and the hawkers – girls in bright skirts or sarongs – cast long shadows as they vied to sell the foodstuffs in trays on their heads, wading into water the colour of milky tea, darting between their customers, swapping cakes, bananas or plastic bladders of drinking water for small change and dirty, diminutive banknotes. Nearby some women scraped clothes on washboards, rinsing them in the brown waters, while men gutted fish, the crimson blood mingling with soapsuds to marble the sloping quayside.

Our commercial *pirogue* was around forty feet from stem to stern, with a shallow wooden hull. The cargo – mostly petrol in orange drums – nestled below, and a subsidiary cargo of sacks of millet had been turned into seating for the lower-deck passengers, surrounded by their personal cargos of chickens and potato sacks.

I climbed an unsteady gangplank onto the upper deck, made from sheets of corrugated iron. Here the lighter-weight cargo had been piled up – a steel door, a stack of folding chairs, two bicycles and a brood of pot-bellied iron cooking-pots in a rope net. To the aft of the boat a soot-blacked funnel protruded; forward, there was a shelter with a bench where the captain sat with his mate. The shelter widened into a space just large enough for three bench-seats – nine people at a pinch. It was to claim a place here that Tin-Tin had been beckoning me.

'You just made it,' he said, as the diesel fired up in the boat's bowels, flinging soot into the evening air. The babble increased as the girl-hawkers urged the passengers to avail themselves of their last chance for days to purchase positively anything, and provoked another round of panic sweet-buying.

I threw my bag into the shelter and introduced myself to this tall, albin northerner. He turned out to be a Flemish Belgian, and well accustomed to schooling new acquaintances in the correct pronunciation of his name. 'X like the ch in loch, then ee as in bee,' he explained pedantically, 'then an r a bit like a French r, and a tuh.'

'K-chee-err-tuh,' I said.

'Not bad.'

As we stood watching the scene at the water's edge, other upper-deck or first-class passengers began to arrive. 'They've been waiting in their hotels until a messenger came to tell them the boat's leaving,' Xeert told me.

The newcomers engaged in an unseemly struggle for space in the shelter. Xeert had tried to claim a bench by spreading out a mat along its length, but a large, orange-robed lady was instructing the man with her to colonize it. When Xeert protested, he was haughtily informed that the lady was ill. As more people crowded in, the territorial spats became rowdier and the shelter so crowded with luggage and spread-out latex mattresses that almost no floorspace remained. A casualty of the reshuffle was my own bag, heaved off the bench where I had left it and onto the floor. I didn't care. The open deck looked more appealing.

The diesel continued to throb, but there was no sign of casting off. Dusk was settling and the washerwomen had been replaced by washing women, performing their ablutions at the river's edge. The mature women contrived to introduce the suds beneath their chest-high *lungis*, but girls of fourteen and fifteen soaped their breasts with indifference to curious male eyes.

It was nearly dark by the time the captain turned on his running lights and the boat shoved off from the now-deserted pier and slid into the broad River Niger. The water's rippled surface had turned bronze, a nuance darker than the sky. Flocks of river birds banked past us, crying. The engine said doog-doog-doog.

After the collapse of the Roman Empire, the eastern Sahara became completely isolated from Europe. It was only during the Middle Ages that trade with North African ports began to be re-established. By the 1800s, caravans were crossing the Sahara with firearms, gunpowder, silk and tin from Europe, traded for gold and ivory, apes and parrots – but most importantly for slaves.

The eastern Sahara was dominated by the Tuareg, who ran a sort of protection racket, with fees based on the value of the cargo or the reputed wealth of the travellers. For this they provided armed escorts who protected traders from – the Tuareg.

In Europe it was widely believed that the Niger River region was the threshold of some interior Africa of fabulous wealth. In particular, Timbuctoo was imagined as a treasure-house of inconceivable riches. Surely you could sail upstream to these fabled cities from the great river's mouth on the Atlantic – but no one knew where that mouth was. In its northern reaches the Niger flows east, so some speculated that it went all the way to the Red Sea. Others conjectured that it emptied into some vast inland lake. None suspected that it reached the Atlantic at the Gulf of Guinea, despite the fact that that enormous delta had been known to Europeans for over 300 years.

In 1788, the Association for Promoting the Discovery of the Interior Parts of Africa, soon known more pithily as the Africa Association, was founded in London with the principal aim of the exploration of the Niger. Its first recruit was the American adventurer John Ledyard, who went to Cairo to find a caravan to carry him and died there. Next David Houghton, an impoverished ex-officer of the British army, set off to explore the Gambia river, wrongly believing it to be the Niger. He travelled 400 miles upstream to Medina, and died on the edge of the Sahara.

Five years after the death of Houghton, one of the legends of African exploration arrived on the Gambia, a twenty-three-year-old Scottish surgeon who rejoiced in the name of Mungo Park. In December 1795, after several bouts of fever, he set off into the African interior. Park was on horseback, his two companions, a servant-interpreter and a personal slave-boy, on mules. In the course of a nineteen-month journey, Park was captured, imprisoned, starved and nearly murdered. He was shown the spot where the body of the murdered Houghton had been thrown. He lost his companions, his equipment and most of his clothing. He almost died of thirst. He received succour from a woman in the jungle, a local king and a slave trader. On 20 July 1796,

I saw with infinite pleasure the great object of my mission – the long sought for majestic Niger, glittering in the morning sun, as broad as the Thames at Westminster . . . I hastened to the brink and, having drunk of the water, lifted up my fervent thanks to the Great Ruler of all things for having thus far crowned my endeavours with success.

But he had not managed to reach Timbuctoo. And it would take him many more months of privation and sickness to get back to the Gambia, and then England, to tell the world his extraordinary tale.

In the top-deck saloon, the first-class passengers were beginning to worry about their dinner. Urgent smells of frying fish reached them, and the rumour spread that a meal was being prepared by the ship's cook. The large lady in orange expressed outrage that the lower-deck passengers were apparently being served first. Adversity drew us together, and bags were opened. My large collection of mangos elicited looks of distaste, as though they were some sort of working-class food or positively unhealthy. Meat or eggs or animal fat was what they wanted, and they consoled themselves with deep-fried morsels of goat and chicken in fat-sodden brown paper bags, bought from those Mopti hawkers who had accurately foretold starvation.

There were nine of us upper-deckers. Sharing food made us companionable and we began to talk. The seven Africans assumed that the two white men, similar in looks and age, were together. Xeert told us he was a journalist. He adored the music of Mali and was planning to visit Niafounké, halfway to Timbuctoo, where the celebrated singer Ali Farkatouré had his home. There were two newly-weds who were both teachers. He was in his mid-thirties, beefy and inert-looking but with watchful eyes, a sort of smaller *Père* Bessita. His wife was a handsome creature in her early twenties, big and well proportioned, her firm flesh exuding health and strength, yet her mouth oddly coquettish, small and pouting and adorned on its downy upper lip with a black beauty spot. Her husband never let her out of his sight.

There was a small and slender man of forty or so who said he was a telephone engineer, going upriver to attend his father's funeral. The fat lady was, she gave us to understand, a Woman of Means, her husband a businessman in Timbuctoo. She was on her way home after a visit to relatives, and the man with her, a shifty fellow who never talked but grimaced awkwardly and avoided people's eyes, was a cousin and chaperone. The last two members of this motley group were a sheepish, silent student and a loose-limbed man in his mid-twenties with an astonishing smile. He told us he was a salesman. Selling what? we asked. He gave a wink and unleashed the dazzling smile. 'Anything!'

By the time the official dinner arrived, I was no longer hungry. It was watery rice and fish lumps deep-fried into brittle scabs. The newly-wed hubby had seconds and thirds, and with paternal concern – another shade of Bessita – urged me to eat more. The lady in the orange robe tucked in with undignified haste, then, sated, felinely licked her slender fingers and, in the manner of gluttons who believe themselves epicures, disdainfully turned down the offer of a fourth helping.

As it grew late, we talked by the glow of a single oil lamp. People became sleepy and the women started to tidy things away. The fat lady lifted the front of her orange robe, exposing enormous, pendulous breasts, and began to rub a fragrant-scented oil into them. Her body was covered in long, fascinating scars. I looked up and caught the eye of the newly-wed teacher, who gave an amused shrug.

Bedding was laid out on and around the three benches in the shelter. The newly-weds and the fat lady stretched out on their enormous foam mattresses and the men spread out around them. People were huddling together like pack animals against the cold and the dark, and something of the collective anxiety caught hold of me, making me squeeze my grass mat and sleeping bag into a narrow space alongside Xeert.

Along the deck, the salesman and the engineer were smoking cigarettes, and Xeert and I went to join them. The salesman grinned at me. 'What did you think of the display?'

'What display?'

'Madame!' He nudged me. 'You know what she's telling us.'

'You're joking!' said Xeert.

'Not at all,' said the salesman. 'She's at an age where she can get away with that. Here women can show their breasts before marriage and when they're old.'

'She's not that old,' I said, 'she's – what – forty-something?'

'She shouldn't be doing it,' the engineer interjected. 'This is meant to be a Moslem country. All this nudity shouldn't be allowed.'

'But she's disgusting!' said Xeert, hotly.

'No, she's a good-looking woman,' said the salesman. 'Look at her. Light, smooth skin, soft hands and feet – obviously never does any work. And – she's excited by having all us around. I get the message, even if you lot don't.'

I stared at him.

'*C'est une courtesane*,' he said.

I woke at dawn and carefully stepped over slumbering bodies. The boat's only head was below, reached, in the absence of staircase or ladder, by swinging yourself over the side, dropping onto the gunwhale and clambering into the lower deck. There the throbbing diesel engine was attended by a grimy young mechanic, behind him the engine's cooling inlet, a well of churning green water, that doubled as the boat's sink and bath. Lastly, behind a rag curtain, was a sloping floor where the canoe stern jutted over the river's surface. Rushing brown water was visible through a small square aperture.

Back on deck, I noticed that the engineer was sleeping alone at the middle of the boat, on one of the metal doors we were ferrying to Timbuctoo. As dawn broke he woke up and lit a cigarette. In silence we watched the sunrise spilling gold into the Niger. It was a fresh, clean morning, with the breeze carrying the black diesel fumes away from us.

One of the boys below was making tea, which was being sent up in small chipped cups. We walked back to the shelter, where our fellow passengers were stirring. The sky became a pale, pale blue, separated by a strip of fawn from the subtly opalescent green of the Niger. On the sandy banks there were ranks of termite mounds the size of tree trunks. 'You take the stuff they're made from,' said the engineer, 'add some water, and it's good for curing fractures – really good. You don't even need a *giri-giri*. *Mais c'est la maison du diable*.'

'The devil's house? The devil lives there?'

'Yes.'

'How can you tell?'

'Oh, we often see the devil. You see him a way off, you can tell it's the devil – you sense it. But if you try to go up close, he disappears. And if you try to find out anything more, he'll make you mad. Mind you, devils are afraid of whites. They're white, too, you know.'

Just before lunch, we reached a fishing village. There were piles of logs on the muddy shore, and blue smoke drifted over rectangular reed huts. Two dozen *pirogues* were moored along the shore like a catch of massive swordfish. Naked ebony-skinned children played in the water. We crowded ashore, stepping past cages of ducks and piles of drying stinking fish – first Xeert, the engineer and I, lastly the fat lady, waddling down the gangplank stabilized by two boys waist–deep in the water on either side of her. There was not much to see, but there was food. Smoke rose from brick grills, and we gathered eagerly around them to buy ourselves fish lunches.

An hour passed, then two, and there was no sign of the boat preparing to leave. I sought out the captain, who told me that the Niger now widened into a lake. The fishermen had told him the wind was rising and he was worried that it might be too dangerous to cross. The wind was blowing like an English summer breeze, warm but assertive.

'*Trop d'vent.*'

'When will it drop?'

'At this time of year . . . could be . . . a day.'

'*A day?*'

'We have to wait and see. If we're lucky, we'll get away first thing in the morning.'

By dusk, it was clear there was no hope of leaving that night. Xeert and I went down to the section of riverbank where the men performed their ablutions and soaped ourselves among the fishermen.

Back on board we found that Mr Newly-wed had bought an enormous catfish and had the ship's cook prepare it for us all. It resembled a Roman fountain, a great mound of rice from which architectural chunks of fish, including its large head and tail, protruded. Once again he was worried by my lack of appetite – or perhaps, not to lose face, he needed to see me appreciating his hospitality – and he kept proffering chunks of ivory flesh. But even after we had all eaten our fill, there was food left over. He threw it in the water. 'This is for the fish,' he said. 'After all, they have to eat too.'

The engineer passed me a surreptitious beer. This was first-class, bourgeois, Islamic Africa, where one did not drink too openly. We lolled on the darkened deck, bottles in hand, watching the flickering oil lamps of the village. Now and then snatches of music reached us. 'As a Moslem,' I teased him, 'you disapprove of nudity, but it's OK to drink beer?'

He gave a sheepish laugh. 'You don't have to tell me, I know it's wrong. But I can't help it, and God knows, there are worse crimes. Every man has a weakness – mine, as a matter of fact, is *whisky*.' He leaned towards me and in the gloom I thought I saw him wink. 'I've got some with me – don't worry, I don't drink alone. I'll pull it out at the right moment.'

Behind us came a sudden burst of laughter from the other passengers playing cards.

'Come,' said the engineer, 'let's take a walk. Do you want to see the skull of the biggest hippopotamus that ever lived?'

We walked along the riverbank. Perched on the roof of a single-storey hut we saw a great skull, vast and concavely hollowed like a Henry Moore sculpture. Spread around it were bones like logs, and its hide, the size of a Tuareg tent.

The man who had fired the fatal shot and earned the privilege of owning the skull emerged from within the hut. There were many hippopotami in the lake, he told us, but this one had become aggressive and had to be killed. He pointed at the alpine molars. 'The front teeth are for eating, the side ones do the damage when they attack humans,' he said.

'Was he difficult to kill?' I asked.

He gave a wry smile. 'Well, it wasn't easy.'

'And it was far too heavy to carry,' enthused his father, who had come out to join us. 'They had to cut it up on the shore and bring it here in bits.'

The hunter pointed to a bone that might have belonged to a dinosaur. 'It took five of us just to carry that thigh,' he said.

'And it yielded so much meat that no one in the village has eaten anything else since,' the old man added, 'God be praised!'

On deck in the dark I heard someone say, 'Me, I want to stay up all night, drinking tea, chatting.' I had to smile at the innocence of it. 'Then tomorrow at dawn,' the voice went on, 'the sun comes up, it gets hot – that's the time to sleep!'

No one was chatting or drinking tea when I woke at 3.30 a.m., but the boat was in motion. The wind had dropped, and our captain, charged with awesome responsibilities over life and property, had taken the decision that it was now safe for us to leave.

The fused grey of lake and sky slowly unmerged until the water seemed oddly brighter than the sky, a lighter shade of satinized steel. Then a black line appeared separating the two: the distant shore. Swarms of birds – black dots – appeared in the sky, weirdly metamorphosing into tetrahedrons and figures-of-eight. It took two hours to cross the mirror of the lake. Sometimes we glimpsed the coral tunnel of an open-mouthed hippo. As we reached the open mouth of the Niger, there were dense reeds on the southern bank, but the northern shore yellowed into almost treeless sand.

'This is where the desert begins,' said the engineer. Soon the boat was sliding past mud-brick settlements and palm trees. By 9 a.m. it was hot.

As I sliced up some mangos for breakfast, Mrs Newly-wed came up behind me, returning from the toilet – one of her rare unchaperoned outings. 'Don't eat mangos for breakfast!' she said in a shocked voice. 'They're bad for you!'

'They're *what*?'

'Mangos give you malaria, they give you yellow fever, hepatitis . . .'

'Excuse me, but that's . . . madness. Mangos are full of goodness. You're a teacher, you ought to know these things.' I cut off another slice and held it out to her.

Grinning coquettishly, she held out her large, shapely hand to receive the golden sliver.

Later, as I sat on deck reading, a boy from the lower deck approached me. 'Big brother!' he said. 'Can you do these sums for me?' He squatted next to me and held out a grubby scribbled list of car parts he had been sent to Bamako to purchase. He could read a little, but only capital letters, and the note was handwritten. The total was what he had expected, but he wanted to be sure he had paid the right price for individual items.

'Would you like to learn how to read joined-up writing?' I asked.

'Yes please, brother!'

He rummaged in a bag for an old exercise book and sat down beside me. In the course of the next two days I gave him regular lessons, and he would go off and practise by himself. He learned in no time at all.

In mid-afternoon we arrived in Niafounké, where Xeert hoped to meet Ali Farkatouré. He had packed his bags and was ready to go ashore, but hesitated. What if Farkatouré was not there? Xeert would be stranded in a very small town on the Niger, waiting for a boat. He had asked the captain how long he would halt in Niafounké, and been told that there was some cargo to unload – so maybe thirty minutes. Surely it would be possible to find the musician's house in ten. To offer moral support, I decided to go ashore with him. If Ali Farkatouré was there and amenable to an interview, Xeert could come back for his bags – for if we left our luggage in the boat, the captain would be unable to go without us. Or so I reasoned.

The boat stood off about twenty yards from shore. *Pirogues* slid out to us with vegetables for Timbuctoo and food to sell to the passengers, and we called for one to take us ashore. The beach was broad, and when we reached the ridge of houses we found a fairly substantial town clustering behind it – but not a soul to ask the way. We jogged through hot, empty streets until at last we met someone who knew Farkatouré's name and led us to his house. A trio of men were relaxing under an awning outside. White, red-faced, breathless and sweaty, we greeted them.

But you do not just breeze into an African village like an urban yuppy with a mobile phone and demand immediate attention. Things are done more circumspectly; one is invited to sit, pleasantries are exchanged, tea or water offered and accepted. There was something undignified about us, but it had taken twenty-five minutes to find the house and we knew that the captain, already behind schedule, would be in no mood to hang about.

The three men were gracious. They regretted our wasted effort, but Farkatouré was in Bamako.

'*Bamako?*' said Xeert, with a note of exasperation. 'I tried in Bamako. They said he was here.'

Deep shrugs and regretful smiles. 'He *was* here. But . . .'

'Who does he stay w— I mean, where does he stay when he's in Bamako?'

More shrugs and smiles. 'That depends, *Monsieur.*'

We withdrew with as much dignity as we could muster, then made a dash for it. We were soon lost. The slumbering town remained devoid of people to ask. Sweating and gasping, we scrambled through alleys and across yards with dogs and chickens, through civic spaces and odd patches of cultivation.

'If the boat's gone, we'll hire a *pirogue* and catch it up,' I panted at Xeert.

'If it's *gone*? You think it might be *gone*?'

'No. I'm just saying, if it has . . .'

We had been away for about forty minutes when we reached the treeline and saw the beach stretched below us. All the *pirogues* were pulled up on the beach and our boat, puffing black smoke, was chugging slowly out towards the deep channel in the middle of the river.

We ran down the beach yelling our heads off and clambered into a *pirogue*. At once our fellow passengers began waving, telling the reluctant captain to wait and urging the man poling our *pirogue* to speed up. As we finally clambered aboard there were comradely smiles from our friends, and a grumpy scowl from the captain.

Mungo Park's second expedition was grander than his first. Instead of three men, one on horseback and two on donkeys, there was Mungo, twenty-six volunteer soldiers, two sailors, four carpenters, Mungo's brother Alexander Anderson, and a Lieutenant Martyn, picked up on the island of Gorée off the coast of modern-day Dakar (where, as things turned out, I would be married). Mungo had been given the rank of army captain, and his mission was to sail along the Niger in search of the river's mouth. At the time, Europe was preoccupied by the Napoleonic Wars, and little energy was expended on exploration. But there was general concern that Napoleon was about to expand his African empire. Imagine – Boney seizing control of those lucrative trade routes, the veins that pumped gold from the heart of Africa. He had to be stopped.

The British redcoats must have been a lively sight in the semi-arid Sahel. But the rainy season was just beginning, and dysentery and malaria began to strike them down. Men too sick to travel were abandoned to their fate. Only ten of them reached the Niger – at Bamako.

With indomitable determination and stupidity, Park drove on. It was not enough to have set off during fever season; he had to shoot at every African who approached the boat. Captain Park had turned into a psychopath. Having built a forty-foot schooner by conjoining two native *pirogues*, the dwindling company approached the falls at Kabara, the port of Timbuctoo. It was still the season of low water that preceded the monsoon, but Park and the remaining four members of his party just managed to float over the falls. The river is some miles from the town, and Park never saw Timbuctoo.

As he must have sensed, Park was no more than halfway through his journey. Most of his company were dead, and shooting at every passing African had resulted in his being pursued by an enraged posse. He had botched it.

There was not enough water to carry the four men – plus a newly engaged translator, and three slave rowers – over the next set of falls, at Bussa. The boat became jammed in the rocks, and there were too many spear-throwing natives for the soldiers to pick off with their muskets. They dived overboard, making for shallower water, and were drowned. A slave lived to tell the world the tale, five years later.

That night we hit a sand-bank. 'We need all the men in the water!' a voice called out. Xeert, the engineer and Mr Newly-wed looked on in horror as I threw off my clothes and, in underpants alone, jumped in. The water reached my chest. There were six of us splashing around in the warm water, while the adolescent girls of the lower deck, on the pretext of helping, shone flashlights at us and giggled. The captain joined us, sounding depths and yelling to his Mate. Shoulders were put to the boat and we heaved ineffectually, our feet sinking into the reedy ooze. 'Watch out!' someone yelled at me. I looked up and saw that I had almost walked into a near-invisible fishing line on which there glittered a dozen hooks. Moving too close to the shore, the boat had snared several of these lines, and soon, with much phosphorescence and splashing, protesting fishermen emerged from the darkness. Painstakingly they unpicked their lines, and were then commandeered to help shove. But we were stuck fast. The cargo would have to be unloaded. Luckily, it consisted mostly of sealed oildrums. One by one, the bright orange drums were heaved overboard, where they floated, upright, just under the surface. I was politely given to understand that I was not really stevedore material, but was allowed, as compensation, to look after some drums. As one-man committees proliferated, all yelling instructions at no one in particular, I delegated the drum-sitting and swam inshore. The shallow water was like warm syrup, the riverbed underfoot slimy and fibrous. The boat became a far-off ribbon of gay lights, a pleasure boat on the Venetian lagoon, laughter dancing across the placid waters. I felt intensely solitary and still, yet connected to some vital human activity. Human beings are nothing without each other. I was not actually needed by these others, yet I felt complementary to them.

By the time I swam back, the boat was afloat and the orange drums were being reloaded. Someone saw me and yelled out, 'Don't go too far from the boat! There are hippos by the shore!'

I helped corral the oildrums towards the boat, and found myself surrounded by half a dozen near-naked bodies, superbly muscled, leaping, lifting and laughing, a scene from the first Olympiad, a celebration of pure physical exuberance. I felt an extraordinary existential freedom that night.

CHAPTER TWENTY-SIX

Altared State

I DID NOT STAY IN Timbuctoo. I had planned to travel north into the desert, but instead I took a taxi – the oldest Land Rover I had ever seen – to Timbuctoo airport.

Air Mali had sold twice as many tickets as it had seats, and I had to fight hard for a place, brandishing the BBC i.d. card I had neglected to hand in when I left the Corporation several months earlier. The plane, a museum-piece Andropov, was crammed with cheerful people in brilliant robes, while the crew, three puddingy Eastern Europeans, looked morosely on. At Bamako I checked into a friendly little Italian-run hotel and waited for the train to Senegal.

Reaching the Senegalese capital felt like coming home to Europe. I arrived in darkness, and woke the next morning to look out on tall buildings and red-tiled villas, a tidy harbour and a quiet sea. I was soon drinking aromatic coffee and spreading *Président* butter on a crusty, fresh baguette.

They say the streets of Dakar are the liveliest in West Africa: the city's energy and money attract migrants from throughout Francophone Africa. Frantic vendors clog the pavements, and perspiring men stand at every traffic light waving their wares at passing motorists – coat hangers, cassettes, carpets; I saw one trying to sell a wardrobe.

In Mali, Xeert had made a gloomy prognostication: 'If you get married in Nouakchott your wife will never forgive you – it's a hot, ugly hell-hole.'

In the event, the problem had gone away – or rather, the pastor had. He was still missing.

Penny had deluged the British consulates of West Africa for

information on matrimonial mores. Morocco, she discovered, forbids Christian marriage (the head of the 'Alawite line is, after all, a descendant of the Prophet's daughter Fatima). Mali was an option, but too far from the sea. We opted for Dakar.

Well-intentioned friends have pointed out that my leaving all these arrangements to Penny was reprehensible. (She also organized the London blessing, the printing of the invitations, the caterers, booking of the musicians, etc). I reply weakly that we both wanted to be married in the desert, and that I arranged what I could before setting off. The fact is, Penny bore the brunt of making our eccentric plan a reality.

After several months of separation, we met again at Dakar airport. Telephone conversation during our long separation had been brief and unsatisfactory, and the tone of some of my letters had been strange – the desert summer, the geographic and cultural (if not always human) isolation had taken their toll. But since Agadez, I had relaxed, and had begun to feel calm, and strong. My emotion at seeing Penny again was one of joy.

We settled into the Hôtel St Louis Sun, a converted colonial villa. It was a charming building, but the big, shuttered, white-painted rooms were getting dingy. The day before Penny's arrival I had bought paint and brushes, and after the chambermaid had changed the sheets, I furtively slapped a coat of matt emulsion on the walls.

In the open central courtyard there were palms and rubber plants and caged parrots. Penny loved to hear the parrots sing, but felt their cage was a bit small. Did I think it would be possible for us to buy a *second* cage and sort of attatch it to the first one, doubling the birds' accommodation?

Yes, yes, I agreed, but didn't we have a marriage to arrange first?

There was a charming little gothic Protestant church in Dakar, a city which is, in so far as it is Christian, predominantly Catholic. Unfortunately, it was locked and barred.

It was a bizarre tale. The pastor, Jean-Pierre Yomo, a Zairean raised in Paris, had not long been in the job. When the previous – Senegalese – pastor had gone on study leave to America, Jean-Pierre had been appointed to replace him. This enraged his predecessor, who rushed back from the States and demanded reinstatement. When the Church authorities refused, his behaviour became increasingly erratic, culminating in the threat that if services were held in the church, he would burn it down with the congregation inside. The Senegalese constabulary felt that the best way to prevent his threat from consummation was to ban anyone from entering the church. Services were currently being held in a suburban schoolroom, which did not seem to us the most romantic place in which to tie the knot. Testing Dakar's ecumenism, we went in search of a Catholic church.

Just off the coast of Dakar is the tiny island of Gorée, one of Europe's first footholds on the west coast of Africa. Gorée's black, sea-battered cliffs face out towards the New World, and it was from the deep-water harbour in the lee of the island that millions of slaves were taken to a life – if they survived the voyage – in the Americas.

Today, Gorée is a UNESCO World Heritage site. It has the somnolent atmosphere of a Mediterranean village, with peeling pale ochre and raw sienna-washed houses draped with billowing bougainvillaea. Its houses and the tall, shady Church of St Charles have hardly changed since they were built in the sixteenth century. It seemed the ideal place to celebrate a wedding. But the church was Catholic; and you needed to take a boat to reach it.

The state of Senegal required us to have a civil marriage before any religious service. Forewarned, Penny had brought with her every conceivable official document pertaining to our existence – all translated into French – but still we arrived at the registrar's office terrified he would ask for something we had overlooked. We need not have worried – he did not ask to see anything. An expansive, larger-than-life sort of chap, M. Sao trusted us.

There were, however, certain important formalities. Did I, Martin Leonard Buckley, wish to opt for monogamous or polygamous marriage (legal maximum, four wives)?

I had not expected this question, and was momentarily confused; I caught something sharp in Penny's eye, and blurted 'Monogamous!' not a moment too soon.

Next, M. Sao read Penny the document she needed to sign. 'Your husband shall be the head of the household, but he shall exercise that right for the good of all concerned. He shall have the right to choose the location of the household. He shall . . .'

Penny's expression made M. Sao laugh out loud. After considerable hesitation she signed, muttering darkly, 'If I'd known about all this . . .'

M. Sao handed Penny a booklet with spaces to write the names of each of our children. There were twenty-one. We posed for photographs under a sign saying '*Ringo-Starr Photographe Professionel*'.

The priest at St Charles had no objection to our making use of his church. A small fee, naturally, was contributed to parish funds. And so, the following Saturday morning, we stood on the deck of a ferry in a new suit and wedding dress, clutching bouquets of flowers. Three Italian desert-travellers we had met in our hotel were joining us as photographers and organist. They were accompanied by Pastor Yomo, his wife Florence, and their three young sons, dandy in little suits and bow-ties, ready to serenade us to the sound of Jean-Pierre's guitar. Our Best Man was Roger Budd, director of the British Council, whom we had met on the beach that weekend. A nice lady on the boat thought it all sounded lovely and decided to invite herself along too. She was very welcome.

It all went wonderfully. The church organist had taken the (electronic) organ home to practise, so Giovanni banged a drum and rang the bell. We exchanged vows in French and received a French copy of the Bible. Everyone perspired freely. As we emerged into the sunlight, the children of the island were waiting and threw handfuls of freshly gathered flower petals over us. It was a delightful surprise, and very moving.

We spent our honeymoon in the desert.

First we drove north to St Louis, a colonial town built on a long boat-shaped island, connected to the mainland by a bridge designed by Gustave Eiffel that had once spanned the Danube and had been brought to Senegal in 1897.

Turn-of-the-century St Louis is vividly and slightly scandalously evoked in the novel *Sahara* by the doyen of French nineteenth-century exoticists, Pierre Loti. It is the story of the life and death of a French yeoman-type who volunteers for service overseas. His downfall is brought about by his love for a St Louis native:

> That little creature with her shock of black sheep's wool, her body of sculptured marble, and her glittering eyes, already fully aware of what they asked of Jean, yet downcast in his presence with a childish pretence of timid modesty.

This highly-flavoured fruit . . . was precociously ripened by the
tropical sun, bursting with poisonous juices, rife with morbid
voluptuousness, febrile and foreign . . .
    Fatou-gaye knew how to soothe her lover with cat-like caresses; how
to clasp him in her black silver-braceletted arms that were shapely as the
arms of a statue; how to lean her bare bosom against the red cloth of his
jacket, rousing in him feverish desires . . .

Naturally, Jean is doomed from the moment he claps eyes on her.
Sahara is unparalleled, as far as I know, in British colonial literature.
The British were presumably no less feverishly aroused, but more
hypocritical about owning it.

The oldest French settlement in West Africa, and the capital of
Senegal and Mauritania until 1958, St Louis shares with Gorée the
status of UNESCO protectorate; what our taxes are protecting is a
boat-shaped island not altogether unlike its namesake in Paris, though
cluttered not with chic apartments but with decoratively decayed
colonial houses. The outstanding interest of the town for me was its
role as a staging post for the French aviators of Aéropostale,
immemorialized by Antoine de St Exupéry. The Hôtel de la Poste is
still in business, the place where he drank a beer and laid his head, he
and the other heroic pilots of those bone-rattling, deafening,
pioneering aircraft – the first men to witness the vast golden Sahara
from the skies.

St Exupéry wrote about the desert with rare beauty. Many people
love his *Little Prince*, but his *Terre des Hommes* (*Wind, Sand and Stars*)
is a neglected classic. Among its descriptions of journeys (including a
near-fatal crash) in the Sahara is the following:

A minor accident forced me down . . . in Spanish Africa. Landing on
one of those table-lands in the Sahara which fall away steeply at the
sides, I found myself on the flat top of the frustrum of a cone, an isolated
vestige of a plateau that had crumbled around the edges . . .
    . . . its sides were remarkably steep, no Arab could have climbed
them, and no European had yet ventured into this wild region.
    I lingered there, startled by this silence that had never been broken.
The first star began to shine, and I said to myself that this pure surface
had lain here thousands of years in the sight of only stars.

But suddenly I kicked against a hard, black stone, the size of a man's fist . . . a sheet spread beneath an apple tree can receive only apples; a sheet spread beneath the stars can receive only star-dust. Never had a stone fallen from the skies made known its origin so unmistakably . . .

I picked up one then a second then a third of these stones, finding them at the rate of one to an acre. And here is where my adventure became magical, for in a striking foreshortening of time that embraced thousands of years, I had become the witness of this miserly rain from the stars.

Penny and I travelled on to Nouakchott, where, contemplating that grey, uncharismatic assemblage of modern buildings that turn their backs on the ocean, we were relieved that we had chosen Senegal for our wedding.

We drove north-west, to the oasis of Chinguetti, once one of the great university cities of the Moslem world, disseminating calligraphed Korans and students of theology throughout western Africa. Today it is a half-forgotten, crumbling oasis. Beyond it lies the ruined town of Tinigui, where legend has it that two dynasties lived in harmony until a feud began. The families bitterly fought until, at last, they had destroyed each other and the town they once jointly inhabited. A few remaining survivors took up their possessions and fled the accursed place.

Deeper in the desert, close to the border with Western Sahara, we reached the Guelb er Rîchât, a vast volcanic site where rings of concentric craters narrow to a central upthrust peak of lava crust. We climbed it, and stood in the hot wind staring at the perfect circles of jagged ridges receding to the horizon all around us. A navel of the world.

Before leaving Dakar we had searched long and hard for somewhere to buy a birdcage, and at last commissioned a craftsman to make one for us. Returning to Dakar for the flight home, we raced to the suburbs to find it incomplete, and with the taxi waiting we stood over the man as he crimped wire and whittled wood. I was uneasy, feeling that the hotel might see our gift as an implicit criticism, but the manager, a bird-lover, was deeply moved. We watched as the parrots explored their suddenly doubled accommodation. Then the love-birds flew back to London.

PART IV

A Thirstland

CHAPTER TWENTY-SEVEN

Nothingnessland

THROUGH THE PERSPEX PORTHOLE the day was being born, in blood and out of darkness. As I sipped my third coffee, the Kalahari slid underneath the jet in hazy ripples, a pink dream-land where sand melted into air. I had reluctantly left my wife of a few weeks to continue my desert circumnavigation and now I was entering a second ring of desert, the ring that runs around the earth's southern hemisphere. The map showed a smudge of yellow covering most of south-western Africa. My atlas described it poetically as a 'thirstland'.

By 6 a.m. I was grounded, one in a queue shuffling through Johannesburg airport. I found myself beside a fat, bearded Italian in a mock-photographer's waistcoat. The customs man searched through its numerous zipped pouches and joked, 'Why do you need so many pockets?'

As we walked on, the Italian leaned towards me and said confidentially, 'It's curiosity with these blacks, you know – they have to know what you've got.' I stared at him blankly. As often happens, the Italian took my silence as acquiescence. As we stood waiting for the lift he went on, 'This country: in four more years—' his hand dropped like a plummeting graph curve '—it will be finished.'

'Even if that were true,' I said, 'it's up to the Africans now to sort their country out.'

He eyed me scornfully and jabbed a podgy finger in my chest. 'Look, the whites never took an inch of South African land inhabited by blacks. This country was empty, *empty*, when they got here.' The lift doors opened and we stepped in. 'The whites civilized South Africa, they *made* it. Now the blacks are turning it into a Third World hell-hole.'

I said nothing, and we stood in silence as the lift lifted.

'So,' he continued with a cheery grin, 'where are you from?'

The doors opened onto the concourse. 'Excuse me,' I said, 'I have to go to the bank.'

The plane to Windhoek in Namibia had left London seven hours late, re-routed via Johannesburg. Now we were told there would be a further delay. I can never relax in airports. Unlike a great railway station with its cosy gloom and newspaper kiosks, and the enticing prospects of a city just beyond its imposing portals, airports have fast-food franchises, and are ringed around by suburban wastelands.

I fidgeted, wrote some premature postcards and studied the marble panels on the walls. In mid-afternoon the Windhoek flight was called, and soon I was again looking down on the drylands of southern Africa.

'What you have to remember, Martin, is that Namibia is a desert.'

Bob Sinclair, proprietor of the Sundown Lodge, was the first of five Namibians to tell me this during my visit. They wear it like a stigma, as though the word 'desert' implied only absence, a *lack* of trees, of grass, of green; in fact, Namibia is a place of intense *presences*; but of course, it is infertility they are ashamed of.

Don, the representative of the company that owned the Land Rover I had booked, was explaining the use of the on-board satellite distress beacon. 'I hope you won't need it,' he said, 'because being recovered by the Namibian Airforce doesn't come cheap.' He tried to demonstrate how to raise the roof tent. It seemed to be stuck, and we both stood on the Landy's metal-plated bonnet struggling with the folding frame. Don was a little the worse for wear, having been up half the night with 'a lady friend'.

Bob called out from the bar, 'Are you two trying to work up a bloody sweat? Or are you going to come and have a drink?'

The sky over the Sundown Lodge turned a livid wound-red, and the horizon was masked by sheets of grey rain. It rained, big drops that plopped in the dust like old English pennies. From somewhere the rain alchemy produced a hissing noise like an electric kettle, and a sweet, damp musk rose from the ground.

The next morning was typical Windhoek weather, the sky cloudless, the dry air over the camel-coloured hills pellucid. Brassy sunlight blared on a modern African town punctuated by Tyrolean towerlets, Bismark Streets and *Biergartens*. Over the years various nations have asserted ownership of the patch of land now known as Namibia, including South Africa, Britain and Germany, but the German imprint remains strongest. The first thing I noticed that morning was the number of wealthy whites about the place, well-padded Germans and Boers pulling their four-by-fours into Windhoek's new shopping mall, while blacks swabbed pavements and filled shelves.

I was looking for a haircut.

I have a weakness for getting my hair cut abroad. I have watched locks of my hair alight on the dry earth of an open-air barber's in Uttar Pradesh, and the marble floor of an art deco salon in Beirut. I never tell the hairdressers too much, but let them decide how I should look. Perhaps I am seeking acceptance, or a disguise.

I found a small unisex salon on the edge of the town centre. The walls were plastered with adverts for 'hair relaxers' – chemicals that overcome the resistance of African hair. I was the only white person there. With the exception of a clutch of wide-eyed schoolboys, the customers were women having their hair done before work, and my arrival caused a ripple of surprise. Women naturally feel exposed by the presence of a man during these intimate operations. One sitting next to me had her hair in an unrelaxed fan like a peacock's feathers, and flashed me a hostile sideways look. I did not blame her.

My hairdresser was a young woman who started hesitantly, lifting my straight fine hair in her comb with exaggerated caution and observing my request 'not to take too much off'. She soon forgot the request, and tufts began to fall. I was quickly given a short back and sides, but with a very long fringe.

An assistant was running an electric razor over the heads of the schoolboys with the glazed-eyed indifference of an army barber working on a squad of new recruits. My hairdresser called him over and explained in detail how he should trim my hair, then went off to heat up some sadistic-looking curling tongs.

After a few minutes my hairdresser wandered back, and in the mirror I saw her eyes widen in horror. She shouted at the man, who backed off with a hurt look. I tried to glimpse the back of my head in the mirror over my shoulder. It seemed to be patterned with a series of intersecting diamonds, not unlike a coconut. Meanwhile an extended Napoleonic forelock covered my nose.

I was a source of embarrassment, and, after some swift remedial snipping, was bustled out of the shop.

I mussed my hair as untidily as I could and edged sheepishly into a coffee shop. Only in an African coffee shop would people come in for breakfast with an eighteen-day-old lion cub, abandoned by her parents.

'Here,' they said, 'hold her.'

I took the lioness in my arms. They handed me a milk bottle and I held it to her lips. She pawed me with her tiny paws, her eyes closed tight, and fell asleep, snoring happily.

The word Namibia is derived from the word Namib, which in the indigenous Nama language conveys the idea of endless expanse and nothingness. Nothingnessland: it must be the most mysterious and poetic name of any nation.

The Namib Desert stretches up the south-west coast of Africa, a hundred thousand square miles of mountain, scrub and sand dune. The average rainfall here is less than half an inch per annum – in some places, they say, it has not rained for a hundred years. But the icy Antarctic Benguela Current chills the western seashore, causing thick fogs which are blown up to fifty miles inshore before they evaporate. Rarely does the fog condense into rain, but the moisture is sufficient to nourish a range of vapour-sipping plants and animals. It means that air temperatures never reach the hallucinatory extremes of the Sahara. Nevertheless, the Namib is one of the world's driest deserts.

Its treacherous shore is known as the Skeleton Coast, a notorious wrecker of ships. The desert is most famous, however, as the world's largest single source of diamonds. They come mostly from a narrow beach just two-score miles long, known as the Diamond Coast. Over $100 million-worth of diamonds are bulldozed out of the desert here each day. The story of the South African diamond fever is well known, a story of indecent wealth for the few and disappointment and poverty for the many.

In 1920 the de Beers family negotiated themselves an eighty-year concession. And they guard their brilliant blobs of carbon zealously, with force of arms. I had no intention of going near the Diamond

Coast; but forty miles inland are some of the most distinctive and beautiful sand formations in the world.

I drove south-east out of Windhoek, through the hills of the Khomas Hochland, where white Namibians have their huge farmsteads, partitioned off with a geometry as mad as the compass-drawn frontiers of Mali or New Mexico. I lunched by the roadside on tinned sardines, pouring out the excess oil and watching it congeal in the red dust. In mid-afternoon I stopped at a farmhouse where a sign said 'Cold Drinks'. Some well-fed farm boys in denim dungarees looked up suspiciously from hunks of yellow machinery. The girl behind the counter was a shy, pale creature, thick in the hips, with a straight long dress draped over columnar legs. I ordered coffee and 'home-made cake', and went to sit out on the porch. The coffee, Nescafé, tasted of burnt cork; the cake was aerated, pink and flavourless. A fat old dog – was everyone here overfed? – collapsed in front of me and stared, drooling, at my mouth. As I surreptitiously passed him the pink confection, the girl came out to tell him to stop bothering me, and caught me in the act of feeding him the cake she had doubtless made with her own plump hands that morning.

Towards evening I saw before me a broken ridge of sandstone with a talus of sand lapping around it, the classic desert mesa, a pure and elemental sight so fine I had to stop and stand in the afternoon light and take it in. In early evening I parked my Land Rover at the campsite outside the Sesriem Park, and pitched my tent within the stone circle provided. Images of ringed Boer wagons came to mind, but this was just someone's attempt at homeliness.

It felt like an odd sort of desert travel after the Sahara, driving a rental four-by-four to a commercial campsite on the brink of the desert's profoundest mysteries. It was summer in the southern hemisphere, outside the season for desert tourism, and there were few other travellers around. I had travelled painfully into the heart of the Sahara in other men's broken, rattling wagons, only to breeze into the Namib in a brand-new air-con'd Land Rover. As sunset came on I hiked to the ridge overlooking the camp and watched a red stain spread over the sand and scrub. Tourist or no, it was good to be in the desert's silence again.

CHAPTER TWENTY-EIGHT

Solitary Pleasures

THE RANGERS UNLOCKED the gate that leads to Sossusvlei just after 5 a.m. Tourist jeeps impatiently gunned their engines, then raced through the barrier – every vehicle after the first one would have to endure the clouds of dust that rise into the still air and hang there.

Forty miles of hardened corrugated track made for a tooth-loosening drive. As dawn broke, I realized I was passing through a canyon of immense pyramidal sand dunes, black-silhouetted in the east and bronze in the west when the sun's rays touched them. For the last mile the Land Rover was paddling through soft sand, and then I reached the Sossusvlei water hole. This dune-locked pan is a magnet for many forms of desert life, most visibly several giant and evocatively named camel thorn trees. Wader birds circled overhead, disturbed by me and the jeep ahead of me, whose occupants were soon scrambling up the nearest dune. I made for a taller dune, due west of the pan.

At Sossusvlei the local yellow Namib sand is blended with red sand from the Kalahari, washed down the Orange River far to the south, and borne here by the wind. The dawn light deepened it to an intense red-gold. There is life in these dunes. In the Sahara I had seen tracts of desert where no organic life could survive, but here the water hole and the banks of fog that roll off the Atlantic provide enough moisture to sustain plant and animal life. After rainfall, tough grasses spring up in the troughs between dunes, and in the transitory moisture there appear urgent creatures such as the blue weevil, insects which hatch overnight, feed on the springing grass, mate, lay eggs and immediately die. Their eggs will wait months or years for the next rainfall. (This eery patience and brevity of life is not unique. Some desert seeds can wait thirty years for rain.) The desiccating, slowly decomposing grasses and far-blown seeds form a messy detritus that gathers in the leeward sides of dunes, forming the basis of an ecosystem known as 'detritivore'. There were lizards and many breeds of beetle. There was an incredible and rarely-glimpsed mole, a near-blind creature that swims through the dune, breathing the air trapped between grains of sand and surfacing only to claim its prey – mostly insects, but also a breed of legless lizard that likewise swims under the sand-surface. It was unlikely that I would glimpse a jackal, unlikely too that I would encounter a sidewinder adder (cousin of the American sidewinder rattlesnake), but I kept my eyes open, just in case.

With many stops to gaze at the delicate traceries left in the sand by lizards and beetles, it took me over two hours to climb to the highest point. There was an archipelago of false summits, and the sun was high by the time I reached the top. I sat and contemplated views thirty miles out across the sand-sea of towering dunes. The Sossusvlei dunes are of a rare form geographers describe as parabolic or multi-cycle – the prevailing winds blow equally from every direction, fashioning sinuous star-shaped entities, crumpled like bed-sheets except for their razor-edged apexes. We are told that these giants move, a yard a year. They were, it seemed to me, the loveliest natural phenomenon I had ever witnessed, seen now at their bosomy gentlest.

The gentleness was illusory. By ten the sun was beating down, bleaching the colour from the dunes and grilling the surface of the sand. In the west a layer of soft grey fog hovered over the dune horizon, towards the sea. In the pale blue sky there was a whirring, and I saw a small plane flying up the valley I had driven through. It banked and passed over me and I waved my scarf as I sprawled, a human dot on the crown of a vast sand star. No one saw me. Dwarf sand-dervishes whirled on the dune surface, sucking in fragments of dried grass and making them dance. A bright green gecko darted out and I tried to pursue it, but it dived into the sand. I scooped my fingers into the dry orange grains but no good, it had swum too deep. There was a high-pitched whine and I jerked my head round to see – incredibly – a mosquito. I lifted my canteen to drink and a few droplets fell, formed into sand-encrusted globes and tumbled away. At once a black beetle appeared to harvest the water. As it came to grips with a water-boulder, I watched to see it drink, but instead it earnestly rolled its sisyphian burden back up the slope. I was entranced by this mountain-top hermit, a fellow creature.

The whole point of desert travel is solitude. As our ever-multiplying human numbers crowd the globe, a few who can take it no longer break free and run for the mountains, the oceans, the deserts. I am enthralled by many of the attractions of metropolitan life, but in a corner of my heart I keep the image of a silent, spacious place. A space, perhaps like this . . .

Reluctantly, I decided I would soon need shade and allowed myself the childish pleasure of rolling down the dune. But as I rolled down the steep eastern slip-face I realized how hot it was, yelped, and scrambled back in a series of undignified hops.

I spent hours scaling the sandy slopes. At midday I slept under a tree beside the flooded pan. Towards evening I climbed another peak to watch the sun sink into the Diamond Coast, the dunes grow coppery and the shadows lengthen and blacken. Only those who have spent time in a desert can understand its powerful, mystical attraction. It is a solitary pleasure.

Solitaire is a lonely name for a town, but an apt one. It is the epitome of the dusty desert halts in Arizona or the Outback that have become a staple of TV adland – the ramshackle filling station, a few tin shacks, some scrubby trees and gutted cars, surrounded by the desert void.

A satellite dish jutted off the wall of Solitaire's solitary shop. A jovial, bearded man stood behind the counter, passing crates of beer to the big, hairy men who came in to stock up. 'Can I help you?'

'Er – do you have any combs?'

Next door there was a restaurant offering sausages and *sauerkraut*. A foursome of German tourists sat opposite me. One of the women looked anorexic, rope arms, prominent veins on her legs. How odd she seemed in this land of fat health. A blond, pretty creature, she wore a twisted grin, as though permanently conscious of the irony of her illness.

Opposite the restaurant was a small Boer chapel. I asked a waitress if she knew who had the keys. A local farmer, she said. And did she go to the church?

'Ach, no – us blacks are not allowed.'

'Are you serious?'

'Oh yes.'

There were five black staff, but the manageress was white, a chubby young woman with an attractive, cheerful face. 'They're die-hard Boers around here,' she told me, 'and they think it's still apartheid. I'm very unpopular. They hate the fact that I employ blacks, and that I treat them like equals. In little towns like this you're expected to go to church, but I stopped going because they won't have blacks. You know, before the service the chapel is cleaned by somebody's servant. One woman then allowed her servant to sit at the back during the service – and several of the men walked out! I ask you, Christians . . . I told them that they ought to paint the chapel in zebra stripes – black and white – now that everyone is allowed in; you can imagine how that went down. Anyway, I can't take it any longer. In a few weeks' time I'm off.'

The chapel windows were of green frosted glass. Green, the colour of fecundity, in this barren place. And frosted, to stop the sinful Sunday-morning eye from wandering, or idle eyes from peering in. The chapel was locked, but a broken pane allowed me to open a window and see the severe interior with its pulpit, pews and harmonium, all covered in fine, Solitary dust.

CHAPTER TWENTY-NINE

Deutschland Über Alles

FROM SOLITAIRE I DROVE WEST under a tall, severely blue sky. The road wound and dipped through the canyons of the Kuiseb River, the barrier which marks the northern frontier of the vast Namib dune-sea. The Kuiseb does not flow every year, and although its mouth gapes at the Atlantic, it has not flowed into the ocean since 1933.

The road straightened out, and bisected a flat, dun dullness. Occasional die-straight tracks erupted from the dust and struck the tarmac at right-angles. They led to white farms nestling in the green highlands.

In late afternoon the desert stopped. The high clouds that had been tantalizingly over the horizon for hours were suddenly drawing closer, and with them the ocean. It flared for an instant as the road crested a hill – a far-off glimpse of acid-yellow, like light bounced off brass – then vanished again. The road descended, dipping under power cables. Two plumes of white smoke billowed from factory chimneys. A chain of giant dunes appeared, peaking and troughing their way north. Then, suddenly, the whole horizon was ocean, a levitating golden bar.

The air, like a Nordic seaport or an Alpine village, was spare and brisk. There were reedy, pea-green ponds dotted with white waterfowl, then palms, tall ones sprouting from the hub of a roundabout, newly planted babies like giant carrots.

This was Walvis Bay, seaside oasis, a Namibian Miami. I passed proper-looking homes with breeze-block walls and sprinklered, palm-shaded gardens, embalmed in calm late Sunday afternoon sunshine; municipal bungalows, toy-set boxes with tin roofs; thatched cottages in spruce plots, their garlanded picture-windows staring out to sea; an esplanade, with pink and turquoise bench-seats; Pikkie's

Chips and Russians, Engen Hamburgers, Grobbies Estate Agent, Beco Plumbing.

Walvis Bay is a Dutch corruption of the Portuguese name *Bahia das Bahleas*, Bay of Whales, given to this place by seventeenth-century whalers. In 1487, the Portuguese Bartholomeu Dias was the first European to land here, searching for a sea route to India. He found no fresh water, and unaffectionately named the area Sands of Hell. Later it was recognized as a valuable deep-water harbour, rich in fish (eventually fresh water was discovered too), and Walvis Bay was to pass through many colonialist hands. Even after Namibian independence in 1990, South Africa clung on to it for four years. Now it is being developed as a major port linking land-locked Zambia and Zimbabwe with the Atlantic.

I wondered how far south the road went towards the Diamond Coast. It passed low dunes and mud flats, and suddenly there were flamingos in their thousands, dabs of white and pink picking their way across the mud or wading in marzipan-blue shallows, or ribboning through the sky like streamered kites, their wings blood-red as the late light streamed through them.

The flats had been put to practical use. There was a saltworks, and a fifty-foot salt cone glowed blue. The road twisted past miles of salt pastures and finally reached a beach of dirty-looking sand furrowed by the tyres of four-wheel-drives. Haloed fishermen cast lines at the setting sun.

I turned back through the town, between those menacing dunes and the seashore ('Duneboarding Here! Take a Camel Ride!'). Twenty miles north of Walvis Bay is the even odder town of Swakopmund. I pulled off the road to prepare myself for human society. My skin was sun-raw and sand-caked, my clothing filthy. I climbed into ironed clothes, and in their stiff cotton sheath felt cleaner. I found a bar. After three solitary days, I had emerged from the desert into a packed tin-roofed beer cellar resounding with voices and German drinking music. I ordered a beer and felt uncomfortable in the throng.

Back in the street, there was a deep, seashore silence. I walked down to the beach, where breakers crackled and hissed. Two men walked close by me, and I realized I felt mildly alarmed. How strange to leave the enwombing security of the desert for the unsettlement of human company. Human beings are social creatures, but unsocial too.

The next morning I looked for somewhere to catch up with my writing. What someone recommended as 'the best café in Swakop' turned out to be part of a grey resort hotel, with a Day-Glo sprinklered lawn, palm trees, sunbrellas and an occluded view of the turquoise sea. The menu promised *Kuchen und Torten Nach Ihrer Wahl*, Cakes and Confectionery of Your Choice, served by cold, polite black waitresses who had had their natural vivacity bullied out of them.

The mock-Bavarian streets of Swakopmund end abruptly in prospects of open desert. German tourists predominate, and German-run shops sell souvenirs proclaiming **Deutsches Südwestafrica**, Namibia's German colonial name. I tried to think of comparisons; 'British Empire' tea-towels in New Delhi; '*My forefathers wiped out the indigenous people of Tasmania and all I got was this lousy T-shirt.*' What was at the root of the desire to perpetuate a German Africa? A need to believe that in the colonial scramble for world domination, Germany had, after all, made its mark? German is widely spoken in Namibia, but in declining numbers, so that the German state is eager to pour in money to provide the language with intensive care. During my stay in Namibia the German president was on a state visit, and facing demonstrations demanding an apology for massacres carried out in German times.

North of Swakopmund are miles of flat coastline with occasional rod-fishers, and seedy seaside townlets. This is the start of the notorious Skeleton Coast, feared by mariners for centuries. The ship-wrecked faced a trek through uninhabited, waterless desert for hundreds of miles in any direction. Death was as good as certain.

Even today, the coast is littered with the skeletons of ships, some recently wrecked. The only people able to see them are the annual 1,000 wealthy tourists who pay the sole concessionaire for an exclusive 'fly-in safari'. This 120-mile-long strip of shifting sand-bars, sand dunes and salt flats is now the Skeleton Coast National Park. Anti-tourism pressure groups believe that high-cost, 'high-quality' tourism is the only way to protect vulnerable environments and human cultures from destruction. Make a place so expensive that it becomes the preserve of the very wealthy, and disruption will be minimized. Namibia plasters photographs of glamorously wrecked ships all over its tourist publicity, but the wrecks themselves are out of bounds to all but the rich.

Disappointed as I was to hit the No Entry signs, I felt a sort of relief. After the Sahara, I was finding even the lightly touristed Namib overcrowded. The free market covers anything beautiful in squalid holiday homes. If exclusive tourism can protect the seashore, then perhaps the rich are serving a purpose.

I did see the bones of the *South West Sea*. The wreck of this small coastal vessel has been picked clean by many a souvenir hunter, but its wooden spine and ribs are still visible, buffed smooth by the wind and sea, its engine block slowly returning via rust and corrosion to base chemistry.

I had a flat tyre on the road east out of the Skeleton Coast, in the hilly hinterland of the Kaoko Veld. It was just after midday, and the heat was intense as I began working my way through a pile of tools that might have been designed for a tank. So far on my journey through the deserts I had been in the fortunate position of having other people do this sort of thing for me.

I was just discovering that the foot-pump was broken when another Land Rover pulled up behind me – the second vehicle I had seen all day. Two men climbed out, and, staggeringly, one of them was wearing a T-shirt saying 'Land Rover Service'.

'Can we help?' he asked, in English.

The second man handed me a can of iced lager. 'How do you do?' he said urbanely. 'My name is Marcus, and this is Roland.'

Roland was a fanatical member of the Austrian Land Rover Appreciation Society. The two of them were on a trans-African trip to test-drive (demonstrate, prove, *assert*) Roland's old (but highly maintained) Land Rover, once owned by the BBC and equipped for wildlife photography in Africa. It would be a positive pleasure to help me change my tyre, he insisted.

Roland's absorption in Land Rover lore was quasi-religious. Its claims to superiority over every other type of four-wheel-drive vehicle had the status of Divine Revelation. He pulled out a felt roll containing tools as glittering and sterile as a surgeon's, and set to work.

Marcus pointed at a pennant fluttering on the radio aerial. 'That,' he said, 'is an *Austrian* flag. So is that, and that.' The vehicle was plastered with Austrian insignia. 'And why do you think we have so many Austrian flags on the car? Because we *hate* being taken for Germans.'

We eased my Rover up on its big jack and released the spare tyre from its place on the bonnet. Marcus and Roland told me that a German-dominated European Community was going to destroy everything that was individual about Austria. Britain, by contrast, they loved. It was anti-EC. It was the home of pop music, *Monty Python's Flying Circus* – and the Land Rover. 'But,' Roland put in tartly, 'it took a BMW, a *German* company, to realize the potential of the Land Rover marque and begin to market it properly.' The utopian creed of Land Roverism, I saw, transcended merely national boundaries.

'Anyway,' Marcus added ironically, passing me another chilled Tafel Light, 'we are all Germans – er, I mean, Europeans – now.'

I spent several days driving through the Kaoko Veld, a land of abandoned mines – though a few gemstones still emerge from the reluctant earth. I saw what looked like trees struck by lightning, mounds of fleshy leaf apparently dissolving into ash. They were Tumboa plants, *Weltwitschia mirabilis*, the miraculous dwarf trees of the Namib. In the course of geologic history the surface of the earth has cycled through phases of moisture and aridity; incredibly, the Weltwischia has survived from a period of global aridity several phases back, perhaps hundreds of millions of years ago. It is known that individual specimens can live hundreds – some scientists say thousands – of years.

The terrain was barren and harsh, words so axiomatic in a desert context as to be almost meaningless, yet here real because of the signs of human struggle. Under steep table mountains I passed villages of impoverished shacks, and shabby clapboard farmsteads with a broken truck on blocks in the yard and a few scrawny animals gathered in the shade of a thorn tree. I stopped at a graveyard where wrought-iron crosses stuck out of hard, rocky earth. The ground was strewn with malicious burrs. I saw the grave of a young man who had died on his twenty-sixth birthday: '*Hier Rus Blasius Florry + 29.7.1971 + 29.7.1997*'. Bricks and plastic flowers were piled on the grave-mounds, and here and there lay empty whisky bottles.

I camped at a luxury 'lodge' deep in the bush (though 'bush' belies the scorched nature of the terrain). A former farm had bagged the best land for many miles around. Nearby blacks lived in rural poverty, but the hotel was built beside a spring on a raised bluff, with distant, gorgeous views. Irrigation had created an oasis where foaming foliage embraced a turquoise swimming pool. But I stayed at the gravel campground, segregated from the luxury rooms.

I saw few black people at the lodge. You glimpsed them on the edge of things, at dawn or dusk, moving silently through your peripheral vision with mops or brooms. They never spoke or smiled. The camp was orderly and attractive. I saw the camp's German owner that afternoon, berating one of his black staff, his blunt face screwed up and red with rage, the black cleaner flinching and humiliated. It was a sight I was to see repeated in both Namibia and Botswana, both countries liberated from the yoke of apartheid.

CHAPTER THIRTY

The Romantic Belgian

ERIC VAN VELDT IS A man in love, but also a man divided, because the woman he loves lives in another century. Mbimba belongs to a semi-nomadic tribe called the Himba, who live in northern Namibia near the Angolan border. She is extraordinarily beautiful, and, as Eric's photographs show, very photogenic. And she loves him too.

But I must admit impediments to the course of True Love: Eric's Belgian girlfriend is insanely jealous about Mbimba. And Mbimba is pregnant by another – unknown – man.

I had met Eric in the dusty northern settlement of Sesfontein, in a hotel still being converted from a ruined German fort. We would sleep in our jeeps at night, but have the run of the hotel by day, with its delicious pool and shaded verandas. The owners were tangentially in touch with the world via short-wave bush radio, the machine kept permanently switched on so that a continuous trickle of life in the bush – births and accidents, deaths and scholarships won – seeped into the subconscious.

Eric was a man in flight from the modern world. He worked as a freelance guide, taking tourists around the Namibian north, cursing them for their facile questions and their childlike desire to peer at the wild animals he felt should be left in peace. He claimed to despise tourism, but accepted that it allowed him to live the way he wanted, out in the bush.

He was sitting opposite me in the bar. Strongly built, not tall, blond, about thirty. His scarred face gave him a dangerous look.

He laughed. 'Dangerous? Yes, I know people think that. I suppose I use it to my advantage, somewhat. But I'm not dangerous at all, you know. I didn't get scarred fighting, I had an accident as a child. Well, actually, I was making a bomb and it blew up in my face.'

'But you're not violent.'

'No.'

'Building a bomb isn't an act of violence?'

'I was adolescent. You know how boys are at that age.'

I do not think Eric would have taken me to meet his friends among the Himba if we had not happened to share an enthusiasm for deserts. He had travelled in the Sahara, and had just returned from a trip through the uninhabited desert in Kaokoland.

'I got stuck in loose sand, I had to dig myself out. Stupid – I mean, I take risks; I couldn't have walked out of there. And I really wasn't sure I was gonna get the car out of that sand. Look at my hands.' The palms and fingers were swollen and chafed, covered in fresh scabs. 'Rubbed all the skin off digging. Still, made it.' A grin. 'So you love deserts too?'

We sat around, talking and enjoying the shade. Eric was going north to spend another week with the Himba before heading back to Windhoek. Yes, he would take me with him. If I behaved the way he told me to.

'And don't think you're going to meet Noble Savages. They're not superior to so-called civilized man – or inferior. They're just human. But their lives are simpler than ours, so generally, I think they behave better. Maybe it is a kind of nobility. If nobility means being far from civilization. Yes, perhaps there's something in it after all. We need a word to replace "savages".'

He pulled a postcard from the book he was reading. It showed a half-naked Himba woman standing in a supermarket, a wire basket in her hand, stacked shelves behind her.

'It's a striking image,' I said.

'Yes, well. That's how people here see the Himba. Anachronisms. Run up with your camera, and take a snap of the naked tribal. A bit ethnographic, a bit sexy, you know. They are sexy, these chicks, it turns men on.'

'Yes, it would.'

'So, what's *your* interest?'

'Curiosity, I suppose. Nothing noble.'

I followed him north, his four-wheel-drive flat-bed bumping over dry hills dotted with thorny trees, an apocalyptic white cloud ahead of me on the ghats, past Opuwo and into the forest towards the Angolan border: Himba land.

It was dusk when we reached the Himba encampment, a circular stockade made from broken branches, with several conical huts and a central corral for livestock. In the dim glow of fires I saw serious faces looking at me. 'I don't usually bring visitors,' said Eric. He greeted people and laughed uproariously, but there was a tension. No one offered the weary travellers food, or even water. I presented gifts – millet flour, salt, apples – to the chief's wife. She accepted them with a curt '*dangee*' – Thank you. 'Don't worry,' said Eric, 'they'll relax when they get to know you.'

Later, as we cooked our meal on a campfire outside the village, Mbimba came out to join us, standing uncertainly, unspeaking. I could not tell whether she was fearful or aloof. Perhaps she was both. Eric had barely spoken to her since he came back from Christmas in Belgium with his family and girlfriend, and found her with child. She was now very visibly pregnant – around five months. Everyone at the village was convinced the child was Eric's.

'I respected her, I never touched her,' he told me. 'Wait till it's born, it will be black.'

The Himba practise ancestor-worship, and keep a sacred fire in their encampment. This foyer is where the women spend much of their time, and the men, who tend the animals, usually return in the evening. The next day Eric took me out to where some of the men of the village were grazing their cattle, a mile or so from the main camp. I was introduced to the chief, who asked Eric, by no means for the first time, if he intended to take Mbimba as his wife. He pointed at a distant hill.

Eric smiled. 'He's offering me land over there. He says it's good land, green. He says I should come and live with them.'

As we walked back, he explained, 'The old chief is canny. Times are changing for the Himba, and he sees me as a way for them to mediate with the modern world that threatens them so much. It's understandable. And from their point of view, I'm rich. Of course, he thinks Mbimba is pregnant with my child, like everyone else. So he's got a surprise coming, too.'

Mbimba passed her days with her sisters, herding cattle, feeding children and grinding millet. Spare time was spent in self-adornment, making bangles from bark and fruit stones, grinding red sandstone and mixing it with fat to smear over her body, giving her the distinctive copper glow of a Himba woman. Her hair was plaited and congealed with mud into a glorious crown of copper ropes. She wore necklaces, and long bracelets at her wrists and ankles, and a short skirt of meticulously pleated goat hide. Nothing more.

'If you really love her,' I suggested to Eric, 'you'll accept her, you'll find it in yourself somehow to forgive her.'

'No. No, you don't understand me.'

'Are you an absolutist, Eric?'

'*Yes*. Her being pregnant changes everything. I asked her how she feels about the baby, and she says she wishes it had never happened. Well, it's too late for that now. It's all over.'

But the next day, his mood had softened. 'How could she do it?' he asked me. 'But I still love her. My God, how I love her.'

'What understanding did the two of you have, Eric?'

'That we were in love. Look, I'm not proud of what has happened. I have a girlfriend in Belgium, we have been together ten years. We had been talking about her coming here to live. But I fell in love with Mbimba. We didn't have sex, I told her I had to talk to my girlfriend in Belgium. She knew I was coming back. She should have waited.'

'For what?'

'To see how things worked out. Christ, now she's pregnant, and she says she's not *sure* who the father is. Is she protecting the man? Or is she really not sure; in other words, is it more than one man?'

I felt slightly irritable with him. 'You know very well that life is different here, Eric. Romantic love and serial monogamy are not the rule all over the world. She isn't a nice suburban middle-class girl in Europe.'

'You're a cynic.'

I laughed. 'OK. How are you going to choose between your Belgian girlfriend and Mbimba? What would your life be like if you married Mbimba? Are you sure you're not living out a fantasy here – the rich white man arrives with his gifts and *savoir-faire*, the native girl is swept off her feet. Are you sure you're not exploiting her?'

'The point is, I love her. I don't pretend any of this is easy.'

Slowly, I was accepted by the village. As I came to know Mbimba a little, I found her an intelligent and complex woman, at different moments gay or moody, flirtatious or curt. She did not have a light heart; she frowned a good deal, but Eric laughed and said she always had, it was evidence of her depth. 'Ask any of the women, they'll tell you she has always been like this. She's thoughtful, she asks questions. Maybe that's why she was interested in an outsider like me.'

I could see that that might be true. I could also see that Mbimba understood very well that Eric was slipping away from her.

Late afternoon in a wooded encampment even deeper in the bush: I was sitting in the shade with a slender, lovely fifteen-year-old, who was weaving out of bark a strap on which to swing a calabash and churn milk into cheese. Ineptly I was grinding corn between two rocks. She called out to Eric, then began to laugh hysterically. He did not really need to translate her question. 'Tell her I'm already married,' I said. 'Only just married, in fact.'

One night Eric announced that we were going to have a party. He turned his car stereo up loud, playing Western dance music as well as Himba singing and drumming he had recorded himself. The music boomed into the black, silent bush. Children came out of the stockade and began to dance.

At other times the tiny, big-eyed boys, still too young to be out alone with the animals, would grab our Western paraphernalia and examine it gravely – the deep-rimmed hats, many-bladed knives, cameras with their water-well lenses.

'When they grow up,' I said, 'they'll want a hat like yours and a car stereo. And you won't be able to blame them for running off to the city to find it all.'

He did not have an answer. But when I saw him crouching with the women by the evening fire, I saw that he was happy there. A man in love, not just with a woman, but a way of life.

After several days in the village, Eric drove me north to the Epupa Falls, where the Kunene River plunges 200 feet into a boiling pit of orange alluvial water. There are palm trees, snakes and crocodiles. On the far side of the ravine is Angola.

Epupa is one of Namibia's most glorious natural spectacles, and the most endangered. The government wants to build a hydro-electric dam that will swamp the valley in a reservoir thirty miles wide and sixty miles long. For the Himba, the effect would be catastrophic. They would lose a vast area of their ancient homeland; 160 grave sites, where the holy men commune with the ancestors' spirits, would be swamped; the town that would need to be built in this remote place for the project's 5,000 permanent workers would bring roads, shops, bars, prostitution, disease.

I looked through the feasibility studies for the project, prepared by a highly paid European consultancy. Having observed one of these grandiose disasters at Narmada in India, I was familiar with the pieties that such documents contain about minimal cultural disturbance, compensation and resettlement. All lies.

One night, after we had turned in, we heard the distant sound of women chanting. 'Get up!' said Eric. 'It's a dance!'

We pulled on our boots and walked through the bush for ten minutes, until we saw a fire. Six bodies, six women aged from twelve to twenty-six, were swaying, their limbs and breasts and braids catching the gold light of the fire.

Eric whispered that at mixed dances the sexes line up opposite each other, but he was unsure of the protocol here. For half an hour we sat in the shadows, watching. One woman would begin a chant, which was taken up by the group. Then another would enter the semicircle and begin a stomping, whirling dance, her elbows high behind her. At once the clapping speeded up and the singers whooped excitedly, urging her on. Usually the dance would stop after a while, but from time to time the rhythm caught like a bushfire, and the dancer went into an entranced state, whirling and screaming herself towards exhaustion. Two or three times one of the women ran into the darkness, and we heard her hysterical laughter as she leapt and whirled around the camp.

Eric was becoming more and more excited. Suddenly he leapt up, jumped into the circle and began madly dancing, the women clapping and screaming and urging him on.

When he finally stumbled out of the circle, the chief's wife darted forward and grabbed my wrist and propelled me into the circle. At once the chanting and accelerating rhythm possessed me, and I too stamped and whirled like a demon.

It was early morning when a sound from the corral alerted the women that some animals had broken loose, and they ran into the darkness, scolding the cattle with high-pitched shrieks. The dance was over.

After I had been with the Himba a week it was time to leave, but I could happily have stayed much longer. The people I had met had touched me with a simplicity and spontaneity, a dignity and lack of aggression. These are clichés about 'primitive' cultures, and the Noble Savage is supposedly a long-debunked myth, but Western society has more to learn from such people than it can admit. To admit it would mean slowing down our headlong dash into industrialization, with the promise of suburban refrigerators and televisual ecstasy for all.

Before leaving Himba country I encountered a missionary woman who had turned up to give some women a Bible class. I asked her what she was doing.

'I am teaching them to read.'

I looked at the photocopied extracts the women were holding in their hands. 'You're teaching them to read *the Bible*. You are *changing* them. You are bringing the modern world closer to them, the modern world that will destroy them.'

She looked genuinely surprised. 'Why do you think the modern world will destroy them? Don't you think they have the right to medicine, and education?'

'Medicine, yes. But education means indoctrination, and becoming embarrassed by their own ways.'

'If they can't read medicine labels, voting forms, newspapers, then they *will* be exploited. You want to leave them unequipped to deal with the modern world.'

'I want the modern world to *leave them alone*.'

She gave a superior smile. 'The modern world will not just go away. I am helping them to protect themselves by being able to communicate on an equal level.'

'By teaching them to reject their own religion? By teaching them to be ashamed of the naked body? You do think nakedness is shameful, I presume?'

She hesitated. 'That is not the issue.'

'You're teaching them to read by giving them Bible stories, by indoctrination. It's cultural imperialism.'

'So what do *you* think we should do with them?' I could see she was beginning to get angry too.

'What makes you think we have the right to *do* anything with them? We should just leave them alone! Let them come to us when they want to.'

She looked at me with distaste. 'I don't think this conversation is going anywhere.'

Six Himba women were watching us with perplexed smiles. I turned to one of them and asked Eric, who had been keeping out of the row, to translate a question.

'Do you want to learn to read?' I asked.

One of the younger women spoke up. 'Yes,' she said, 'because some of the men go to school and they tell us things we can't believe. We want to be able to read the books too. Also the lady is nice. The stories she tells us are interesting.'

She meant the Bible stories. I said, 'She will make you turn away from the religion of your ancestors. She will make you dress like her.'

The women squirmed and grinned. 'No, no, we will never do that.'

'Oh yes. Hear what I am saying. This woman wants to change you. Don't trust her.'

There was nothing more to say. The missionary was wearing a smug victor's grin.

Of course I knew of places where Christians had protected indigenous peoples from the rapacity of capitalism – where Jesuit missions, for example, had kept European land-grabbers at bay. But the Namibian government and the missionaries are embarrassed by the Himba's 'primitive' ways. When they have finished, the Himba will be 'civilized': their land will be sold to farmers, and they will be living pointless lives in slums on the edge of towns, wearing third-hand American T-shirts, bleating 'Jesu' and getting drunk.

CHAPTER THIRTY-ONE

Bushmen and Bushwomen

'LOOK!' SAID MR FRIEDRICH. A glossy antelope slipped by. 'That was a gemsbok – and there, a hyena!'

It was dusk, and the sun's last rays were flushing a cumulous cloud the size of Kent. 'I think,' mused Mr Friedrich, 'I'll buy some giraffe.'

Mr Friedrich had a very large garden. It measured about 1,100 acres, was in fact a farm in the process of becoming – like many Namibian farms – a 'game park'. Most of Namibia has been parcelled, fenced, stocked and farmed within living memory. But beef rearing in this thirst-land has always been a struggle. Now, wildlife tourism is coming to the rescue. Game parks are springing up everywhere.

'The trouble is,' he said, 'giraffe are so expensive these days. Everybody wants them.'

I reflected silently on the irony that the grandsons of the men who blew the wild animals off this land must now dig into their pockets for the cash to bring them back.

Friedrich was huge, with legs like oak-hearts, one of them a poignant précis of the man. A few years ago he had stepped on a mine near the Angolan border, had his leg blown into strips, and been urgently helicoptered to South Africa, where military surgeons spliced it back together. Months in hospital. Now the leg carried him again, good as ever, thick purple scars coiling around it like vines. Friedrich wasn't a man to let something like that stop him in his tracks.

He had been walking and loving this land since childhood. He used to disappear for whole days, usually alone or with the Bushmen who taught him their language and much that they knew about the land. Now he puts up tourists in tall versions of traditional Bushman twig-and-grass shelters, and takes them on nature trails, indoctrinating them with a love of Nature.

He showed me dozens of canny Bushman devices for staying alive in the desert. Digging up succulent roots, scraping nourishment out of fibrous leaves, making vacuums with hollow grass stalks to suck water out of deep, moist sand, using ostrich eggs as water bottles, animal bones for arrowheads, crushing beetle larvae to produce lethal poisons.

I was impressed by all this – and by the great variety of trees on his land, wildlife too. There was a manifest absence, however, of Bushmen and Bushwomen (or 'San', as they are known in the clicking Khoi languages of Southern Africa). True, we were accompanied on our walk by a shy, baseball-capped Bushman called Jimmy, who would demonstrate techniques of trap-setting and reed-sucking. But there are only a few San now in this part of Namibia, struggling to adapt to the ways of the immigrants – white and black – who have pushed them off this land and destroyed their way of life.

'Ach, the Bushman Way of Life,' said Mr Friedrich. 'You're like Laurens Van der Post – a romantic. I'll tell you – if you think the Bushmen lived in some sort of paradise, an Edenic state, you should think again. Heat, thirst, starvation, disease, endless walking across inhospitable desert – I tell you, they lived in *hell*.'

I had to disagree. Surely the Bushmen once had free rein over much of Southern Africa, which included many places that might indeed be described as paradisal – land now dominated by far pushier black and white tribes.

There are popular notions about the Bushmen: the romantic and the scientific. The romantic view, most notably propagated by the author Laurens Van der Post, was that the Bushmen were an innately spiritual race, non-violent and mystically in harmony with nature. This contrasts with the dour scientific view, articulated in this instance by Mr Friedrich: that the Bushmen were a primitive people, well adapted to an appalling landscape and climate, but far from happy in their squalid poverty.

I told Friedrich that I had met Van der Post, and that if romanticism consisted in trying to preserve something valuable and fragile from destruction by something stronger and more aggressive, then I was indeed a romantic.

Friedrich was involved in a charity set up to provide these traditional hunter-gatherers with some permanent land and the skills for sedentary farming. (Back at the office, he sold me one of their calendars.) 'Ask them if they want education and so on. They *want* to be more like us.'

'That,' I insisted, 'is because, stealing their land, we leave them with no choice.'

We walked on.

'Look at that,' Friedrich boomed, pointing at a termite nest that rose before us ten feet tall like a fairy fortress, and tunnelled perhaps fifty feet down, mining the damp deep beneath the desert floor. 'Enough termites in there to make New York look like a small town. And every termite knows its role, does its duty. That's what's missing from the modern world, it's all *me, me, me,* no one thinks any more about what they can do for others. My philosophy is, "Only when you've done your duty do you have any rights!"'

I suggested that Nature, being unconscious, is easier to love than Man. Nature is simple, human beings complicated. It was easier to preserve a tract of land than a way of life, but weren't there certain ways of life that might be worth trying to preserve?

'But what is it that's destroying everything of value? It's the modern world!'

I did not voice the thought that came to my mind: that in Namibia, it was the white farmers who had destroyed the African tribes and the wildlife, and made the semi-arid land susceptible to drought. At what cost did a few hundred thousand whites make a living, the tenuousness of which they constantly complained about?

Mr Friedrich was clear who were the villains of the piece: the manifestly foolish and wicked assembly of politicians, journalists, anthropologists – and Walt Disney: 'I was a consultant for them on a film they made with Bushmen characters. They ignored everything I told them, the thing was a farce.'

We passed a tall cactus-like shrub. 'In Europe they keep these in tubs in their offices, I hear, ho ho! And your UN is handing out eucalyptus trees by the score for people to plant. It's madness! They kill the undergrowth!'

'*My* UN?'

It was Friedrich versus The Rest.

Two nights later I was in Botswana. I had looped out of Namibia north-east towards the Victoria Falls, via the pan-handle of the Caprivi Strip, and south into Botswana. In this remote north-western corner of the country there is a row of puddingy quartzite hills that stand alone amidst thousands of square miles of level Kalahari. Here the San left a treasurehouse of paintings of antelope, lions and even fish – like finding the Louvre in the middle of the desert, said Van der Post. A shrine, and an epitaph to an all-but-vanished culture.

I missed the track to Tsodilo, and when I turned west after Sepupa I had to endure an unremitting three hours of bone-shaking ruts, and what they say is the softest sand in Botswana. I could see the hills ahead, but they refused to come any closer, the vehicle screaming and wallowing in Low Range through the sand. But gradually the hills took shape before me, an archipelago of domes known in Bushman legend as the Male, Female and Baby Hills.

A man in a village nearby the hills told me I was the only foreigner there. Then, walking in the midday heat, I stumbled through some thorn trees into an oasis of green sunbrellas, canvas chairs and cold beer.

Andy MacGregor is a young Scot who is building the first-ever lodge at Tsodilo. After years of buying and selling African handicrafts, he had fallen in love with Tsodilo, and decided to stay here.

'Tourist numbers have been increasing in recent years, most of them flying in to the airstrip for a quick shufti. Some have been deliberately damaging the paintings. Some of the designs are surprisingly modern, and we found one well-intentioned ignoramus scraping them off in the mistaken belief that they were recent graffiti! Then there were the Happy Clappers.'

'The what?'

'Fundamentalist Christians, up from South Africa. They thought these designs were evil – animist, pagan, ungodly – so they defaced them. This is a national monument, so theoretically entrance is restricted, but in effect anyone can wander around unmonitored. There's a little camp of government archaeologists, but almost no control of who's coming up here. This place should be a UNESCO World Heritage Site, with international funds to help protect it.'

I asked Andy if, for all his concerns, he wasn't exacerbating the problem by building a lodge here.

'Fair point; but no. The people I cater for are the top end, who are coming anyway. The problem is with tourists wandering around unsupervised. There's going to be a new tarred road that will give easy access to Tsodilo from the outside world for the first time. There'll be tour buses rolling up from South Africa, and then who knows what we'll see. Fast-food outlets, souvenir touts – and damage. I'm against the road, I don't think Tsodilo is ready for it. But it's part of the tourist infrastructure, you see. I've been petitioning the government to try to ensure that no one can get in to see the paintings without a guide. Otherwise, vandalism and theft are inevitable.'

That night it was too hot to sleep in a tent, and I chose a camp bed.

'Just keep your arms on the bed,' said Andy, 'there are scorpions everywhere.' He pointed his flashlight at the ground, and I saw seven or eight slits in the sand like flaps of flesh.

Andy is, I think, a romantic. He told me I must climb the Male Hill at dawn. At quarter to five we set off, scrambling up the rocky slopes by torchlight. After a few minutes the sole peeled off one of my hugely expensive, specialist 'desert boots', and I lashed it back on with a bootlace.

It was steep and rocky. 'Watch out, and listen out, for snakes,' said Andy. A few minutes later he said, 'Look.' A rudely awakened three-foot puff adder was curling into a shallow crevice. We scrambled on. 'Oh, and if you find anything interesting, my advice is, leave it alone. There are angry spirits in these hills.'

The rest of the climb was uneventful. Exactly as we reached the summit, a saffron sun bisected the sky and poured light onto the crest of the Male Hill.

I saw something glinting at my feet, a fine piece of white quartz, and picked it up. Soon the light was ebbing across the vast 360-degree plain of the Kalahari. In moments, night turned to day.

The summit was marked by a metal post with some words scratched onto its black surface: 'Big Jo, RIP'.

'He was an SAS soldier,' said Andy. 'He fell to his death a few years back.' He led me to the edge of a precipice and we stared into its depths. 'That was the spot. These hills get used a lot for training by special forces. But something went wrong that day. It seems his body was quite messed up, the soldiers were pretty shaken by the accident.

Then they were attacked by wild bees on their way down. Maybe they'd done something to upset the spirits of the hills.'

I looked at him sharply. But he was serious.

'Like I said, you disturb the spirits at your peril. You probably think I'm barmy, and I don't care. I've experienced the power of this place. You don't muck about with it.'

In the shadows a thousand feet beneath us, the fire at Andy's camp was burning brightly. 'That's odd,' he said, 'I raked the embers when we set off.'

On our way down, I was stung by a bee. It was nine-thirty when we reached the camp. Smoke and a stench of burnt feathers hung in the air, and four Bushmen were crouched around a sinister, black, smouldering mummy. It was all that remained of my sleeping bag, which the wind had apparently blown into the fire.

I reached into my pocket, and felt the sharp edges of the quartz I had stolen from the hilltop.

Early the next morning Andy and his San assistants were at work slicing tomatoes and salami. At lunchtime, people began to walk into the camp. They had flown in that morning, and San guides had led them the mile or so from the airstrip to the rock paintings. By lunchtime, gasping and perspiring, they were staggering into the bottle-green shade of Andy's elegant sunbrellas, sinking into the canvas directors' chairs with groans of agony and relief, and calling for drinks. One minute they had been roughing it, out of condition and without adequate shade and water, the next they were barking orders for beer and iced water like tourists on the waterfront at Cannes.

I was joined at the table where I was writing by a German in his fifties, wearing a dirty, crumpled T-shirt. He began to roll a cigarette, then hesitated. 'Do you mind?'

'Not at all.'

He smiled. His crooked teeth were nicotine-stained the colour of teak. 'Most people do these days . . . Writing a travel diary?'

'That's right.'

'I'm a computer programmer. I work from home. So I don't have to stand in the rain every time I want a cigarette.'

A San girl from the local village who helped out in the kitchen brought over his beer. She was beautiful, with golden skin and full lips, and I saw the German admiring her. Quickly he turned to Andy. 'Can I take a picture of her?'

'I suppose, if you ask her permission,' Andy said hesitantly. 'But why do you want to?'

'Well, er . . . She's another race, you see, not black but . . . you know. It's interesting, no?' He became uncomfortable.

'She's not an exhibit in a museum, you know. Take a picture of the old man if you want to, he's your guide, and you're with him. He's a member of the same race as her.'

'Yes, yes, quite so . . .' Embarrassed, the man nodded vigorously.

Andy turned to the pilot. 'Any rain your way?'

'Hey, the Delta's dry, man.'

'Did you hear it rained in Etosha, though?'

'Yeah, in Angola it's good, too.'

In the desert, people get excited about water.

Ghanzi, the desolate urban heart of Bushman life, is around 200 miles due south of Tsodilo. I was thirty miles from Ghanzi when the sky curtained over, darkness at noon, and bolts of pencil-straight lightning stabbed the desert scrub. By the time I reached Ghanzi a storm was raging, water lying unabsorbed on the hard-baked earth and spurting as though it were being strafed. I pulled over to pick up a soaked man holding a briefcase over his head. He clambered into the cab and sat beside me, steaming.

I pulled in at the town centre service station and we watched sheets of solid rain striping the streets, drenching anyone who ventured out. Some young men were taking cover under the petrol-pump shelter, but the high winds drove the rain at them. A girl in a thin dress made a dash across the street, but slipped and slid in the mud, her dress riding around her waist. There was ribald laughter from the watching boys, and the embarrassed girl dragged her hem back down over her knees and raced for the shed of the female pump attendants.

Ghanzi has the ramshackle, untidy feel of a frontier town, the sort of place where dreams meet an early death. The Kalahari Arms Hotel in the town centre is a blockish pre-independence building to which age has brought no charm. The tin-roofed rooms have cavernous bathrooms with bathtubs so long and deep you feel guilty about wasting so much water. Yet there are no showers, and I wondered whether this was evidence of a more leisurely age, or just that Ghanzi was so dusty that only a good long soak would get a chap clean.

In the bar, the mounted heads of reebok and gemsbok eyed me glassily. Two slick Barclay's bankers sat at the next table, all shirt pleats and aftershave, enthusing about investments and returns. I wondered what mischief they were up to in Ghanzi. Nothing that might help the San, that was for sure. The indigenous inhabitants of this place got by as cheap labour on the mostly white-owned farms. It almost amounted to indentured labour.

I hoped to meet Roy Sesana, a Bushman who was campaigning against the eviction of the San from their last stronghold of traditional life, the Central Kalahari Game Reserve.

It was still raining when I found him in his mud-walled one-room house on the edge of town. Water was pouring through a hole in the corrugated-iron roof into a galvanized tin bath. Roy's wife, Bulanda, was dishing up a lunch of rice and boiled beef. They welcomed me, and she put the kettle on the paraffin stove for some tea, while I dashed out to the car for a tin of evaporated milk and some ginger biscuits.

Roy was in a genial mood. A Land Rover, a vital tool for anyone trying to get around this tough country, had just arrived from South Africa. 'It is a gift to our organization from your Prince Charles,' he told me. Roy had a broad, infectious grin.

Sir Laurens Van der Post had been a friend and mentor to the Prince, and was associated with the Bushman cause from the 1950s, being instrumental in having the central Kalahari declared the Bushmen's home in perpetuity. But the Bushmen have no friends in the Botswanan government, and their fate is as grim as that of any tribal people. The government of Botswana claims it is evicting the San from the game reserve for their own good, but I did not meet one Bushman who believed this. They suspect that diamonds have been found there.

The last survivors of the hunter-gatherers of Southern Africa have become – like the Himba – an inconvenience. But while the Himba still live on a large territory they can call their own, with their culture substantially intact, the San, who once had the unrivalled freedom of all of what is now Botswana, have been forced into slums. Their culture has been destroyed by the combined forces of the Tswana – the dominant group in Botswana – and the whites.

'We may have been poor,' Roy said, 'but we had dignity and self-respect, and we could live from hunting. Outside the reserve, without an education, people have no way to support themselves but as low-paid labour. They are exploited. Our traditions are disappearing. We have nothing left. This town is full of AIDS and alcoholism. What hope is there for us?'

I asked him if he would take me into the reserve.

We had to set out at four the next morning to visit Molepo, one of the last traditional communities in the Kalahari. The rutted track made heavy going. At dawn we reached so-called New Xade, the miserable squatters' camp into which the Bushmen from the game reserve are being forced. The evictions are, according to the few independent journalists who have been able to witness them, forced and even violent where necessary.

It was a gloomy sight at dawn, a scattering of tents and shacks on flat, unvegetated land, a couple of dozen people wandering listlessly around, the blue smoke from a few fires drifting across the grey sky. I asked several people how they felt about the settlement. This was bad, dry land, they told me. 'We rely on hunting to live; to be relocated to such a place means we have no way to survive.'

A younger man told me, hot-headed, 'It is a deliberate abuse. We have no option but to drift towards Ghanzi, and what hope is there for us there? There is no work for us. The government wants to destroy us.'

The sun climbed, and Roy, Bulanda and I arrived at Old Xade, the settlement from which many people had been evicted. It had trees, water and a number of buildings, including a school and a dispensary. It was quite a pretty little village.

'Tell me what possible reason there could be for refusing people the right to live here,' said Roy. 'Tell me this isn't genocide.'

We plunged into the sandy tracks of the vast reserve. If we broke down here, we would just have to wait – probably for days – to be found. Wildlife was abundant, the antelope glossy and fat on the rich rainy-season grasses. It made nonsense of the government argument nobody believed anyway – that the San were irresponsibly killing large numbers of game-park animals.

We pulled up at a water hole, and a lion huffily strode back into the shade of some trees.

In late afternoon we arrived in Molepo, the only traditional village left in this great reserve supposedly set aside for the San. There were several acres of widely-spaced huts and acacia trees on thick, granular sand. Slowly people gathered under the shade of some trees to talk to us.

I asked them what was going on.

'The government has told us it wants us to leave this place,' they told me. 'They say we are killing too many animals. But it is not true. We only kill to eat, and we have been controlled by quotas for years. There are nothing like enough people living here to affect the animal numbers. Have you seen how many animals there are?'

I said I had.

Men sucked in air through their teeth and shook their heads. 'If they make us go, we will have to go.'

After we had been there an hour, a jeep pulled into the shade beside us and two South Africans climbed out. I saw Roy and Bulanda flash them angry looks. When they came across to introduce themselves, we asked what they were doing here. 'Repairing the airstrip,' they said.

'For what?'

'For tour groups. Our company does ethno-tourism. We fly people in to see the Bushmen.'

'And then what?'

'Well, we fly them out again.'

'So the Bushmen have the same status as the animals in the reserve, do they?' I asked. 'Something to be gawped at and to have your picture taken with?'

It was a rather frosty encounter.

Later, I asked Roy what he thought abut 'ethno-tourism'.

'It is patronizing,' he said. 'But on the other hand, regular tourist visits would make it more difficult for the government to evict these people. I haven't got time to worry about the ethics of tourism. I just want the San to be given back their land.'

PART V

Between the Andes
and the Ocean

CHAPTER THIRTY-TWO

The Longest Country in the World

IN SANTIAGO, THE LEAVES were falling but it felt like spring. April, they told me, is the kindest month, when balmy breezes succeed the hot southern summer.

The plane had drifted west over the sun-lit desert slopes of the Argentine Andes. We sooner associate South America with jungle than desert, but I recalled that these dried-out salt pans and bleak corrugations had stood in for Tibet in a recent Hollywood movie. There were valleys etched with silver tracks, turquoise-tinted shadows, and wiggles of green where snow-melt ribbonned down. Then the Andes disappeared beneath a quilt of cloud and the plane sank into it, to emerge in shadow with jagged peaks towering all around.

They say that drawings by Chilean children always show a line of mountains, because the Andes are always there. Chile is 2,700 miles long, but never more than 100 miles wide. Today the proximate mountains were invisible from Santiago, the Pacific too – as the jet dropped I craned my neck to glimpse sapphire water, but could see only fog.

The airport bus passed through hard-working suburbs where starched washing hung out to dry. I had an impression of brilliant paintwork, of cupreous greens and marine blues, of walls graffiti-daubed and overpainted, the old messages bleeding out as the newer layers peeled away. There were orange and pink hand-painted shop signs, a line of livid plastic bowls dangling outside a hardware store, ochre-roofed taxicabs, fat old Chevrolets and petite new Asian hatchbacks, Oasis Watercoolers, Asia Auto, *Ferreteria, Plastificacion*. A bus driver's eyes, coal black, met mine. At the Mercado Central I saw Indian and Indo-Spanish faces, well-fleshed women in tight dresses. Santiago was unmistakably Latin, but calmly so. 'The Chileans are reserved,' someone had told me, 'they're the British of Latin America.'

The architecture became grander, with wrought-iron grids guarding secretive colonial windows. Handsome boulevards were divided by strips of spongy, irrigated grass. There were street cafés, date palms, giant cacti. Schoolgirls sipped fast-food drinks through straws and flirted with boys. Two conical green hills probed the foggy sky.

It rained. That night I sheltered in a plastic-tablecloth eating joint where overalled men yelled cheerfully across empty beer bottles, and cigarette smoke twisted under the ceiling. A television and an antique jukebox strained separately to defeat the hubbub, but only increased it. As I ate *empanadas*, pastry envelopes filled, in this case, with seafood, I became aware of a group of middle-aged men whose drunken voices were rising above all else. As I rose to leave, one of them intercepted me.

'You are American?' he slurred.

'No, English.'

'Never mind. Come and drink with us!' Comm an dreeng we doss.

'Ingleeeesh,' chimed his friends, 'Arsennall! Manchayster United! Leeverpool!'

They were Chilean football referees, celebrating the selection of several of their number for the World Cup. They splashed out huge beakers of *pisco*, a kind of brandy, and pushed them into my hand. They sang, misty-eyed, of football, and toasted the beauty of Chilean women.

'You speak Spanish?' one asked.

'*Un pocito.*'

'It doesn't matter. There are only two words of Spanish a man needs to know . . .'

As they became increasingly raucous, I wondered if in Chile it was the refs rather than the supporters who were the hooligans. I could see the newspaper headlines: 'COPS BLOW WHISTLE ON REFS' NIGHT OUT'.

One of them rose swaying to his feet, and asked whether I knew Chile's outstanding quality.

I groped for superlatives. The world's best wine? Its most beautiful women?

Referees always get the last word. He leaned unsteadily forward until his nose was an inch from mine.

'Chile', he announced portentously, 'is the longest country in the world.'

On the map, Chile is almost comically thin. If it were flat, it probably would not exist at all – it would merge with Bolivia and Argentina. But the great wall of the Andes plunges down the western flank of South America, isolating a narrow strip of land. In the south, the strip is moist; but northern Chile is the world's driest desert – and this narrow desert continues north for 2,000 miles, to the border with Ecuador, my ultimate destination in South America.

No one I had met in Santiago had travelled to their country's north; the desert was a sort of mythic realm. 'Oh yes, I know the Atacama,' said one, who had visited La Serena. But La Serena, a holiday resort 400 miles north of Santiago, is on the desert's southern extremity. Covering the 700 miles before Altiplano, the Peruvian high plains, is the Atacama, a huge basin 7,000 feet above sea-level, with sandy alluvial fans at the feet of the Andes, and salt pans, the remnants of ancient lakes.

La Serena, then, is where southern Chile ends. It is a quaint colonial coastal town sliced around by multi-lane highways, its beach a burgeoning nightmare strip of high-rise holiday blocks. A rusting wreck sits on the beach, and the waves, unable to reach the tower blocks, vent their fury on it.

I stayed in an imposing colonial hacienda run as a boarding house by a tubby old matriarch. My room had shiny dark floorboards and carved colonial wardrobes, and smelt aromatically of dried wood. It let onto a courtyard of golden Hispanic tiles with a knot of luxuriant potted trees. In the morning they were watered by serene, pre-pubescent girls with long black hair and chaste white blouses.

For breakfast I went to an open-air café and ordered coffee. Chilean coffee is among the worst in the world. Chileans have not heard of the trees in neighbouring Brazil where coffee beans grow. They believe that coffee is a brown powder that comes in a sachet marked 'Nescafé'; in cafés they bring you a cup of hottish water and a sachet, and charge you for the privilege of tearing it open and stirring it in.

In the world's poorest countries (Chile is not one of them) there is blanket advertising for Nestlé products. Even today some governments still find it necessary to run public health campaigns to persuade mothers that breast milk is better than powdered milk. As for Nescafé, it runs seductive ad campaigns all over the world for its products. It has persuaded Indians, who once had access to real, home-grown coffee, that Nescafé is better . . .

From La Serena the desert coast has a few large towns, tied to the south by plane, and by buses that make the long journey overnight. Further north, there is a handful of fishing settlements crammed into the chinks in the desert's armour, where water trickling out of the Andes meets the sea. The Humboldt Current is to Chile as the Benguela Current is to Namibia, its upwelling cold waters creating fog – *camanchacas* – onshore. But inland of the coastal mountain range, the Cordillera de la Costa, there is no moisture. The road north ascends by twisting ghats into increasingly dry, searing, uninhabited mountains, dotted with sage scrub. Chileans call the first tract of desert the Little Atacama. Later, the real Atacama begins, markedly more barren. Unencumbered by anything but the meanest vegetation, it feels like an illustrated geology lecture gone mad. The sharp-toothed Andes are a constant presence in the east. Sometimes they advance on the coast, rearing up like furious, naked gods, forcing the road to mount their moulded shoulders as they dive into the chilly Pacific. I passed red and gold volcanic cones, and green cliffs layered with black, like some poisonous Battenburg cake. Then there were vast, tranquil sandy plains where nothing happened, disused tracks wandering into the distance past battered signs saying 'To the Mine'. The mines, or *oficinas*, once hauled chile saltpetre, a nitrate-rich natural fertilizer, from under the desert crust. Elsewhere the desert is plundered for copper ore, but here it is silent, except for occasional trucks that thunder down the grandiosely named Pan-American Highway, the thread of tarmac that tenuously connects northern and southern Chile.

As night fell, I left the Pan-American and followed a winding road down through the Cordillera to the coast. Tal Tal is a declining fishing town, far enough from the capital to be immune from holiday developments. It was built, like all the settlements on the Atacama coast, from Oregon pine, brought as ballast in the ships that sailed here to carry away the saltpetre. Now the paint is peeling from its clapboard buildings, many of them derelict. The few shops around the main square are ordinary homes where the front room is the shop, and the slightly-larger-than-usual window onto the street displays shiny plastic shoes, or a bouffant dress confected from nylon lace. They reminded me of the pebble-dash shops you find in small towns under the murky skies of rural Wales. I saw the shabby socialist party building, and felt a grim satisfaction that in the aftermath of the totalitarian regime that lasted until 1990, it could at least exist.

The Hotel Verdy was a clapboard house in which all the rooms felt like seaside chalets − barely big enough for a single bed, but painted candyfloss colours. The courtyard had a naïve mural showing the church, and fishing boats on a miraculously placid sky-blue sea. The floor was paved with shiny brown tiles that appeared freshly daubed with blood. A window gazed through the branches of a pomegranate tree, past terracotta-tiled roofs, to the Pacific.

Every Chilean provincial town has its Placa de Armes, and on that warm night, it being Easter Sunday, a band was playing in a corner of the square. Tubby matrons sat on the low park walls and clapped, while maidens, sleek and glossy, refused to display anything so embarrassing as enthusiasm. They were *jeunes filles en fleur*, possessed by mysterious and alarming forces of life. The prowling suitors impressed me less; moustachioed, wearing tight nylon shirts and pointed shoes, they looked like cartoon mafiosi.

One girl was exceptionally beautiful, ethereally desirable, and perhaps because she was aware of this, and perhaps to accentuate it, she had enclosed her syrupy skin in a dress of virginal white. I looked at the circling sharks with their self-conscious avoidance of eye-contact and their matey facetious sniggering. Which of them, I wondered, would she give herself to? In a few years, she would be a silver-toothed mother sitting on the park wall, balancing a child in a flouncy dress on her podgy knee, holding its wrists in her hands and encouraging it to clap to the music.

CHAPTER THIRTY-THREE

Ice Cold in Anto

NORTH OF TAL TAL a rough, little-used track runs beside the ocean. At 10.30 a.m. I pulled off the road, and straight into soft sand.

I had arrived in Santiago before a four-day public holiday, and this two-wheel-drive pick-up was all I had been able to find. It was either take it or wait five nights in Santiago to find a four-by-four. I had taken it.

This was not the first time on my desert journey that I had got stuck in sand. But previously I had always been in a four-wheel-drive vehicle, with sand-tyres, sand-ladders, shovels and so on. Père Bessita had taught me the techniques. And I knew that you could use branches instead of sand-ladders, so I found some flotsam to put under the wheels; they span the brittle wood to dust in instants. Then I tried flat rocks.

After an hour I had managed to reverse a yard, and was still some way from the road. At 1 p.m. I saw that the rear axles were now beneath the level of the sand. As I took a deep swig from my water bottle and contemplated the burrowing axle, I recalled how many other cars I'd seen that day: none.

It would be pleasant to report a heroic outcome. But by three o'clock I was only two yards further back. Then, to my unspeakable relief, a truck came by, and three hearty fellows jumped out. They pulled various planks and tools from their truck, but it still took the four of us half an hour to get the pick-up back onto the road. For the hundredth time on this trip, I was grateful for the kindness of strangers.

As I continued north, the sun sank towards the Pacific. In the late afternoon light the colours on the beach glowed – the delicately yellow sand, the blue stones and the fibrous dark-red seaweed. From time to time I saw impoverished huts made from jetsam bound with scraps of nylon fishing rope, a few clothes on a line, a truck jacked up on bricks. It was hard to imagine how anyone made a living on this desolate shore. Piles of the dark-red seaweed had been bundled up, and days later someone told me it was harvested and sold for fertilizer.

By dusk, I was in trouble again. Despite being alone on this track, I had automatically pulled over to consult my map, and was now again marooned in soft sand, this time on a slope. I had seen no human life for many miles, and this far from Tal Tal I might not see another vehicle for days. I cursed my impatient folly in renting a two-wheel-drive jeep, and cursed my ineptitude in getting it stuck again.

Then I remembered my frying pan.

As darkness fell I dug, using the pan as a shovel. Its soft metal handle quickly snapped, but the pan held up. It took me ninety minutes to scoop out of the slope behind the vehicle two horizontal tracks that I could reverse into. My intention was to provide enough grip under the tyres to let me accelerate towards firmer, gravelly sand ahead.

There is a classic British film called *Ice Cold in Alex*, the story of a tough wartime journey in the North African desert. The travellers fantasize about arriving in Alexandria and ordering an ice-cold beer. At the end of the film they go into a bar, and a barman sets up glasses. The ice-cool lager foams and John Mills, the reformed-alcoholic British officer, fervently raises it to his lips. As I dug, I thought of that film, and of the beer I was going to drink that night if I ever reached a town.

My situation was fairly serious. The only escape from here would be to drive down the sandy slope towards the ocean, trying to pick up enough speed and grip to climb back towards the track. If the jeep got stuck further down, I would have to walk back to the last shack I had seen with any kind of roadworthy vehicle parked by it. Maybe a day's walking, and in the sun; if I walked at night, I might overshoot.

I threw the frying pan in the back of the pick-up, and climbed into the cab. I really ought to wait until morning to try this. Tentatively, I reversed back into the grooves I had dug. They seemed to be firm underwheel: the vehicle did not sink any deeper. I revved the engine wildly, let out the clutch and rocketed forward, jerking the gear lever into second, keeping the engine screaming. Bessita had taught me that driving on soft sand you have to keep turning the steering wheel from the eleven to two o'clock positions. I got about a hundred yards across mixed gravel and sand drifts, and began to climb. I drove as fast as I could, still in second, trying to keep to the firmer ground which was recognizable in my headlights from the detritus of small black rocks on its surface. I was approaching the track, its edge marked by a ridge of sand. I accelerated, furiously jerking the steering wheel from left to right. The vehicle shuddered and whined as it met resistance, then after a moment's hesitation, surged free.

I felt elated, but deeply stupid. I drove on in blackness, with the phosphorescent breakers of the Pacific on my left, seeing not a single vehicle or hut. The road left the ocean and climbed high into the desolate Cordillera de la Costa, with moonlit views onto arid valleys. The air was cool. It was nearly midnight when I regained the main road and found a truck-halt on the outskirts of Antofagasta. Exhausted and filthy, I walked in and ordered a beer.

CHAPTER THIRTY-FOUR

Copper Country

SAN PEDRO IS THE NAME of a hallucinogenic South American cactus, and of a small town in the Atacama Desert. Eight thousand feet up in the Andes, I came to an unspectacular settlement of the mud-brick construction known by the Spanish-American word *adobe*. At first glance San Pedro seems similar to many Saharan oases, but it is set apart by the presence in the central square not of a mosque but a church. It is an oasis, however, and was once a staging-post for cattle driven from Argentina to feed the nitrate miners. I had seen animals collapsing from thirst in the Sahara, and I could imagine the spectacle of 10,000 cattle on the 14-day drive from the Argentine province of Salta, over the lowest passes of the Andes to San Pedro de Atacama.

Today it is gringo tourists who come to San Pedro, for the desert scenery around it, and for the small museum of archaeology. The oasis has always attracted visitors, including Graham Greene, famously a Catholic, drawn here to meet the solitary Belgian missionary who made the first serious archaeological study of the region. *Padre* Paige's collection is replete with the paraphernalia of hallucinogen use: wooden tablets and snuffer tubes decorated with angelic creatures – winged men and human-headed birds.

The desert cherishes its dead. Over 6,000 years ago the people of northern Chile we now know as the Chinchorro mummified their dead. They kept the mummies about the house, presumably for reassurance, for contact with the supernatural world. Archaeologists have found in adult corpses signs of a syphilis-like disease probably caught through cuts on their hands as they excavated the stomachs of dead relatives to pack them with mud and straw.

The Chinchorro had abandoned the practice of mummification

before the Egyptians, a people on the shore of a far distant desert, even took it up. But many peoples felt the need to preserve the earthly remains of their dead, interring their kings in cairns – as modern Britons still do, cloaking their monarchs in mystery and burying their bodies in the cairns of millennial cathedrals.

Bodies interred in the sands of the driest place on earth suffer extraordinarily little decomposition. Skin still clings to hands and faces, and there is a collection of these well-preserved bodies displayed, or desecrated, at the museum. There is also a recreation of its founder's monastic bedroom, and if the *Padre*'s spirit hovers around, it is able to peek across the hallway at one of Chile's most iconic images: the mummified figure of a young girl, gruesomely but affectionately known as Miss Chile.

A further 6,000 feet up into the Andes are the El Tatio geysers, which boil and gush each day at dawn. At 3 a.m. I set off to see them. The narrow road twists up towards the border with Bolivia, through arid passes with few habitations or signposts, and I had been warned that I would get lost. I did not, but twenty miles from the geysers my brakes failed. There seemed little point in turning back so close to my goal, so I carried on, mostly in second gear.

I reached El Tatio at dawn, and the geysers steamed impressively, but did not spurt. It seems they are temperamental.

Descending without brakes proved more difficult than climbing, but I was most put out by an attack of *soroche* – altitude sickness. I should have stayed a day or two longer in San Pedro to acclimatize. My head throbbed, and my eyes seemed to be bursting. All around me pink volcanic cones were levitating in skies of immaculate blue, over meadows of corn-yellow grass. I popped some painkillers, slept for two hours, woke up feeling god-awful and inched on down the winding mountain ghats in second gear. The camelids of South America, wild guanacos and domesticated alpacas, scampered over the rocky slopes. I drove out of the mountains, found a rare stream, washed myself in the icy water and fell asleep in the sunshine. When I set off again, the brakes were working. I remembered that I had forded a shallow stream in the early hours – they had been soaked, and had taken this long to dry out. For the rest of the journey, I never really trusted them.

I was amazed to see a small white dog running along the verge. I thought it must have fallen from a vehicle, and stopped, but the dog refused to come near me. Worried that here in the desert, miles from any water, it would quickly die, I followed it. The dog trotted steadily ahead of me until it saw a plastic bag and tore it open, revealing sandwich crusts. I realized it was a scavenger, anxiously patrolling the verges until it found scraps discarded by passing trucks.

I stopped again, and called to the dog. It kept its distance, watching me warily. I keyed open a couple of cans, and put down my pan with a heap of tuna and bread.

'Come on, boy,' I said.

He would only approach the food when I retreated about twenty feet. He gulped it without even chewing, and retreated when I went to fill the pan with water. We repeated this ritual until he had drunk his fill, several pints. Again he backed off suspiciously. And I drove away.

Near the middle of the Atacama Desert is Chuquicamata, the world's largest copper deposit. At the Visitor Centre I watched a video proclaiming that cyanide poisoning of local streams by the refinery process had been considerably reduced in recent years. Well done, I thought. How proud that must make the Guggenheims, who made their art-philanthropizing fortunes here.

A tourist bus carried me to the crater rim, and I stared a half-mile down into the world's largest man-made hole. Yellow trucks the size of family houses rumbled past, corkscrewing their way into the pit's depths until they were mere insects, then, charged with 225 tons of ore, crawling back out. It was all so dwarfing that I felt my spirits dampened, and I was still feeling dazed from the *soroche*. Then something happened to cheer me up.

The tour was over and I was standing near the guide. A tall, scholarly looking German, a member of a tour party, asked in a low voice, 'It seems that with the mine's vast male workforce, there are quite a few strip clubs and brothels to, er, look after their needs?'

'Well,' the guide began, but was distracted by an elderly couple searching for the toilets. 'Over there, on the left!' he yelled. Turning back, he remembered the man's question and distractedly went on in his megaphone voice, 'For strip clubs, try Calle Vargas.'

Several other members of the German party looked up to see who had asked this interesting question.

'And if you want a woman—' boomed the guide.

'It is not for *me*!' The man hissed, blushing furiously. 'I was only asking out of interest!' He scuttled out to the waiting coach.

CHAPTER THIRTY-FIVE

Blood and Sand

I WAS WADING into the cool Pacific, on a golden beach in the palmy town of Iquique, halfway up the Atacama coast. Iquique is a holiday resort and – the reason for its contemporary success – a free port. For all its isolation, Iquique is Chile's fastest-growing city.

In 1835 Charles Darwin described a wretched and remote port that

> *stands on a little plain of sand at the foot of a great wall of rock two thousand feet in height, here forming the coast. The whole is utterly desert. A light shower of rain falls once in very many years; and the ravines consequently are filled with detritus, and the mountainsides covered by piles of white sand, even to a height of a thousand feet.*

The town was transformed when chile saltpetre was discovered in the mid-nineteenth century. It was here that enterprising investors and entrepreneurs came, laying the foundations of the Chilean economy. The most notable member of this band of entrepreneurial buccaneers was John Thomas North, the self-styled 'King of Nitrates'. North, an Englishman, managed to establish a virtual monopoly in nitrates, and became fabulously wealthy, a powerful public figure in both Chile and England. He succumbed to hubris, however, when he used force of arms to try and crush Chilean government opposition to his power. When he died in 1896, his nitrate empire was in ruins.

José Garcia, a local teacher I had met at the beach, took me around the town. As in Tal Tal and the other Atacaman ports, every building in Iquique was made from Oregon pine. But Iquique's clapboard structures are uniquely ambitious in scale – there is even an opera house with a classical façade and an ornate, Wedgwood-blue ceiling, entirely carved from pine. As I walked across a stage once graced by the stars of European opera, cleaners were swabbing it with a familiar-smelling product.

'What's that smell?' I asked José.

'It's kerosene – I think it keeps the woodworm away. Don't even ask about the fire implications – this building is a bomb. The fact that it's lasted this long is nothing but a miracle.'

He led me past ropes and pulleys (the brass plate on a piece of Victorian stage-machinery gave the name of a Glasgow manu-facturer), through a door and up a narrow staircase to the flies. It was a gargantuan loft of outflung beams and the cobwebbed, tinder smell of dry wood. We scrambled under a sloping roof, where old chairs, bolts of cloth and props from a hundred years of performance lay in dusty heaps. We stepped through a door, and Iquique was laid around us, the ocean and the bright-painted clapboard houses crisp under an azure sky.

I asked José if the wooden buildings were protected by law.

'Yes, but Iquique is booming, and this land is getting valuable. People aren't allowed to demolish the wooden houses, so every now and then there is an "accident", and one of them burns down, and a tower block comes up in its place. Right now the city centre is maybe 90 per cent complete. I don't know how much longer it will remain this way.'

Until recently, the best-known contemporary resident of Iquique was General-Senator Auguste Pinochet. It was Maria Martinez, a journalist I met at Iquique's radio station, who told me this.

'He's based in Santiago, but Pinochet likes Iquique so much he keeps three houses here. And we've been told that he's in town today.'

'What for?'

'Military stuff. He's giving prizes at a passing-out parade. And – if you believe this you'll believe anything – he's making an inspection of rural poverty.'

I smiled. 'I take it you're not a fan?'

'I hate his guts. But he's quite popular in Iquique. The fact that it's Chile's fastest-growing city has a lot to do with Pinochet's patronage. For many Chileans, recent history is exactly that – *history*. The Disappeared are an uncomfortable subject, and besides, a lot of

Chileans supported Pinochet. You Europeans think all the time about these relatives of the Disappeared who make demonstrations in Santiago, but it's disproportionate to their significance. A lot of Chileans wish they would just shut up, and the young are bored by it all. They'd like to put it behind them and get on with being mindless consumers.'

As the major industrial centre of the nineteenth century, Iquique was the birthplace of the Chilean labour movement. This combination of socialism and severe right-wing authoritarianism struck me forcibly. I wanted to see a place they do not talk about much in Chile, even in school history books. I asked Maria to take me to the site of the 1907 massacre.

The nitrate miners had gone on strike for higher wages. Four-and-a-half thousand workers, their wives and families, walked all the way down to Iquique from the arid mountains and camped in the yard of the Santa Maria school.

An eyewitness wrote:

On the central balcony . . . stood 30 or so men in the prime of life, quite calm, beneath a great Chilean flag . . . They were the strike committee . . . All eyes were directed on them just as all the guns were directed on them. Standing, they received the shots. As though struck by lightning they fell, and the great flag fluttered down over their bodies.

Next, the army was ordered to flush the strikers and their families right out of the area.

There was a moment of silence as the machine guns were lowered to aim at the school yard and the hall, occupied by a compact mass of people . . . There was a sound like thunder as they fired. Then the gunfire ceased and the foot soldiers went into the school by the side doors, firing.

Hundreds were murdered. Witnesses said that for blocks around, the normally dry and dusty streets of this desert town turned into a bloody swamp.

The road out of Iquique is forced up the wall of sand and rock that Darwin described. Thirty miles inland is perhaps the most fascinating of the scores of deserted mines and ghost towns that litter the Atacama. Humberstone was founded as a nitrate town in 1875 by James Humberstone, and the last residents left only in 1961. The church stands empty and desecrated, but the theatre's seats are still poignantly in place, as though awaiting the evening's performance. The town square is spookily alive, its cosy park benches shaded by lattice-trained bougainvillaea. I expected that at any moment an old man would appear, and sit down with a contented groan.

Most bizarre is Humberstone's swimming pool, made out of dozens of sheets of iron plate from a ship wrecked in Iquique harbour, carted into the desert and riveted back together; there is even, under the ladder, a welded-up porthole. The rickety diving board offers views over the dead wooden town, which waits under the desert sun for the spark that must, one day, reduce it all to ash.

A little to the north of Humberstone, a road peeled off into the sandy hills now concealing the ocean. As though some Moses had commanded a mountainous dune to part, the road bisected two vast, velvet-smooth sand-alps, and at length reached a barren clifftop. Two thousand feet below, beyond a strip of sand, the ocean glittered. At the foot of the cliff was a large graveyard; through my binoculars I made out a forest of wrought-iron crosses. A narrow track zig-zagged down the cliff-wall, my jeep's wheels scraping past vertical drops unobstructed by any barrier. I nervously imagined the jeep tumbling down, coming to a burning rest conveniently close to that necropolis. I still did not trust the brakes.

It almost defies belief that a place like Piscagua can exist, an oasis caught between ocean and desert, a town built of timber with a third of its buildings decaying into orange matchwood. Piscagua, population 200, looks out on a wide and placid bay, where fishing boats rock on scintillating water. A small square on the waterfront has at some time been beautified, corniched, but now its lamps are broken and the wrought-iron balustrade twisted shapeless. Some fishermen were sitting near-naked in the shade of a tree, smoking cigarettes, and I asked them where I could find some fish for lunch. They pointed to a doorway in a clapboard building across the street.

It was a small eating house, a woman sitting breast-feeding her baby, her mother-in-law knitting. They split a fish in half and pan fried it, and tossed it on a plate with tomatoes and raw onion.

There was little to see in Piscagua; in the shop, a gloomy cave with no electricity for its 1950s fridge, I bought some cheese; I took photographs of the timber church that showed a fine face to the street, but behind its façade was a shed surrounded by rubble and the stink of urine; I walked to the top of the town, where a white clock-tower was dated 1887.

Remoteness can be romantic, or transcendent, or sinister, and I found something sinister in Piscagua. I knew that a mass-grave had been found, of people murdered by the Pinochet regime. But it was only when I returned home, and Pinochet was arrested in Britain, that I saw an extraordinary black-and-white film shot illicitly by an East European film crew in 1970s. Now, in my memory, the sapphire-blue, dust-brown of Piscagua is fused with the grainy black-and-white images of young men and women – union organizers, socialist party members and vaguely left-of-centre teachers and secretaries – staring at the lens, obliged to perform group exercises in the dazzling sunshine outside their rows of identical new prison huts. Many of them never left Piscagua. The huts have been demolished, and only the flat concrete foundations remain on the seashore outside the town.

CHAPTER THIRTY-SIX

To Peru

THE PAN-AMERICAN HIGHWAY smacks of can-do technical supremacism, aspiring to be a multi-lane freeway arrogantly spanning two continents, lined with fast-food outlets. Climb into your Chevy Transam in LA, and burn south all the way to the ice cap. But no: in parts of the desert in Chile and Peru, the highway is no more than a narrow, switch-back country lane, and in poor repair.

I was about 60 miles from the frontier with Peru when I noticed grey smoke pluming from a lorry coming towards me. The driver flashed his headlights wildly as I passed, and jerked his articulated lorry with its load of brand new Chevrolets off the road. I braked and reversed, but already he was running towards me.

'Motor's dead,' he panted. 'I need to get to Arica fast, before someone tries to steal this lot. Can you take me?'

'Of course,' I said.

'But fast?'

'Get in.'

I drove fast – but not fast enough for Javier.

After a few miles the road dipped and there opened before us a fantastic gorge, one of the loveliest desert spectacles I had seen. The Rio Camarones has cloven the sandstone into two smooth cliffs that drop thousands of feet like bolts of golden silk. The road carved into those cliffs, a jagged run in the silk, has few crash-barriers. When someone dies in a road accident Chileans always plant a cross at the site of the crash, and there were plenty of crosses stuck into the crumbling edges of this road. And Christ, it was a long way down.

Conscious of Javier's impatience, I drove as fast as I could, even overtaking a bus on the inside, but, as I nervously braked before each vertiginous hairpin bend, I could feel him chafing beside me.

When we reached the bottom of the gorge, I went briefly insane. Javier casually asked, 'Shall I drive?'

And I, naïve visitor from another planet, said, 'Yes.'

Delirious, irrelevant thoughts had flashed through my head: after so many miles, it would be nice to be driven for a change. I could enjoy the view of the gorge – perhaps take a snap or two . . . Milliseconds later, the penny dropped: this guy was going to drive like a bat out of hell. But it was too late to change my mind.

Javier left rubber on the road as he pulled away. He exploded past other vehicles. He did seventy, eighty on the short straights, and cornered with the jeep leaning, tyres shrieking, back tyres sliding. As we hurtled into every bend I saw eternity beckoning in gorgeous, bottomless vistas . . .

'Javier, slow down!'

'Do you have vertigo?'

'No, I have a desire to stay alive,' I said, pointing at the cross that marked someone's descent into eternity.

Javier thought this was hilarious. '*Yo con ojo,*' he said, pointing at his eyes.

What the hell did it mean? I am with the eye? I have the eye? What eye?

The evil bloody eye, I decided, because every time we approached a cliffside cross, seeing its morbid effect on me he would point and laugh uproariously. '*Yo con ojo!*'

At last, thank God, we reached the top of the gorge, and the road straightened out to cross a broad plain. Javier accelerated to top speed, supersonically passing all other vehicles. 'Are you afraid?' he asked, with an insincere show of concern.

On the contrary, I felt great. We had left that bloody gorge behind – I was going to live.

We passed a huge construction site. 'The new prison', said Javier. 'There's a lot of drugs here these days, marijuana and cocaine coming down from Peru and Bolivia, going out through the port at Arica. So they need a new prison here.'

'Uhuh.' My attention was distracted: the road was dipping again. 'What's that ahead?'

It was another gorge. 'Javier, how many more of these things before Arica?'

'Three', he said.

'For Christ's sake slow down!' I yelled.

He grinned. '*Yo con ojo*,' he said.

At last I understood. He had been pointing at his forehead, not his eye. He was saying '*Yo conosco*' – I know the road.

Well, that was all right, then.

CHAPTER THIRTY-SEVEN

The Nazca Lines

FOR MUCH OF THE LENGTH of southern Peru the grey cliffs of the Andes dive straight into the sea, an elemental clash that forces the ribbon of the Pan-American Highway inland in search of gentler topographies. There are a few green snow-melt valleys, but mostly it finds itself in volcanic desert, amidst the geologic display of crystalline, pale pink and ochre andesites, pahoehoe and scoria. These are mountains that wear their hearts on their skin, gashed and veined, and mined by man for iron, copper, uranium, molybdenum, tin, silver, gold.

The road descends eventually to a level grey plain dotted with perfect white crescent-shaped barkhan dunes, like a giant tray of unbaked croissants waiting for the oven. Then it returns west, cutting into cliffs high above a pounding, unpacific ocean.

I saw few other vehicles; they were not the neat little south-east-Asian hatchbacks common in prosperous Chile, but chugging, gas-guzzling 1970s Dodges and Oldsmobiles; and coaches, driven recklessly fast along the ledge-like road, and tall, wood-flanked, galleon-like trucks with one of South America's oddly conjoined names painted on the sides in gothic lettering – 𝔉𝔯𝔞𝔫𝔷 𝔏𝔢𝔢, 𝔄𝔩𝔣𝔬𝔫𝔰𝔬 𝔖𝔠𝔬𝔱𝔱 – or else 𝔍𝔢𝔰𝔲𝔰 and 𝔐𝔞𝔯𝔦𝔞.

Here and there a fishing hamlet had been levered into a cleft in the cliffs, and I stopped in one for some lunch. There were several grim eating shacks of the type I was used to – windowless for the shade, unlit within, a Madonna clock made in Taiwan mounted next to beer ads of half-naked blonds (the indigenous Peruvians I was meeting were short and heavy-set, but in the ads it was always lissom blonds). One restaurant stood out, whitewashed, with a jolly mural of a fish. Inside it was white too, with white plastic chairs and tables. It smelt sweetly of fresh cement. The woman in charge listened intently to my order, and disappeared into the kitchen – an unwhite, soot-blackened box – to gut fish. Her two young children sat near me, intent on their Spanish homework.

The fish took time to cook on the paraffin stove, and I walked across the sun-struck road to watch the insistent waves. I returned and asked for the toilet, but there was no toilet, just a brick wall behind which you crouched, the hygienic sun burning your waste to harmless ash. There were no flies. There was no running water.

The fish arrived, succulently fresh. I ate, I sighed. I felt like a siesta, a rest. No days off on this desert trip, I just kept moving. Only when some impediment put itself in my way did I stop.

I ordered coffee and enjoyed the long wait as water was heated on the paraffin stove. The sweet-natured children looked up from their school-books and giggled.

Towards dusk the road turned inland once more, and another broad plain led me towards Nazca.

Sometime after five in the morning, a chilly desert dawn, I stood on the tarmac at Nazca's small aerodrome. The volcanic Andes, tall over the town, were turning red, including the sandy face of Cerro Blanco, claimed to be the tallest sand dune in the world. But I had come to see the Nazca lines.

Carlos, my pilot, fired up the engine of his Cessna. The Nazca lines can only really be appreciated from the sky, and dawn is the best moment, when the shallow ridges are shadowed by the low sun, and the air is clear and vapourless.

We flew up and out over the pampas north of the River Nazca, and soon saw them, hundreds of straight lines radiating across the desert as though from a compass rose, and animal shapes – monkeys, fish and the harmonious, somehow emblematic, image of a humming bird hovering with wings outstretched.

'What do you think all this was for?' I asked Carlos.

'Magic, no doubt about it! When you go out there on foot you can feel it – the power of the place.'

'I thought you weren't *allowed* to go out there on foot.'

'You're not, but it doesn't stop anyone.' He winked. 'Most of those lines were made by motorbikes.'

No one knows why, more than 1,500 years ago, people engraved these lines into the desert crust. The most notorious Nazca explanation was propounded by the Danish writer Erich von Daniken. When I was a schoolboy, von Daniken was firing millions of imaginations with his book, *Chariots of the Gods*, which found evidence of extra-terrestrial interventions in human history all around the world. Sometimes the evidence amounted to little more than a rock-carved figure with some sort of halo – according to von Daniken, a space helmet. As for the Nazca lines, they were landing-strips for alien spacecraft.

Entering Nazca from the south, I had driven to the hotel where I had been told Maria Reiche lived. Now in her nineties, this German woman had devoted her life to trying to protect and decode the lines. She believed them to be some kind of celestial calendar.

'She does not live here any more, *señor*,' the man at the reception desk told me. 'Miss Reiche is an old woman, and for years she has not been strong. Her illness has forced her to leave Nazca. She is receiving medical treatment in Lima.'

But the hotel did possess, he told me proudly, a video film of the lady. Perhaps I would like to see it? He pressed the buttons on an ancient video recorder, and a hazy recording of the film appeared, showing the old woman poring over her alignments and calculations.

Alternative theories of the origin of the lines abound. The one I find most powerful is that they played a role in hallucinogenic rituals. Andean museum collections groan with the paraphernalia of shamanistic powder-snorting, and it was – is – common for shamans in these drug-induced trances to identify with some symbolic creature. Often they believe their spirits take to the sky and soar. Each of the creatures on the Nazcan crust is etched in a single line. Perhaps they were paths, walked by the shamans during their possession by the animal – spiritual, not alien, runways from which the shaman's soul took flight.

From the air, the animals seemed smaller than I expected – one of the birds was supposed to have a wing-span of 300 feet.

'It's because of our height,' said Carlos, 'they're a long way down.'

'Well, can't we go a bit lower?'

'We're not allowed to. And it uses too much fuel.'

'Oh, go on.'

He looked around, scanning the horizon for traffic. Then we

dropped out of the dawn-pink sky to buzz the drawings. As we banked tightly they whirled around us – monkey, dog, 'owlman', spider.

'Do any of the local people ever get a chance to see the lines from up here?' I asked.

Carlos gave me a withering look. 'Of course not. Flying would be far too expensive for them.'

As we came in to land, the engine cut out and the propeller clunked to a halt. 'Gas,' said Carlos, in an 'I told you so' sort of voice, tapping his fingernail on the gauge with its needle showing Empty. We glided. 'But it's OK, I've got some more.'

He flicked a switch to the second tank and pressed the starter. The engine coughed and fired, the plane accelerated, and I watched our shadow race across the sun-stained sand to meet us.

Leaving Nazca I stopped at the small, newly opened museum of the lines. It includes, like the homage to Padre Paige in San Pedro de Atacama, an evocation of Maria Reiche's monastic workroom, scattered with rulers and compasses and charts. Her sincerity and doggedness made one want to believe that she had been right about some mysterious relationship between Nazca and the stars, but for all her years of painstaking research, she had been unable to prove her theory.

A few days later, I read in a newspaper that Maria Reiche had died in Lima.

CHAPTER THIRTY-EIGHT

Señor Grau

NORTH OF NAZCA the desert moderates into cultivated land, and tractors congest the narrow Pan-American. Then, a hundred miles or so south of Lima, the sands return. It was here that I was pulled over by the police. I had given them a smile and a cheerful wave as I went by – asking for trouble. It was the fifth or sixth time I had been stopped since entering Peru, and these cops were the nastiest-looking yet – facially scarred and pot-bellied – but they proved to be very friendly. It took five minutes of ransacking my car for it to sink in that I had lost my passport – that is, had it stolen when I changed money in Nazca. The cops could have arrested me, but instead they showed solicitude, telling me I must report the loss when I reached Lima. I had not even been intending to visit the Peruvian capital; now it seemed I might be in for a stay.

Lima announced itself with giant billboards advertising cigarettes, beer and cars, their common iconography, as ever, the bikini-clad blond. Like Chile, Peru is dominated by Spanish-descended élites who use television and advertising to proclaim their dominion over the darker-skinned original inhabitants of South America. Billboard blonds, TV-announcer blonds affecting European Spanish accents, TV comedies depicting indigenous Indians as mop-haired bumpkins. One begins to understand the grievances of the *Sendero Luminoso* (Shining Path) terrorists, who wanted to sweep the Western, capitalist influences out of Peru.

South of Lima huge sand dunes loom. Beyond them, in the west, are the beach resorts where prosperous Limans while away their weekends. Shanty towns are creeping up these dunes; Lima is a magnet for people fleeing the impoverished countryside, a poverty ironically exacerbated by the years of anti-rich Shining Path terrorism.

Lima is a grey, flat, uniform city, and, standing as it does on the seashore, frequently fog-bound. But today the sun shone. After my long drive on desert roads from Santiago, the seething traffic was a culture shock.

At the Miraflores police station I joined a queue of crime victims on a wooden bench in an echoing hallway. The man next to me had an impressive selection of facial cuts, and explained that he had been mugged taking a shortcut through an alleyway near his house.

'I should have known better. It's a really gloomy alley, and every time I walk through it I think, "This would be the ideal place to mug somebody."' He grinned ruefully, and ran his fingertips over his swollen forehead and lips, then stared meditatively at the pale stripe of bare skin where his watch had been. 'Thank God they only beat me up – obviously they have to rough you up a bit, to frighten you. Puts you in shock, makes you less likely to resist. But sometimes they kill people, you know.'

I felt grateful to the thieves in Nazca who had merely stolen my property.

A taxi drove me suicidally through hectic streets to the British Consulate, housed in a tower block of brutalistic ugliness. Once inside, however, I was grateful for the cool, and the calm hyper-efficiency. I was issued with a new passport in minutes, then packed off to a central police station to spend several hours trying to get a replacement entry stamp. Amidst the bureaucracy and paperwork, I was awed to see that they had instant access to a computerized record of my entry into Perú at Tacna. Every day the world is becoming more secure, more tightly controlled.

By the end of the afternoon I had a new passport and entry stamp; I was legal. As the sun sank, so did I, into a well-padded chair on the terrace of the *Tienda Cita Blanca*, one of Lima's posher cafés. Here was luxury: tea with real milk, accompanied by a tiny basket in which six chocolates nestled on the linen bed of a crisp white napkin. The evening rush-hour had begun, with soot-belching buses and aged VW-beetle taxis rattling past. Wooden-faced policemen patrolled with sub-machine guns; gold-encrusted, well-padded *señoras* swayed by, beggars shuffled, hustlers twitched; a woman in a nurse's uniform stood on a corner, taking the blood pressure of passing businessmen for a few *soles*; a blind beggar sat beside a begging bowl, playing a flute – a Dickensian gallery of the rich and the very poor, and I confess that I sat behind the plate-glass screen and pot plants of my privileged enclave and ate the chocolates – all six of them.

I strolled. The Parque de 7 Junio is the scene in early evening of a lazy Liman *passeggiata*, where every man is languidly elegant, every woman languidly beautiful. At last, depressed by the sight of kissing couples and feeling insufficiently languid or elegant, I found a backstreet restaurant that seemed popular with locals. I ordered *ceviche*, a seafood stew where fat orange mussels floated in a white fish stock.

'You have made a good choice of restaurant,' the man at the next table said, in impeccable English. 'I have been coming to eat here for over twenty years. Would you care to join me?'

He was gaunt, with pale, powdery skin and facial scarring. 'I must make a bizarre sight,' he said apologetically, his voice weak, almost inaudible. 'It's the chemotherapy. I am supposed to be dying from cancer.' He grinned wryly. 'The doctors told me I would go three weeks ago. But as you see, I did not. So I consider myself twenty-one days into my Renaissance.'

I smiled. I did not know what to say. Señor Grau might well have been eighty, but was, he told me, sixty-five. His great-grandfather, Admiral Miguel Grau, had been one of Peru's military heroes. I knew the name. In Iquique I had visited the naval museum that commemorated Grau's clash with the Chilean Navy in the War of the Pacific. His father, he went on, had died heroically in the struggle for peasant's land rights. But Señor Grau himself had expressed himself with a more dilettante heroism. 'I was a racing driver. I didn't do badly, either. I won ninety-nine races in my time.'

He had always had his engines built in England, by Jaguar. 'The most reliable engines. And the factory was in Coventry – I liked that place. It rose from the ashes of war. I found that very moving. It had a spirit, that city. I especially liked the cathedral, half-ancient, half-modern.'

Señor Grau did not shirk the subject of death. On the contrary, he was philosophical. 'I have had a good life, you know – yes, and a comfortable one. And I have two healthy sons, whom I leave well provided for.'

And a wife?

'Ah no, I never married her. I have been a liberal, you see – too liberal for most Limans. I have had an easy attitude to life. And so I am able to contemplate death without fear.

'You know, as a racing driver you face death many times.' He looked at me unemotionally. His eyes were watery and pink, his skin a yellow parchment. He was ill-shaven, and his raffish 1950s pencil-moustache was prematurely white. His voice shook with a steady, Parkinsonian rhythm.

'Death, my friend, is just another journey.'

CHAPTER THIRTY-NINE

Señor Grau's Great-Grandfather

IN LIMA I DECIDED I had driven far enough up the coast of South America. I parked the jeep securely in a private yard and continued north by bus. Now I was close to the border with Ecuador and the area of shifting sands known as the Desert of Sechura, the northernmost extreme of South America's coastal desert. *El Niño* had been doing strange things, and major floods had transformed the Sechura Desert into a lake, cutting off the road through it. Ill portents, people kept telling me.

The bus was diverted inland into the foothills of the Andes, which, in these tropical climes, were riotously creepered and green. Astounding things had happened here. All along the valley, half the road's width had gone, tumbling into the river far below and leaving a jagged serration where there should have been a white dotted line. A few weeks earlier, northern Peru had been assailed by its worst storms in twenty years. Now the bus crossed a riverbed where twisted girders – all that remained of a substantial iron bridge – were strewn around like casually crumpled cigarette packs, and trees lay uprooted. A tyre-rutted track found a firm way across the riverbed, while on either side of it were the roofs of lorries submerged in quicksand. We were halfway across the riverbed when the bus ground to a halt: it was the gearbox.

Nearby was the workcamp of the labourers who were beginning to repair the road. Seeing us becalmed, children ran towards the bus to sell us unchilled bubble-gum-flavoured Inca Cola and boiled sweetcorn.

As usual, it was all male hands to the pumps. The driver, sweating profusely, ordered us to push the bus forward so that he could try and jerk it into gear. We pushed, and the bus grunted. We heaved, and the gearbox ground horribly. 'Try backwards,' the driver shouted, 'maybe it will work in reverse.' It was hot work. Meanwhile, the women sat in the bus looking bored, yelling unhelpful advice out of the windows and eating sweetcorn.

A caterpillar truck came to give us a hand. It lowered its sand-shovel against the rear of the bus, plumed black diesel fumes into the sky, and shoved. The back of the bus crumpled impressively. Our driver hastily thanked the caterpillar driver, but said it would be better not to destroy the back of the bus.

At last, and quite unexpectedly, the bus started in second gear and progressed slowly across the riverbed, while we men ran after it and piled in. It continued in second gear. The next time we have to cross the riverbed, I thought, we are going to get stuck again.

And so we did.

Piura reminded me of a small town in northern India – not just the parched landscape, the boxy concrete buildings with iron railings and foaming bougainvillaea, not even the press of humanity; it was all the three-wheeled Bajaj auto-rickshaws darting like Drake's little ships in and out of an armada of fuming buses and trucks. The autos emitted clouds of lavender-blue exhaust fumes, sweet and nauseating, the distinctive smell of New Delhi or Ahmedabad. I felt myself nostalgically transported by this Proustian pollution.

The bus pulled into the town centre, with its colonial cathedral and balconied Spanish houses. Piura is the oldest colonial city in Peru, founded by the soldiers Pizarro left behind when he went off to plunder the Incan empire. It has the reputation of making its enforced exile from the mainstream of Peruvian life into some sort of virtue, holding itself aloof. Be that as it may, it is not aloof from that most characteristic trait of South American politics, the assassination. The day I arrived, the local newspaper headlined a story about a local female journalist who had been publicly attacking the landowners for their exploitation of peasant farmers. She had been murdered.

A few days earlier I had eaten dinner in a Liman restaurant with Señor Grau, who told me with as stiff an upper lip as you could hope to meet that he would die any day now of cancer. Now I stood in a cool, high-ceilinged museum devoted to the memory of his great-grandfather.

I remembered the museum I had visited in Iquique that commemorated the Chilean version of this conflict. The War of the Pacific, from 1879 to 1883, was one of the few major international wars in Latin America, and fought for possession of a desert. Deserts are valuable to people only for what lies buried beneath them. This was during the height of nitrate exploitation, and the Atacaman frontier between Chile and Peru was unclear. With vast wealth at stake, the two countries had gone to war.

Admiral Grau's iron-clad *Huáscar* had sunk the Chilean *Esmeralda*, and gone on to have further successes; but eventually Chile captured the *Huáscar* and won the war. It led to the enrichment of Chile, with the political dominance the country enjoys to this day, and left a legacy of bitterness in Peru. (Nor has it been forgotten that Britain assisted Chile in the war.)

The Peruvians compensated somewhat for their loss of the war with this shrine to the hero Grau, glorifying in lurid and badly executed canvases his destruction of Chilean tonnage. The Admiral gazed at me from innumerable busts and portraits, a tubbier version of his terribly wasted grandson.

I returned to the room containing a model of the *Huáscar*, the Admiral's flagship, which had been built in Britain (the former pride of the Peruvian Navy is now on exhibition in the Chilean port of Talcahuano). Trying to take a picture of it, I contorted myself to minimize the reflections on the glass and mahogany cabinet. The caretaker appeared, and I cowered guiltily, expecting to be told that photography was forbidden, or possible only on purchase of some costly permit. Instead, he offered to take off the case so that I could photograph the model more easily.

The Victorian case was heavy, and we struggled together to lift it clear of the boat's rigging, then carefully posed it on the stone-flagged floor. I took my snaps. When we came to lift the thing up again, it wobbled perilously, and for a moment it seemed that we were going to drop it. In my mind I saw it shattering on the stone floor, the scandal, the expense . . . We managed to keep our grip, but as we unsteadily lowered it we knocked the miniature pennant off the Admiral's flagship.

I broke out in a sweat. We grunted as we heaved the miniature Crystal Palace back to the ground and attended to the blow struck against Peruvian national honour. With trembling fingers I reattached the little flag to the rigging. Hastily lifting the case, we nearly dropped it again.

At last it was in place, and with heart still pounding I tipped the caretaker generously and ran for the door – only to find my path blocked. It was the museum's curator, a slender and elegant woman of a certain age. 'Stop,' she ordered, 'wait here,' and she disappeared into her office.

I waited like an errant schoolboy, hoping the caretaker would not lose his job.

The curator came back, and with a benign smile handed me a brochure. 'It is a wonderful exhibition of religious art,' she told me. 'It's on just up the road. I'm sure you won't want to miss it.'

PART VI

Sonora

CHAPTER FORTY

In Which I Plumb the Depths

THE STEEP AND SINUOUS road inland from the Pacific coast took me through craggy peaks covered in virgin forest. It all seemed dreamily unfamiliar – so much water, so much verdure. Here in north-western Mexico, high in the Sierra Madre Occidente, there is a network of deep gorges, deeper even than Colorado's Grand Canyon. Chilly and alpine at their rims, they descend through several microclimates to become dry and cactus-strewn deserts in their depths.

The town at the centre of this system of *barrancas,* or canyons, is Creel. It is a ramshackle, Wild West kind of town, slowly being domesticated by tourist hotels and log-cabin boutiques selling Tarahumara knick-knacks.

The Tarahumara Indians are the indigenous people of these mountains, said to be descendants of the great Indian nation of the Apache, who came here over a thousand years ago. They are known as phenomenal runners, and as a silent people; while the Yaqui, for instance, violently resisted the demands for workers in the white man's gold and silver mines, and were almost wiped out as a result, the Tarahumara responded by withdrawing ever-deeper into the mountains. But today *mestizo* farmers and logging companies are flushing them out. They are paid little for the handicrafts they make in their isolated *ranchos.* Unintegrated into mainstream Mexican life, they are poor, and like so many indigenous people whose culture has been engulfed by an alien invasion, they are prone to alcoholism. I saw several Tarahumara men stumbling through Creel in a drunken stupor, or simply gazing around in boozy incomprehension.

South of Creel, the road switchbacks ever higher into glossy mountains. This is remote territory, but enterprise has got its teeth into it, and forests were being cut everywhere. By dusk on the third day after leaving the Pacific, I was close to one of the deepest canyons, or so I thought. All I could see through the fir trees was mist. A gang of Tarahumara women were mending potholes in the dirt track, and I asked for directions.

'It's somewhere over there,' they said vaguely. 'You need a guide.'

I drove on for a few miles, down a deep-puddled track. The earth, or mud, was orange. I passed picturesque stone buildings, then the trees opened onto farmland. A tall and wiry man was walking across the fields, carrying a hoe on his shoulder. I told him I was looking for the canyon.

'The canyon? Well, it's just over there.'

'Could you show me?'

He had an extraordinary face – bulging eyes, gap-teeth, a loopy smile. 'Why not? I was only on my way home.'

His name was Hector, and he farmed fifteen hectares of maize, he told me, just over the way. He led me through the level maize fields and into a pine grove. The land inclined upwards slightly, and we scrambled over a mush of copper-coloured pine needles. It began to rain. All at once, the trees and the earth parted and I felt my stomach tighten. I was standing inches from an abysmal gash in the earth, a chasm many miles wide, in which clouds drifted beneath our feet. Cascades of pink rock plunged what seemed an impossible distance, to a river that marked the depths of the gorge, a slender thread of silver. We were standing in drizzle, but down there the sun was shining.

'I'd like to go down,' I said. 'Do you know anyone who could guide me?'

He pulled out a packet of cheap Pharos cigarettes and offered me one. Then he looked comically at my legs (I was wearing shorts) and flashed me his gappy grin. 'Are you sure they're up to it?' He giggled, and I knew I was going to like Hector. 'No one knows the *barranca* better than me,' he said. 'It's my family who live down at the bottom there. I've been going up and down all my life. *I* could take you.'

It was dusk, and as we walked back to the car we agreed a fee. I asked Hector where I could pitch my tent for the night, but he would have none of that: I was going to spend the night in his house, as his guest. We ate a dinner of cheese, tomatoes and tortillas, by the glow of a pot-bellied stove. Then Hector and his wife retreated into their young daughter's bedroom, and I lay in the tall matrimonial bed, securely surrounded by saints in glass boxes and teddy bears in lace dresses.

At six the next morning we were climbing towards the rim of the *barranca*. We passed a level, grassed clearing. 'An airstrip', Hector said. 'Drug running.'

'Really?'

He winked and touched his nose. 'We sometimes hear a plane in the night. I don't ask any questions. It's safer that way.'

Soon we were clambering down precipitous paths through the eroded table formations at the canyon's edge, following a long-established track.

'We'll take the steep track down,' said Hector. 'There's a slightly easier way out, but it's much longer.'

'Which way would you come out?'

'This way, but . . .' He flashed his loopy grin . . . 'I don't think you'd make it.'

Pine needles lay on the pink earth, and there were clusters of mauve and yellow alpine flowers. Quartz crystals littered the hillside. Sometimes the slopes were gradual enough for us to trot down, other times they were so steep that narrow, zig-zagging paths were scored into them. Giant buttes and rock pedestals loomed over us, hundreds of feet high and bigger than cathedrals, yet minor details at the edge of the gorge. There were mossy caves, waterfalls where you filled your hat and drank, and ledges over fearful voids. Stopping for a break, we perched on the edge of a mushroom-like pedestal. There was a vertiginous drop to a carpet of treetops.

Hector would exuberantly call out the names of trees and birds as we went, and taught me a whistling dialogue with one particularly chatty bird. We disturbed two red and turquoise parrots, and the echo of their cries was so loud rolling among the mesas that I thought we had disturbed a whole flock of birds.

Hector told me about his small-holding. 'Fifteen hectares, it's nothing. My neighbours have two hundred. Fifteen hectares doesn't earn you enough to live.' He laughed incongruously. 'I'm a poor man. I was better off before, and now I'm saving again. Next year I hope to buy some more land, and maybe an old truck.'

'Why were you better off before?'

'My father had cancer. He was in the hospital at Chihuahua. The doctors just kept saying, "He needs this treatment," and what is a son to do when he hears that? It cost thousands. We're eight brothers, and we all had to give, we all had to sell something. I lost my truck, but only about a third of my land – I wasn't very well off to start with.'

Slowly alpine vegetation gave way to plants of a spikier, desert variety. We made many halts for photographs and note-scribbling, and Hector warned me that by midday the sun would be intense – we would be out of the moist central area of the canyon and into the desert. The air grew hotter, less humid. By 3 p.m. we were tired and hot, with some way still to go, scrambling along dusty paths among giant cacti. 'I normally do this in four hours,' said Hector. So far it had taken us eight. We had descended 5,000 feet – the equivalent of three Empire State Buildings.

The rushing alluvial waters of the Rio Verde were now clearly visible, and at last we saw the roof of Hector's uncle's house in a cleft beside a mountain stream. The abundance of fresh water had made a miniature tropical climate, and for a few irrigated acres around the house there were orchards instead of thorn trees.

'We'll soon be eating mangos!' said Hector. 'You see the trees? Oranges, figs, grapes – you name it, it all grows here. This used to be a big village, people lived all along the river. But these days it's too cut off, nobody wants to live where you have to climb a mountain to buy some sugar or see a doctor. So there's nobody to sell the fruit to. It falls on the ground and rots.'

Hector's uncle was sitting by the stream, underneath a mango tree. He was the most laid-back man I ever met. As we burst sweaty and bedraggled from the trees, he looked up as though he had been speaking to Hector just five minutes earlier, and mumbled '*Hola*' without even taking out his cigarette.

Hector was his usual manically enthusiastic self. 'This is an *Englishman*,' he announced. 'I promised him mangos.'

The old man grinned, and pointed a thumb at the laden branches above him. 'Help yourself.'

The next morning a 'neighbour' wandered into the compound, a one-eyed man from what they called the next village. I expressed an interest in seeing the village, and he said he would be happy to take me. I presumed I was letting myself in for a short walk along the river bank. But first we had to cross the Rio Verde, a hundred feet of not green but brown water.

'How do you get across?' I asked Hector.

'You just hang on to that wire over the middle and wade.'

It was all right for him; he was staying behind.

I tied my boots (along with my dictionary and money belt) around my neck – a mistake, because the riverbed was ferociously rocky. The river was running a strong current – it was the rainy season – and, unlike the man leading the way, I had no staff to lean on. I hobbled after him, somehow managing not to stumble. When I was halfway across, he called back that the wire crossed the river at its deepest point: if I wanted to keep my head above water, I should follow him. I let go of the wire and tottered towards him through bubbling, chest-high mud.

On the far shore my purblind guide set off briskly, pausing only to point out a dead snake. The path wound into boulders high over the river, narrowing to a rough-hewn ledge eighteen inches wide. In places there was no real ledge, just a series of oval footholds hacked out of the cliff-face. A hundred feet down, the river seethed and spangled. I found it hard to believe that this could be their only path to the outside world; and yet it was.

It took two hours to reach the 'village', which turned out to be a single, half-derelict house on a shield-shaped terrace of alluvium on a bend in the river. Everything was dry and wretched. 'A stream cuts through the middle of my land, but it only flows at this time of year, in the wet season,' the man told me, explaining the absence of crops or fruit trees. He lived by raising a few animals and fowl. I was amazed: a broad, year-round river ran nearby, but he did not use it to irrigate his land. He was content to live in a desert.

He and his wife and the eleven children in their ragged hand-me-downs were the most reticent people I met on this journey. They looked up apathetically when I arrived. Cut off from human contact, they had even forgotten how to say 'Hello'. Theirs was a material and a social poverty.

My guide led me a mile or so across the seasonal stream that divided his land, to show me 'the old houses'. Within a grove of thorn trees we saw a dozen old stone dwellings – ancient, I judged, from the size of the doors and windows – small forts, really, with tiny turrets set into the walls. I felt sure that no archaeologist's trowel had ever disturbed the place. I wondered if the people who had lived here had known how to irrigate the dusty soil.

As we walked, grey veils of rain drifted across the canyon peaks. When we tried to retrace our steps, we found that the small stream that crossed his land, carrying water out of the heights, had turned into a deep and violent torrent. 'It's dangerous to cross now, there's too much pressure,' he said. 'We'll have to wait a couple of hours for it to subside.'

Two hours later the water was again low enough for us to wade in, waist-deep. We reached the house in darkness. He owned no electric torch, so we could not contemplate trying to return to the home of Hector's uncle. I was stuck here for the night.

The small guide's fee I had paid was, to him, a substantial sum, and he generously announced that a turkey would be slaughtered in my honour. By firelight I watched two of his youngest daughters, perhaps seven and five, cut the bird's throat and hold it upside down for two long minutes as its purple blood throbbed into the dust and the life jerked out it.

No one spoke to me; nor, beyond immediate necessities, did they speak to each other. There was no laughter and no play, no toys and no books, no writing of any kind other than a hardware store's free calendar pinned to the wall. I sat at the table with the two adults and the others waited their turn to eat, the oldest boy fingering the handgun he normally kept tucked in his waistband.

After dinner I was given a wooden cot in the ramshackle outhouse. Slowly, the joyless little children climbed into nooks in the wooden structure around me. By moonlight I saw a thousand holes in the roof. Rain came, and fell everywhere but on my artfully placed bed. I slept, and dawn arrived with someone banging pans together just behind my head: it was the clanging of goats' bells, accompanied by the shrieking of a regiment of cockerels. There was no tea for breakfast, just bread and turkey-gravy.

Politely, but gratefully, I departed, one of the elder sons leading the way.

Hector had been up half the night calling my name into the darkness, sick with worry. He was angry with me when I arrived, but so relieved that joy quickly overcame anger. We wallowed for another day in his uncle's orchards, gorging on oranges and mangos. After the deprivations I had witnessed in the next 'village', I knew that this was Eden. At five the next morning it would be time to begin a 7,000-foot climb, with rucksacks loaded with water, to the canyon rim.

Over dinner I asked Hector's uncle his opinion of the one-eyed man. He was contemptuous.

'He's a lazy bastard,' he said promptly.

'He says he's very poor.'

'Poor? Hah! He's never done a day's work in his life! You think he couldn't irrigate that land if he wanted to?'

'Well, how on earth does he support eleven children?'

'If he's poor, why does he *have* so many children? The man's a goat. He's poor in money, but rich in the testicles. He's got one eye but two dicks! The randy fucker should leave his wife alone for five minutes!'

CHAPTER FORTY-ONE

Salvador and the Meat Market

A DAY'S DRIVE EAST brought me out of the western Sierra Madres and onto Mexico's central plateau. As I arrived in the drab and practical town of Ciudad Cuauhtémoc, I realized that with the dousings and temperature changes of the last few days, I was coming down with a chill. My temperature soared, and I started to shiver. I pulled on my thermal underwear and several layers of clothes, piled on extra blankets, and lay in bed with my teeth chattering.

But by morning, under the influence of extra-strength paracetamol, two coffees and two portions of a popular local breakfast dish – porridge – I was strong enough to carry on. As I ate, a flat-bed Chevrolet pulled up in the street outside and a blond-haired, pink-skinned family climbed out, the women in broad straw hats and long skirts, the men in denim dungarees. They were Mennonites from Belgium, Holland and Germany, whose grandfathers came to Mexico in the early twentieth century.

'*Los Mennonites*', the old man sitting next to me said. 'They're excellent farmers, you know – better than us Mexicans. They've brought a lot of prosperity to this region.'

'What do they produce?'

'Oh, cheese, ham,' he pointed at my bowl, 'oats – all good stuff, they know all the modern techniques. Wherever you go in Mexico you'll see their cheese for sale, look at the label – it's always from Chihuahua.'

'Are they friendly?'

He did a comic double-take. 'To *us*? They don't know us! Some of them even pretend that they don't speak Spanish, but I don't believe them. No, they're not friendly, they keep themselves to themselves. But so what? That's their right.'

A few miles further east was the city of Chihuahua. Feeling weak, I searched for a hotel, and found a sunny guesthouse in a colonial hacienda. The owner was a very round lady with bulging calves and perilously thin ankles, peroxide hair and carmine lips and nails. She was followed everywhere by two of the miniature dogs this town is famous for – Chihuahuas. I had read that they are well-adapted to the desert, having a constant body temperature of 104°F. Rodent-like and permanently hysterical, they seemed unpleasant creatures, but their owner evidently adored them.

I slept through the midday heat, then walked to the *Museo de la Revolución*, a dignified colonial edifice devoted to the assassinated bandit and revolutionary hero Pancho Villa. The most popular exhibit is the open-top Dodge in which Villa took his last ride, its shiny black coachwork riddled with bullet holes.

Back at the guesthouse, my landlady informed me that I 'should buy a Chihuahua doggy to take back to England'.

'They would die in the cold,' I replied.

'No, no, there is a long-haired variety.'

'I'll think about it.'

As I drove north the next morning I passed a huge Ford plant. Chihuahua is a down-at-heel industrial city, whose future largely depends on the US corporations who are exporting jobs to Mexico to benefit from the lower wage costs. Detroit and Pittsburg's loss, Chihuahua's gain.

The principal barriers between Mexico and the US are not mountains and deserts; they are not physical, but financial and racial. You can easily drift across the dividing line that separates the mean Mexican desert from its prosperous Anglo-Saxon twin. Mexican motels, dentists and plastic surgeons are cheaper than American ones, and their signboards crowd the streets.

The motel I reached late at night was a dump, though the proprietor drove an attention-seeking Cadillac. Young men lolled around looking heavy, and in the room next to mine they threw a noisy orgy. They seemed to be modelling themselves on favourite characters from Orson Welles's *A Touch of Evil*. I had been looking for a place to spend a couple of quiet days writing; this obviously was not it. I left before dawn, to try and cover some miles before the heat of the day.

At the next town, I stopped for breakfast and met Salvador. He had been hired to manage a soon-to-open hotel, and had only just arrived in town himself. But he already knew his way around.

'The dump where you stayed belongs to a mobster,' he said. 'Those grease-balls are drug couriers. They have shoot-outs in the streets there, you know, people get killed. It's a dangerous town.'

This, by contrast, was a haven of tranquillity, 'though they got some mobsters here, too.' He took me on a tour of the town, past an ugly ranch-style mansion that lurked behind tall gates. 'That place belongs to a mobster. Some out-of-town guys thought he was muscling in on their territory, so they flew over in a helicopter and dropped grenades; bombed the fuckin' shit out of him.'

He pointed out the town's 'only halfway OK motel – until my place opens, which will be the best in town, naturally. See you for lunch.'

I set up an office in my motel room. Even with the air conditioning on full, the heat was terrible, and I sat naked in a pool of sweat, banging out words on a typewriter borrowed from reception, breaking off every half-hour for a not-very-cool shower.

I lunched with Salvador in the same workers' café. He was ostentatiously a Man of the World; he wore smart trousers and good shoes, and an expression of cheerful contempt for everything around him. He had just returned to Mexico from thirteen years in Chicago, and told me over the tortillas a string of lurid and possibly true stories about the mafias he had encountered there – ' "Look, Salvador," says the Old Man, "normally we couldn't let you get away with insulting a member of the Family like that. Because we like you, we're gonna let you live; but don't let it happen again." '

Now he was a Reformed Character. 'I'm thirty-eight,' he told me. 'I gave up drink and, you know, women. Fuckin' around. I met a nice Mexican girl. I'm gonna settle down and lead a small-town life.'

'So what's small-town life in Mexico like?'

'You wanna know? I'll show you tonight. It's the Festival of Grapes. They have a procession and chicks throw grapes from pick-ups at passers-by. Then they elect the Queen of the Grape or some shit. Anyway, you'll see.'

It started late, when the air had cooled, with a lacklustre grape-tossing cavalcade. We followed the thin crowds to a half-empty fairground on the outskirts of town.

'Apparently there's gonna be some dancing,' said Salvador. 'You're gonna see some Mexican culture.'

First of all there was a sort of beauty pageant, with the girls in their white dresses and red sashes parading past the judges. A fat man in a wheelchair had been pushed to the front row. 'Look, the only guy wearing a tie,' Salvador whispered sarcastically. 'He's the Mayor, Mr Feel-sorry-for-me-I'm-a-cripple-family-man-butter-wouldn't-melt-in-my-fucking-mouth. He's the bastard I just paid a twenty thousand *peso* bribe for a liquor licence.'

The girl who won was not, we agreed, the prettiest. 'You see,' said Salvador, 'in Mexico even a little girls' beauty pageant is rigged.'

The lights went down again, and the tannoys crackled: there would be a hula-hula dance. Eight girls aged twelve to fifteen trooped onto the platform, wearing peach-coloured bikinis. The theme song from *Titanic* began to play, and they began to wiggle.

'Jesus, I didn't expect a meat market,' said Salvador.

Most of the girls danced badly. But two of the older ones, fleshy creatures with rather mature smiles, wiggled well, relishing the attention. They turned and stuck their bottoms out, and wiggled some more.

A strangled silence had fallen over the men in the audience. 'Salvador,' I hissed, 'this is *erotic*.'

'You bet. I think I'm turning into a Dirty Old Man.'

'I mean, is it *meant* to be erotic? They're just young girls.'

'They don't look so young from where I'm standing. No, of *course* it's not meant to be erotic. I mean, I don't think it's hypocrisy. But we Mexicans are pretty confused about sex. Blame the Catholic Church.'

As we walked out of the fairground Salvador said, 'You know what we should do now? Go to a cathouse. But in this dump, forget it, the average age of a hooker is fifty. Jesus, *Christ!*'

I grinned at him. 'Salvador, are you sure you're ready for small-town life?'

'You got it', he said. 'I'm depressed in this two-bit shit-hole, and I've only been here a month. I think I'm gonna move back to Chicago.'

CHAPTER FORTY-TWO

Snakes, Teddy Bears and Desert Rats

*Most would agree . . . that the ultimate among the
various provinces of The Great American Desert is
Sonora's Pinacate region, at the head of the Gulf of
California. This region is the bleakest, flattest,
hottest, grimmest, dreariest, ugliest, most useless, most
senseless desert of them all. It is the villain among
badlands, the most foreboding of desert realms . . .
Pinacate desert is the final test of desert rathood; it is
here that we learn who is a true rat and who is
essentially a desert mouse.*

Edward Abbey

AFTER READING Edward Abbey's words, I could not pass Pinacate
by. So today I was the only human being in the Pinacate
National Park, a vast and remote wilderness of volcanic rubble and
organ-pipe cacti. It is not often that you get 4 million acres of land all
to yourself.

But to hike through Pinacate in July, when the average
temperature is over 106°F? The park rangers, crouched in a silver-
painted air-conditioned cabin at the park's entrance, clearly thought I
was crazy – all except one, Conrado, a recently arrived young ranger
full of puppyish enthusiasm for the desert.

'Nobody comes out here in summer,' he said. 'Do you know it hit
117 degrees last week? *That's hot!*'

But being a desert rat himself, Conrado did not try to talk me out
of my mad scheme. 'If you're not back here by 5 p.m. tomorrow,
though, I'll come looking for you!'

That was reassuring, because I had a tough climb ahead of me. Edward Abbey, doyen of American desert writing, scaled the Pinacate Peak one May day in 1968. At 3,957 feet, Pinacate Peak is not much of a mountain, and the entire hike should take only 8 hours. But I had to try and complete it before the searing midday heat set in.

I climbed up to the mountain's lower slopes and camped there, then set off again before dawn. Slowly the sun revealed evidence of Pinacate's explosive relationship with the landscape – fields of basalt, jagged laval boulders lying where they were blasted millennia ago. It is a positive feast for volcanic *amateurs*, of large and small craters known as fumaroles, calderas and maars, a black garden of lava hurled, squeezed, extruded and drooled, ridges and grottoes of smooth or clinkered or carious rock, flows and pools and solitary fragments that possess dainty scientific names – porphyry, xenolith, scoria, pahoehoe, tephra and lapilli.

On the shallower slopes, where they could gain a foothold, organ-pipe cacti towered twenty feet and higher. Tiny, long-beaked birds sipped from the crimson flowers at their heads. I saw prickly pear – cascades of jointed peardrop-shaped pads – bulbous barrel cacti, low-growing 'hedgehogs', and teddy-bear cacti, whose profusion of scalpel-sharp spines makes them seem cuddly only at a distance.

A leafy plant with 100 times the surface area of a cactus loses water 6,000 times as fast. In order to live in the desert, most cacti have abandoned their leaves, letting their chlorophyll-green trunks and branches do the work. Waxy skins prevent water loss, and spines protect their moist, pulpy cores from predators. Cacti are mysterious organisms, their origins unknown. The Mexican Sonora Desert has the greatest number and variety of them in the world.

I had seen a surprising profusion of birds and geckoes, and all around me crickets were whirring like electric saws. Summer is also known to bring the Sonoran snakes out, and it was not long before I encountered a rattler. It must have frozen when it heard me, because at first I thought it was dead, and since there were no carrion birds hovering around it, long-dead.

But when I was just a few feet away it suddenly reared to face me, coiling its length behind it in a tight spring.

I stared in fascination, happy to have been granted this encounter with nature. The snake stared back. We watched each other for about twenty seconds. Then it slowly turned and began to meander away. And I picked up a stone and threw it.

'English, eh; are you a hooligan?'

A barman had asked me this the previous day, with just a touch of facetiousness. Everyone in Mexico loves football; football teaches you about the world, and the English have the world's best-known football hooligans.

'Yes,' I replied.

His eyes widened, and for a moment he looked unsure. Maybe hooligans did not just haunt far-off football tournaments, maybe he had one in his bar.

I had thrown the stone in front of the snake to make it turn, to prolong our encounter. It did not know that. It turned sharply, reared again and made two or three strikes in my direction, its forked black tongue flickering in and out of its thin mouth in its bald little head. In his poem 'Snake', D. H. Lawrence recalls throwing a stick at a snake, 'one of the lords of life', because he was offended when the magnificent creature crawled into a hole. 'And I had something to expiate,' Lawrence writes; 'A pettiness.'

I had to expiate a self-centredness. Now, as the snake and I regarded one another, I had the absurd, irrational thought that I was going to be in these deserts for a long time, and that this snake would have a lot of cousins. 'Go in peace,' I said out loud, in an ironic voice. Of course, snakes don't get irony. As I walked away, I looked over my shoulder: still erect, alert, tensile, the snake was watching me go. 'You're talking to snakes now, Martin,' I said, as though talking to snakes was worse than talking to myself.

I climbed on. I had no problems until I reached the base of the main volcanic cone, a near-vertical mound of cinders, ash, lapilli. I thought I saw a track, and decided to follow it. I was mistaken. It petered out halfway up a steep slope, too late for me to double back. I had to carry on this way, even though I was clearly on the steep eastern face of the cone. It was 9 a.m., and starting to get hot.

My right boot came lightly into contact with a baby teddy-bear cactus. The size of tennis balls, teddies lie around waiting to hitch a ride and take root elsewhere. It clung on to my boot with the microscopic barbs on its needle-like spines, and as the boot swung forward, embedded itself in my left calf.

Long trousers should be worn for this sort of hiking, principally because of the risk of disturbing a snake, but for other reasons too. The little brute had to be cut off, twenty or thirty of its tenacious spines pulled out with tweezers, and the bleeding wound dressed. Larger cactus clumps are man-traps; I have heard stories of men falling into them and, being immobilized, dying there; I have no difficulty believing that it could happen.

I scrambled on up the steep face of the Pinacate cone, wading calf-deep in ash and cinders. I reached the summit at ten-thirty, an hour later than I had planned, and much more tired.

But God, what a view. On the western horizon was the broad azure band of the Sea of Cortez, the gulf that probes the desert behind the thin, long finger of Baja California. Between me and the sea was an erg – a forty-mile-wide expanse of perfectly barren sand dunes. To my right was a 750-square-mile volcanic field, one of the finest volcanic landscapes anywhere on earth – lunar, black, prickly with cinder cones and craters and extinct geysers. It was here that NASA astronauts came to practise for the moon walk.

Incredibly, Indians once lived here, the Pinacateño Areñero, a sub-group of the Sand Papago, a people rumoured to be proficient in the arts of magic. Proficient anyway in the arts of survival, for they lived on cacti and the occasional jackrabbit or bighorn sheep. When white men – missionaries, traders, gold-panners – appeared in the desert, the Pinacateño ambushed them at water holes and killed them. Naturally the white man could not tolerate such a state of affairs. He came back in murderous force, and since 1912 the Pinacate Indians have been, like the Pinacate volcanoes, extinct.

I added my name to the register kept in a steel ammo box by an Arizona climbing club – where it joined Edward Abbey's – and started down again.

Of the eight pints of water I had started out with, two pints remained as I began my descent. I should have sat out the midday heat under a bush and left at about 4 p.m.; ironically, I was now hurrying only so that Conrado would not hit the panic button. A slog through midday heat was exactly what I had wanted to avoid.

I followed the established trail down, a less direct route, but much less effort. The afternoon heat was furnace-like. The surface soil in Pinacate can be fifty degrees hotter than the air temperature, and by mid-afternoon it is like walking over coals. My feet and head were both extremely hot, and I had no water to cool myself. I was dehydrating rapidly. Whenever there was an organ cactus beside the track, I would crouch for a minute or two in its shadow.

It was like those cartoons we see in the papers. My dehydrating body filled my head with fantasies of cold drinks, and I talked about them out loud. Stumbling along the track, every step an effort, I constructed detailed and highly specific liquid feasts. That night, I told myself, I was going to create cocktails of – oh, pomegranate juice, pear juice, grapes, and oranges, oh, oranges. I remembered a visit to a garden of fountains in Italy, and laughed at the plashing, wanton wetness of it all.

At last I tottered into my campsite. I fell into the meagre shade of some thorn bushes, laid shirts across myself and doused them with water. I shuddered with a sort of exquisite pain as the evaporation quickly cooled my body.

It was late when I reached the park rangers' cabin. Conrado, who had been starting to get worried, greeted me like a conquering hero. 'I don't think anyone's ever climbed Pinacate in July before.'

'Well, why would they? I'm obviously mad.'

'But it's exciting, no? A challenge?'

There is a line somewhere in the writings of Dag Hammarskjöld that talks about the path to a summit lying between two abysses, the first being physical fear, the second the death instinct, tinged with narcissistic masochism. There are questions to be asked about the motivation of people who walk around deserts. I do not think we are maniacs half in love with death. The experience of being in the desert – not the struggle against the desert, but simply *being* – has a quality of ungraspable mystery. Edward Abbey wrote that

> the desert, any desert, suggests always the promise of something unforeseeable, unknown but desirable, waiting around the next turn in the canyon wall, over the next ridge or mesa, somewhere within the wrinkled walls . . . God? Perhaps. Gold? Maybe. Grace? Possibly. But something a little more, a little different, even from these . . . The desert rat loves the desert because there is something about it he cannot explain or even name.

PART VII

Americana, Australiana

CHAPTER FORTY-THREE

On the Road

Today a person with fair skin can be exposed to the
sun for 10 minutes before burning. But if today's
projected high of 118° holds true, then it's starting
to cool off!
The Desert Sun newspaper, Palm Springs

NO ONE PAID ATTENTION as I walked out of Mexico, but over the border a US customs man inspected me carefully. He had a sculpted Indian face, his leather holster gleamed, and I could see my own face in the polished lenses of his sunglasses; even his smile – when it came – was clean-cut.

'Welcome to America, sir,' he said.

A quarter of California is desert, and down on the Mexican border it is a series of scorched, hyper-arid basins and mountains and sand dunes. I had seen the volcanic fields of Pinacate; soon I would be approaching another part of the same geological system, the San Andreas Fault.

The bus headed north, between orderly phalanxes of palms and the blue gash of the Pacific. A broken sign said 'Desert Convenience Stores', and another, 'Mecca Beach Trailer Park'. We passed used-car lots and cut-price stores, and an abandoned service station with a rusting water tank riddled with bullet holes.

The freeway opened up, a river of tar as broad and placid as the Ganges. The morning rolled by, but I had no sense of progress. Travelling on American highways one feels motionless, due only in part to the softness of the suspension systems and the plastic-surgical smoothness of the roads: America's vastness induces in the traveller a kind of helpless placidity.

In Mexico the desert is a fact of life, but Americans live *despite* it. With tarmac and air-conditioning, the arid enemy has been defeated, even commoditized. Billboards proclaim

Heritage Palms Active Adult Community
Desert Hills Country Stores
Fantasy Springs Casino
Sun City Model Homes
Oasis Realty

New houses everywhere, God knows where they get the water from; to the right and left mountains tower, snow-capped in June; perhaps from there.

The bus pulled over at a roadside diner and the driver said, 'Ten minutes'. A sign told us that the diner was 'independently owned', though it was surrounded by Big Brother food franchises with their towering, metal-stalk-mounted signs, the kitsch logos reminding customers of an earlier time when all in America was well. The diner offered several varieties of mostly deep-fried food, all of it, as far as I could tell, inedible.

I had taken the driver's 'ten minutes' literally, ordered a sort of coleslaw, and sat myself at a turquoise melamine table. Suddenly he was honking the horn, and passengers were trooping back aboard with their meals in polystyrene boxes. Ten minutes was apparently shorthand for the amount of time a guy needed to take a leak and smoke one cigarette.

The bus passed hundreds of slow-turning windmills, entire fields of them, the crop being electricity. They were duralium and fibre-glass, high technology, but there was something inadvertently beautiful in their crazy cartwheeling.

In late afternoon we pulled into downtown LA, its deserted Sunday streets sun-spangled and squalid. Black men sat splay-legged on pavements, staring disgruntledly into the distance, like the Tarahumara drunks I had seen in Creel. Asians stood behind the counters of tight little shops where signs said 'We Accept Food Stamps'. Only the Mexicans, tubby and quick-grinned, grandmas and aunties, bundles, picnics – only they seemed to belong. I was in the United States of America, and suddenly I realized that I had not seen a pioneer-type all day. Black, Mexican, Chinese, Vietnamese faces – but no Anglo-Saxons.

The next morning I acquired, like all self-respecting Angelenos, a car.

Think of Sal Paradise and Dean Moriarty, Humbert Humbert and Lolita, *Zen and the Art of Motorcycle Maintenance*, *Easy Rider*. Think of Thelma and Louise's doomed drive across the Southwest, or the journey to Death Valley in Antonioni's *Zabriskie Point*. The American desert is one of the most mythologized landscapes of the modern imagination. And Hollywood's children are so fully fused into the collective imagination that they can only achieve meaning through these shared mythologies, being reborn in filmic incarnations. Everyone exposed to Hollywood has had implanted in their subconscious the desire to get behind a wheel and *become* a road movie.

You need a convertible. You may say I was weakening, that I started out all manly in a goddam jeep crossing Saharan ergs and degenerated into a suburban poser, but remember: Thelma and Louise had a 1966 Thunderbird convertible, and Hunter S. Thompson and his attorney Raoul blasted up to Las Vegas in their exotic open-top Shark, Raoul coughing and a cloud of cocaine lost to the sky and the cacti.

The cowboy film was the staple of my parents' generation, greed and genocide rewritten as the extension of wholesome Protestant values into lands God had unaccountably, but wrongly, placed in the hands of savages. In the second half of the twentieth century the cowboy film, with its tamed and clean-living Wild West, was subverted into the individualistic and usually anti-social road movie.

For European directors like Wenders and Antonioni, road movies were expressions of the colonization of the European imagination by American images. The images of hard blue skies, desert buttes and fetishized automotive chromium were plundered by the makers of pop videos and TV ads. Pop stars endlessly drive convertibles across the Southwest, and Marlboro cowboys and the Camel camel employ the iconic emptiness to commercial ends. The family saloon car is reborn as existential hero, its natural habitat a cracked Nevada lake-bed where its shiny surfaces miraculously reflect and resist the searing heat.

To drive into the Sonora Desert, or the Mojave or Colorado, is to become drunk on light and space. You are insignificant, yet exhilarated and alive. Freed from the limitations of streets and people, you are more fully you. You seduce yourself into a state of euphoric self-absorption. The car is a mobile altar, the desert the bread and wine of pagan communion.

Life must end in death, and road movies are always tragic: the pursuit of freedom is doomed, the spark of individualism must inevitably be snuffed out. The American deserts inspire the same Romantic hero who, two centuries ago, was awestruck by the lakes and mountains of Germany and Italy. Today, this is the 'live now' philosophy, the existential consumerism, of all Western youth; the irony is that it is sold to them not by visionary artists but by media corporations. When I came home from the Sahara, I found Western consumer fantasy entirely unreal – I was dazed for days. It took time to adjust from Saharan heat to a post-modern environment that manipulates images of the desert to sell goods to consumers.

There are other realities never glimpsed in the media: the indigenous people of North America are living in sullen poverty on reservations; in Africa, I had seen how desert dwellers like the Tuareg and the Bushmen were affected by the modern world – they were being culturally exterminated. But the American Indians, having been internally exiled to the deserts, are now at least left pretty much to themselves. New laws and land rights are slowly letting them take control of their own destinies – while at the same time their absorption into the American mainstream continues apace. I wanted to discover what stage this process had reached.

And there was Nature, a genuine wilderness untamed and, from a human perspective, brutal; you could still go alone into the desert as the writer Edward Abbey had done, grabbing his rucksack and canteen and disappearing for days on end.

Driving away from LA's megalopolitan tangle, I came to the Joshua Tree National Park, with its giant out-thrusting yuccas. They were filming a car ad, and under a red sky a reflective box was racing for the cameras along a sinuous desert road.

CHAPTER FORTY-FOUR

Koyaanisqatsi

> *This is the only country that gives you the opportunity*
> *to be so brutally naïve: things, faces, skies, and deserts*
> *are expected to be simply what they are. This is the*
> *land of the 'just as it is'.*
>
> Jean Baudrillard

IN AMERICA, EVERYTHING is possible.

Roswell is a nondescript Main Street kind of town in the middle of a bleak, flat desert, a place that would normally deserve and receive utter neglect.

But you've heard of the Roswell Incident? Well, apparently in 1947 this flying saucer crash-landed in the desert out there. No, *really*. There were these dying aliens in the wreckage, the government kept them clandestinely on life support for years.

'Well, sir, you either believe it or you don't,' said the waiter in the corner diner, a friendly cove with bouffant hair and an immensely camp manner.

'You mean it's like a religion?' I asked.

'No, sir, I certainly do not. I'm a Christian, and I don't believe in the aliens because then I'd have to believe that Jesus Christ came to save little green men as well as us, and I don't want to believe that. *Is that Mr Smith or Mr Jones?*' he yelled suddenly.

'Mr Jones!' a waitress called back.

A man of at least ninety summers, with sallow, liver-spotted skin, tottered through the swing door.

'Morning, Mr Jones,' the waiter sang out. 'Your usual, sir?'

'You got the noospaper?' muttered Mr Jones, almost inaudibly.

And with the *Albuquerque Journal* clutched in a wrinkled claw, he shuffled to a booth.

The waiter turned back to me, dropping his voice and giving my arm a friendly squeeze. 'That coffee is on the house,' he said. 'Then when you've been to the museum you have to come back and tell me if you believe it or not.'

It was 10 a.m., and the heat was already fierce. The 'museum' was a few yards down Main Street, distinguished by the model of a flying saucer displayed outside – two satellite dishes lashed together and sprayed silver.

At the entrance an elderly lady with red beehive hair greeted me. 'Oh, from England? We've had two Germans and a Danish person this morning, but you're our first English person so far!' I felt very special.

It is not so much a museum as a progeny of 1950s sci-fi B-movies like *Them!* and *I Married a Monster from Outer Space*. If you did not already know the alleged sequence of events, you would struggle in vain to discern them from the random assemblage of yellowed newspaper clippings, curling photographs and 'reconstructed' silver foil. In one corner, a café sold 'alien cappuccino'. There was an exhibition of art 'inspired' by the Rockwell Event, of which the quintessential example was a silver flying saucer painted on black velvet. A pair of booths showed Rockwell Event videos, attracting more attention than the display cases. A kiosk sold alien fridge magnets and key-rings.

I stepped back into the white-hot street and returned to the diner.

'Well, I'm *not* a believer,' I told the waiter, a little too loudly. 'In fact,' I spluttered, 'it's obviously complete and utter garbage, I mean, I, I . . .'

The waitress came over with a mildly reproachful look in her eyes. 'We had a woman in here last year,' she said, 'and she'd been abducted by aliens. She had scars and everything.'

'I hope she didn't try and show them to you.'

'Oh, no. Ha ha. But you know, I think *she* was from England . . .'

The inference was plain. My scepticism was as suspect, as exotic and extreme, as that woman's looniness. Far better to maintain a polite indifference. That was the American Way.

Old Mr Jones had worked his way through the introverted pages of the *Albuquerque Journal*, and was trying to pay his bill. Gazing at his shrunken bones and wrinkled skin, I wondered if he was one of those life support aliens.

'I like it here,' he suddenly boomed at the waiter.

'Well, I like it here too, sir,' the waiter replied in a tenor chant. 'I've lived in Chicago, and I've lived in Saigon, but I still prefer Rockwell.'

'Gloria subberful garble,' yelled the old man.

'Pardon me – Gloria?'

'Gloria the waitress. Wonderful girl.'

'Oh, I see. Yes, she's going out with my son, you know. Gloria the blond girl?'

'No, the other one.'

'Oh, that Gloria. Yes, sir, she's also a very nice person.'

Jones shuffled towards the door, and I followed him out.

'Have a very pleasant day,' the waiter called after us, 'and you all be sure to come and visit us again soon.'

> *Farmers on the southern plains gathered to pray for*
> *rain today in the midst of a heatwave already*
> *blamed for 110 deaths.*
>
> The Associated Press

God bless New Mexico, blue-skied paradox of deserts, orchards and grassy plains. Cowgirls with tans, blue-check shirts and pink cheeks say things like, 'I'm kinda tired, I put up 500 bales of hay today.' They are making hay while the sun shines, and the worst heatwave since records began indiscriminately kills crops and people. 'The heatwave has so far claimed over one hundred lives,' the radio intoned, 'including forty-two illegal immigrants who were *literally* baked to death in a sealed railway car. The authorities are appealing to potential illegal immigrants to think again.'

The American desert has been irrigated, greened, domesticated, it has had its teeth pulled; but it can still manage the occasional snarl. Its victims are the unirrigated and unair-con'd, the poor, the huddled, the illegal.

CARE – DEAF AND BLIND CHILDREN CROSSING
STATE PRISON: DO NOT STOP FOR HITCH-HIKERS

I pulled up at a roadside halt. The tarmac was black and fresh-painted, and the buildings gleamed, planted solidly in the ground with white cement foundations that shone like Carrara marble, the grey galvanized metal of their tall security-light pylons still unsullied by traffic soot or graffiti. Everything was big-bolted, massive. The parked trucks were titanic, new, chromium-plated, steam-cleaned to immaculacy. Men in check shirts with silver monkey wrenches dancing on their cowhide belts stood by, ready at any sign of deviation to get the show back on the road. Can-Do: America is a land of possibility, expressed essentially as movement. It is this forward motion that above all is America's *raison d'être*, its spirit, its hunger, and it endlessly requires new land to digest and excrete as suburbia. Houses are spreading like a disease. When America has used up all the land, it will grind to a halt. But for now, there is still all *this* stuff – barren desert under carcinogenic skies – to be turned into some folks' homes.

Once this dusty plain had been an undisturbed home to volcanic gravel, a few creosote and mesquite bushes, some snakes and jackrabbits. Then had come a road, the first wooden shacks, a petrol pump, a soda fountain, electricity and phone lines. Now this latest road, dried black tar like hardened lava, and houses, houses, cheap and ugly boxes for men and women to buy with the fruits of their labour and make the developers rich.

Why, when they made roads and fast-food restaurants out of concrete and steel, did Americans make their houses out of plywood and plastic-coated cardboard? Perhaps because it was enough that they looked good; or for reasons of built-in obsolescence – long-lasting houses did not make economic sense.

How I resented the space these trashy homes took up, with their TV aerials and satellite dishes, and wires for their telephone conversations, internet access and interactive, online home shopping: a perfectly good desert, spoilt. When would America wake up to the despoliation of the land? Probably it would not – too many people stood to lose too much if it did. Anyway, the destruction of the land is just one element in the delirious consumerist carnival which

America leads. *Koyaanisqatsi* is the Hopi Indian word for it: a world out of balance, a life in turmoil.

I reached the site of the first atomic bomb blast. The spectacle of an incandescent cloud mushrooming over the New Mexico desert is one of the indelible images of the twentieth century. The Manhattan Project physicists did the research at Los Alamos, a few hours north of here, then looked for somewhere desolate to detonate the thing. They called the place Trinity, invoking the most sacred Christian doctrine with a machine designed to cremate living millions.

Driving through flat and dusty scrubland just beyond the perimeter of the bomb-site, I came into a two-horse hamlet called Bingham (Bang 'em?), and pulled into the driveway of a shop selling rocks, gems and 'bomb souvenirs'.

The owner, Tom Nelson, stepped out to meet me. He was tall and thin, and packed a revolver on his hip. Tom did not look made for the desert. He had pink skin, albino-blond hair and white-fleshed Mongolian eyes that narrowed when he smiled as though they were going to disappear. And smiling was something he did a lot. Perhaps sales of bomb souvenirs were booming.

He showed me, amongst the glistering chunks of quartz and amethyst, a piece of gen-u-wine 'trinitite', taken from the bomb-site.

'Getting very hard to find, this, yes sirree.'

It was sand, fused at Ground Zero into something resembling a minute green sponge – pocked, porous and bizarrely light in weight. There were little pieces of trinitite at $20, and larger bits for $100. And trinitite set in 22-carat gold, to be worn as jewellery.

'A bit ghoulish, isn't it?' I suggested.

'People wear it mainly as a joke,' Tom said, 'just to watch their friends back off when they tell 'em what it is. Actually, it's hardly radioactive at all.'

Trusting as ever, I believed him; I even handed over twenty bucks (well, it did have a certificate of authenticity).

Tom was a mine of bomb lore. 'This old guy that still lives twenny miles from here, he had the side of his face to the blast – and d'you know, half his beard went white! And they didn't warn the folks round here about fall-out or nothin' like that. Right here where you're standin' is where they got the biggest amount of it.'

I looked around nervously. It seemed safe enough now.

He invited me into his house to see a rarity, a gobbet of melted metal from the pylon that supported the bomb. 'I'd sell it for five thousand dollars,' he ventured, rather misjudging my level of interest.

'Could I have a glass of water?'

'You bet.'

It came from his well, the only one for many miles around.

'Does it glow in the dark?' I asked, ho ho ho.

He gave a polite smile. 'Haven't heard *that* one before.'

I winced. 'So tell me, Tom, why are you wearing a gun?'

'Well, you seen the gems in the store. You find semi-precious rocks all round here, and I own several mines. And some o' the local folks ain't so friendly. Just last week I put in prison two guys who were trying to rob me. I need to carry a weapon. In fact, some o' the folks round here are tryin' to persuade me to stand for sheriff. But I ain't so sure. Could be more trouble than it's worth.'

'And it seems so quiet out here.'

He grinned. '*Looks* quiet, don't it? Well, things ain't always what they seem.'

Conversation drifted towards the desert, which was, we agreed, a Good Thing. Tom said he loved the isolation, the remoteness from city life. And his wife Allison, a Californian by origin, was a fervent convert to the desert wastes. Deserts, she told me, are character-building.

'For example,' she said, 'take the old guy who owned this place before us. You see the porch out there?'

I turned and looked where she was pointing, through the mosquito screen at the concrete steps in front of the mobile home.

'Yes?' I said.

'Well, the old man discovered he was fatally ill, and there was nothin' he could do about it. So he sat right there on the porch. And he shot himself in the head.'

Tom nodded approvingly, and Allison gave me a broad smile.

CHAPTER FORTY-FIVE

Abbey Country

The time passed extremely slowly,
as time should pass.

Edward Abbey

THE ARCHES NATIONAL MONUMENT is the most remarkable
rock formation I have seen. Aeons of geological effort left fine
sandstone fins standing on the barren landscape, and a peculiar form
of erosion slowly caused their insides to cave in. The ragged hoops
that remained were wind-smoothed over more millennia, resulting in
the stone arches we see today, looping across the desert like so many
Loch Ness monsters. Up to 100 feet high and 150 feet wide, these
majestic and serene works of nature stir the soul.

I was amazed to see that tourists are allowed to drive within a few
feet of most of the arches. The US National Parks Service believes
that its educational mandate requires people to be able to park within
wheezing distance of every natural phenomenon – in fact, cars are fast
clogging up the national parks. In Arches, for God's sake, I saw signs
warning 'Congestion Ahead'.

It seemed to me that those who wish to gawp at a natural wonder
from their car seats – or stagger around for twenty-seven seconds
complaining of the heat, then climb back into their motorized air-
conditioning – should be invited to relax in city-centre cinemas and
watch IMAX films of the wonder in question. But if you want to
set your own eyes on the golden cliffs, the giant mesas, the cacti, the
kangaroo rats, the elegant rock arches of this desert national park –
then *walk* in. There would be fewer tourists, less litter and pollution,
more solitude. It will be better for the wild plants and animals, and

better for the soul of the tourist.

My anti-touristic ire was fuelled by the fact that my journey to the Arches was a pilgrimage. This national park has a special place in American desert literature. While I was being born in London's East End, Edward Abbey was a young ranger living in a caravan in this park, an experience which resulted in the classic of modern American writing on deserts, *Desert Solitaire*. Abbey described Arches as the most beautiful place on earth, though he conceded that

> *There are many such places. Every man, every woman carries in heart and mind the image of the ideal place, the right place . . .*
>
> *For myself I'll take . . . the canyonlands. The slickrock desert. The red dust and the burnt cliffs and the lonely sky − all that which lies beyond the end of roads.*

Desert Solitaire has passages of desert-drunk serenity, and passages of outrage. A novelist and social critic as well as a naturalist, Abbey raged against the exploitation of the wilderness by the government, ranching interests, the mining industry and the tourist industry. The Park Service has a policy of driving highways past any obstacle into the heart of the national parks. Abbey writes of an isolated beauty spot that is under the protection of the Park Service, but

> *who can protect it against the Park Service? Powerlines now bisect the scene; a 100-foot water tower looms against the red cliffs; tract-style houses are built to house the 'protectors' . . . an artificial steel-and-asphalt 'campground' . . . historic buildings razed by bulldozers to save the expense of maintaining them . . . and hundreds of thousands of dollars spent on an unneeded paved entrance road. And the administrators complain of vandalism.*

The Park Service was established by Congress in 1916 to administer the parks, and 'provide for the enjoyment of same in such manner and by such means as will leave them unimpaired for the enjoyment of future generations'. Abbey observes that the service is split into two factions, the developers, who stress the words *provide for the enjoyment*, and the Preservers, who stress *leave them unimpaired*. The developers are in the majority.

In Arizona I had driven along a truly outstanding piece of vandalism, a new highway cut through the middle of a vast virgin forest to save motorists a few miles of extra travel. America's huge highway programme started with the 'transcontinental highways' constructed after World War II as part of a defence and mobilization strategy, and now has a runaway logic of its own (mirrored by EC policy in Europe). Roads *must* go everywhere. Abbey observed how frequently local commercial interests influenced the cutting of roads through publicly owned, 'protected' wilderness. He was outraged too by the ways in which industry was given permission to dam rivers and strip ore out of these same areas. His 1975 novel, *The Monkey Wrench Gang*, is dedicated to Ned Ludd, the eighteenth-century weaver who smashed two mechanical looms, and inspired the Luddite weavers who tried to destroy the machinery that was replacing them. Abbey coined the term 'monkey wrenching' to describe a similar form of protest, the destruction of the machines being used to dam rivers and mine deserts. His picaresque novel presents an eclectic bunch of characters, with a variety of pyrotechnic skills, setting out to blow to bits a vast new dam complex. Abbey hoped his novel would inspire real acts of copycat destruction.

In *Desert Solitaire*, he described how industrial tourism ensures that every wilderness is opened up with paved roads, and encrusted within and without with shops, RV sites, hotels and gas stations. The rangers of the Arches park, he said, go quietly nuts answering the tourists' three basic questions five hundred times a day: (1) Where's the john? (2) How long's it take to see this place? (3) Where's the Coke machine?

At the main ranger office at the entrance to Arches, I met a sympathetic ranger who, by coincidence, was preparing a thesis on tourism impact in the national parks. I asked her whether anything had changed since Abbey's day.

'Oh no, it has all just consolidated. On holiday weekends we have nose-to-tail traffic through here.'

I wanted to find the location of Abbey's little housetrailer, the place where he stood and made his morning coffee and watched the sun rise over the rocks like hobgoblins, where he observed the jackrabbits and the deer, and trapped a gopher snake and brought it to live with him, because gopher snakes eat mice and have a reputation for keeping rattlesnakes away.

I found the spot. It is a sort of service area now, with heaps of gravel for road-laying, and a yellow mechanical digger. Industrial tourism. The summer heat was incredible. I lay down on some of Abbey's orange dust under a juniper bush, ate an orange, drank a pint of water, and slept.

That evening I went for a hike to a remote corner of the park where, the sympathetic ranger assured me, I would be at no risk of encountering a human being. It was a seven-mile hike to this particular arch, and when I got there I had an urge to throw off all my clothes and scamper naked on the rocks. You get great *grip*, bootless, waterbottle-less, unencumbered, scrambling naked on rock. It was as pink as a baby's bottom, and still very warm from the day's heat. I felt *essential*, and that is what wilderness is for – especially the desert, so raw, so elemental.

Naturally, I left it too late to leave, and soon found myself wandering in pitch darkness along unknown tracks. I made it back, though, having in my back-pack all the ancient and traditional tools of the true nomad – map, GPS satellite navigation system, compass, torch. Still, it was a tough climb up rocks and down dunes in the dark, and it gave me a bit of a shock. A pleasant shock.

Every time I hear about some schoolkids getting lost on some outward-bound trek, I think, 'Good: they're meeting nature. They're getting its measure, and their own.'

Deserts are good for that.

CHAPTER FORTY-SIX

Indian Country

'INDIAN COUNTRY' was the name on my Rand McNally map, a patchwork of Indian reservations covering a chunk of northern Arizona and scraps of Utah, Colorado and New Mexico. The largest reservation was the 'Navajo Nation', with the island of the Hopi reservation within it.

Indian Country is a patch of desert no whites had wanted to take away from the Indians. Never mind that much of it had been better land 100 years ago, that it had been degraded by over-grazing, and that rainfall had fallen by perhaps 40 per cent in the last 50 years; the original occupants of North America, the people dubbed 'Indians' by sailors who thought they were Hindoos, were squeezed into reservations like these, several hundred square miles of harsh, dun land, while the surrounding billions of moist, lush, verdant square miles were farmed by the white man (when he was not covering it with plywood houses). The Navajo land is not fertile enough to support the 170,000 or so Navajo who live there; thousands are forced to earn a living as transient workers.

Southbound on US Highway 666 I stopped for a young Indian hitch-hiker. He looked a little wary as he climbed in – perhaps he had expected a lift in the back of a pick-up – but he soon relaxed. His name was Danny, he said; he was half Navajo.

The road was straight and empty. I put my foot down and the Chevrolet ate the fawn, white-and-yellow striped tarmac. Danny and I sat back, talking little, sun on our grinning faces and wind roaring in our ears. Car, desert, sun, speed. Barren mesas shimmered in the grey haze. Evening came on, and towering, inky, gold-haloed clouds loomed in the west.

Danny asked me to drop him in a small town near the edge of Indian Country, and I offered to run him home. Thank you, sir. His girlfriend worked in a gas station, could he look in on her? Sure, I said, I needed gas anyway.

His girlfriend was six feet tall, her face a softened version of the sculpted, hawk-nosed Indian look, with full lips and skin an extraordinary, pale copper colour. There was no one in the shop, and she was standing near the till, loading packets of cheese biscuits on a shelf. I pretended to browse while she and Danny talked. When they had finished, I asked her where I could get a meal in town.

'Well, there's a Dunkin' Donut and a McDonald's, up on the Gallup Road.'

'Is there anywhere they serve salad? Fresh vegetables?'

She looked thoughtful. 'No, there's nowhere like that in this town.'

As we drove away, I said to Danny, 'She's a very interesting-looking woman, your girlfriend.'

He smiled, and nodded. 'Uhuh. She's not pure-blood Navajo. Her daddy was army, big white guy.'

'I've never met anyone before who looks anything like her.'

He laughed shyly. 'Yeah. No Indians in England, huh?'

I dropped him at a gate on the outskirts of town, and he walked across sunburnt grass towards a mobile home.

I continued south. Signs said

Bingo
We Buy Rugs
Fat Sheep $85
Rosebud Optical is an American-Indian-Owned
and Operated Business.

Acoma Pueblo is one of the oldest settlements in North America, a mesa-top village that has been continuously inhabited for over 800 years. The Acomans have taken advantage of the opportunity nature offered them to live isolated and protected from other people, and they still have a reputation for aloofness. I had read that they shun contact with outsiders, striving to keep their culture alive and distinct. After seeing the struggle for survival of tribal culture in Africa, I was interested in seeing how the subsidized, politically conscious American Indians preserved their identities in the commercially driven melting pot of the United States.

I stopped in a Grant's gas station, and an old Indian man heard me asking for directions to Acoma.

'Give me a ride, and I'll show you the way, sir,' he said. He had been bent on the antiseptic tiled floor, fumbling with beer, cigarettes and crackers in a large carrier bag, to the amusement of the two female cashiers. He had thick grey hair, a round face and a noble hooked nose that had sometime been bent sideways. He was wearing an old white Stetson and a faded denim cowboy shirt. The beer was at last hidden underneath the other items – on purpose, I presumed, since alcohol was banned on many reservations.

'I'll wait for you in the car,' I said, catching the smiling eyes of the cashiers. As I sat outside, the old man bumbled comically across the forecourt, with bow legs and a rolling gait. 'I'll be right with you sir, don't go away!'

'Which way is it?' I asked, as he settled in beside me.

'That way,' he said, vaguely waving at the windscreen. As a result, I nearly drove down the wrong slip road.

'No, the *next* turning,' he said irritably, turning to look at me. 'Doncha know the way tah Acoma?'

'Never been there in my life.'

'Well I'll be . . .' He shook his head in puzzlement.

I drove towards the hills in the south.

'Lloyd's the name, sir,' he said.

'Martin,' I said.

'Uh-huh.' I drove on. He went on conversationally, 'I just said to my brother, "I'll go get some crackers." So got me a ride over here. Lucky to meet you, now ah got me a ride back, hee-hee!'

The road ran alongside the freeway, where vehicles sped east and west, across New Mexico and across America. 'See that hill, the last one? I got land yonder, cattle – 'bout a hundred fifty.'

'That's a good-sized herd, isn't it?'

He turned and scrutinized me. 'Say, you got a strange way o' talkin', where you from?'

'England.'

'They speak English there?'

'Yes, they do.'

'Well, you got some kinda weird dialect, ain'tcha?'

Lloyd had been born right here, he told me, and here he had lived all his life. But he travelled, all right – he had served in Vietnam.

'None of that shootin' stuff,' he added, gesturing with his hands as though he were holding a rifle at fixed bayonets. 'I worked for them – whatever they call 'em – gold brass, ain't it?' His hand sketched an officer's epaulettes. 'I was in come-ooni-cations. Pluggin' in phones and stuff.' He simulated the switchboard-plugging of an old-time telephonist. His hands began to tangle spaghetti, and he giggled. 'Ah didn't have nothin' to do with no war though, no sir, hee-hee. All ah did wus eat – an' drink that whuskey, hee-hee-hee!'

I dropped Lloyd under a rocky bluff just outside New Acoma, and drove on towards the original, hill-top pueblo.

The road climbed across greenish semi-desert, then wriggled through some hills which suddenly opened onto a vista of Acoma Pueblo. It was a thick knuckle of brown rock, isolated on a broad plain. The village sprawls organically over the flat top of this mesa, 350 feet up. A natural fortress, it was described in 1540 by a Spanish soldier, one of the first Europeans to see it, as 'the greatest stronghold ever seen in the world'. (The Spanish admired good military efforts – it intensified the pleasures of destruction.) As befits an age of pithy slogans and mass tourism, Acoma has adopted the name 'Sky City'.

The road dropped out of the hills, and an elevated causeway led across the flat valley floor, as though the mesa were indeed an island. Access is by guided tour only. Just before the mesa is a car park with a museum-cafeteria-bookshop, where you buy a ticket for the minibus that carries you to the top of the mesa. The attitude of the Acomans towards the tourists seemed evasive to the point of hostility. Was this the aloofness I had read about?

Our guide was an obese young woman with a curious, unpunctuated way of speaking. She warned us that photography was forbidden; then pointed to my notebook, and brusquely added that I had no right to take notes.

'You're *kidding*,' I said.

'I'm not kidding, sir. That's the rule here. I didn't make it and you will have to respect it sir.' Reluctantly I put my notepad away. 'Also we ask that you do not talk to any of the people here. You will meet potters and you may buy their potteries if you wish but we ask that you respect the privacy of the citizens here and not approach them.'

The minibus carried us up the cement ramp that has come with the advent of tourism – access used to be by the narrow, easily defended path that still snakes around the sides of the rocky bluff. The guide led us to the San Esteban del Rey mission, a sturdy adobe church first completed in 1640 (thereafter the ungrateful Acomans burned it down from time to time).

'This church was built by the Acoma people they brought the stones up from the valley floor many tons nobody knows how many,' she said, in her unpunctuated drone. 'The rafters you see above were brought from Mount Taylor it is over there in the north they were transported without being allowed to touch the ground. Here you see the graveyard. Please do not approach it and remember no photographs. Follow me.'

The hilltop village was larger than I had expected, with two- and three-storey buildings of stone and adobe. Here and there were open-air cisterns to which water was carried from the plain below.

'Why are the doorways so small?' asked an ageing female tourist, grotesque in a turquoise nylon leisure suit and jangling bracelets.

I knew why; I might have been in the Sahara or the Chilean Atacama, among the peculiar and wonderful architectural solutions human beings devise for living with the desert.

'People used to be smaller because they carried a lot,' our guide stated, 'and it stunted them.' I stared at her in disbelief. 'But they also used to live longer,' she went on. 'My grandfather said the Acoma people will live less long now because of the artificial products in the food we eat because it is not our traditional food.' The personal reference made her blush, and she gave a funny little shame-faced grin, conscious perhaps that she was the physical embodiment of her grandfather's prediction. She quickly rattled on, 'On the right here you see the potters. You can buy potteries from them if you want.'

They stood behind rows of trestle tables loaded with Acoman earthenware, white clay decorated with fine brush-strokes. I looked at some of it, but found the potters unfriendly. In the shade to one side stood a very old woman, with some pots spread out on a low wall. She had an open, unguarded, toothless grin, the sort of grin you see in poor countries, the secret of which has been lost in the West.

GRAINS OF SAND

Her pots were of appalling, almost cretinous, crudity. 'How much?' I said.

'A hundred fifty dollars,' she replied.

We moved on. I asked the guide a few questions, and she tried to avoid answering them. Yet something about this edgy, fat woman suggested to me that she would have liked to have been friendlier. Was it that she was not *meant* to, or that she did not know how to be?

Returning to Grant's, I stopped at the Acoman gambling den, a metal shed erected just off the east–west freeway. If the casino was externally an eyesore, its interior was even worse, like the inside of a black dustbin liner, and crammed with garish chromium slot machines. In recent years, Indians have discovered that their limited rights over their land exempt them from state laws on gambling. Casinos are springing up wherever major highways cross reservations, the indigenous people's contribution to the contemporary landscape.

I checked into a motel, 'The Cactus' or 'Sands' or 'Oasis' or 'Desert'. I was never depressed when I was alone in the desert – *in* the desert, that is, under the stars – but I was often depressed on nights like this, when I slept in some low-price dump. (I tried to go for the *slightly* better-class joints, the ones where they welcome travelling salesmen with free synthetic morning coffee and a defrosted patisserie, rather than the truly desperate joints, the ones with original 1950s plumbing and holes kicked in the cardboard bathroom doors.) On nights like this, I felt lonely and empty. I imagined myself as an old man dying in one of these places. Climbing into my car and driving out to Motel Desperation, lying on the bobbled nylon bedspread with a booze bottle and the TV turned up loud to shut out the sound of the AC (too cool with it on, too hot with it off). Suicide or natural causes? Aerial shot: dawn; the bedroom door; pull back and tilt up to those purple desert mountains.

The next morning, I had an appointment at the tribal headquarters with Petuuche Gilbert, an outspoken advocate of Indian rights.

The government complex where he worked housed the police station and court-house, and as I walked in an Indian wearing bright orange overalls came out. I assumed he was a maintenance man, but when he raised his hands to greet me I saw handcuffs.

'How are you, sir?'

'Fine,' I smiled.

'You have a very good day, now.'

There were two other prisoners behind him – chained to him – and two cops. The convicts docilely climbed into the back seat of a police car, as though they were all going out for a nice drive.

Gilbert was a compact man in his late forties. He looked up politely when I entered his office, but when he realized who I was, his smile vanished. He pointed at a seat.

'I don't have much time,' he said. 'What is it you want?'

I said I was hoping to learn something about the ways in which the Acomans were trying to preserve their culture. I would like to spend some time at the mesa, and to get to know some Acomans, old and young.

'We can't allow that. Even if we considered it, it would take weeks to arrange, consulting the tribal council and so on.'

'I see. So I absolutely can't visit Acoma.'

'No.'

'Would you be prepared to talk to me for a few minutes?'

'Well, a *few* minutes, yes.'

'Thank you,' I said coldly.

'Look, er, Martin, we sometimes help people, but we have to be very sure of who they are and what they want from us. We were too welcoming in the past. Anthropologists and ethnographers came and lived amongst us, and stole our secrets. Today we have nothing, we're prisoners, prisoners in occupied America. We *refuse* to be exploited any more.'

He paused. I said nothing. He gave a slight sigh, and began the lecture.

'The first outsiders to come here weren't the most recent wave of Europeans, you know. Our stories talk of others, who came from the west. There's a peaceful climate here because of the ancient people's prayers, which they still do today. The Navajos, the Apaches, the Yute all learned a lot from us in relationship to the land. This is what people come to see. People like you.

'We say man's history goes back aeons, right *here*. Why did we choose this place? It was originally called Acco, which meant "the place that always was, the place prepared for my people". Our ancestors migrated here from the north – near Chaco Canyon and Mesa Verde, north of here, in what the Americans call Colorado. We emerged from the *underworld*, not *Asia*, as your anthropologists tell you.'

I nodded. I had not come to argue. He leaned towards me and looked me in the eye. 'Early this morning people got up and went to our sacred sites to pray for all of us, for you too. For everything, the world, the universe. That's what they're responsible for – the maintenance of everything. This is romanticized by white people, of course, our relationship with mother earth, living in harmony with nature. But I believe we had a spiritual relationship with the earth, knowing that everything is a gift from the Creator.'

He sat back again. 'I'm supposed to do it, too. At sunrise, to go out and say prayers. My father did. You welcome God's gifts, of rain for wildlife, grass, people, the *land* . . . But I don't pray any more.'

After his initial brusqueness, I was surprised by the intimacy of this revelation. He stared at the wall for a moment. I thought it was a meditation; but then I realized that he was looking at a clock.

'At one time, Martin, we indigenous people were one with the white race, but we decided to split away. We chose the oral way and the spiritual relationship with the earth, the responsibility of maintenance and the power of prayer; the white race assumed the power of the written. A lot of white people think indigenous people are ignorant, but it's not true. The point is, the white race took on book power. It was prophesied that they would control nature itself, and the lives of all the people. Well, this has come true. But your dominion is based on an intrusion into our land, the imposition of your ways, genocide. Hah – while keeping what *you* call the rule of law, of course.'

There was something about the way he emphasized 'you' that made me ask, 'You mean *me*, personally?'

'Yes. *All* of you. You are all complicit.'

I looked at him sourly, and said, 'A lot of people would agree with much that you have been saying. But do you think that, for example, building a casino has enhanced your moral stature?'

'So we are meant to be better than the white man, and this moral superiority will sustain us in our poverty? The casino helps us survive economically. There are things we need; how else will we get them? Gaming is a necessary evil.'

'You're saying that it's OK to embrace something you claim to despise, in order to get money?'

'We play the white man's game! How are we expected to survive in the modern world in these – *reservations*? What choices do we have? Whether I can be an Indian or not is decided by the state! North Americans have to be *citizens* before any other identity, even if their ethnic identity means a lot more to them than being a part of this Spanish-Anglo-Saxon invasion! The whites came here and colonized this land and made us their subjects – this is cultural and economic imperialism. I hope it's appreciated that we're not angry for nothing. We are oppressed peoples! We *denounce* the US for refusing to recognize our rights to self-determination. And *why* do they make us swear allegiance? Because they're afraid we'll secede! They have to keep us a subject race, to keep their Union – their money-making machine – in one piece. Great power dominates and protects the Union; it is an immense beast. Fighting it is all but impossible. That's why I personally try to be a thorn in American society, an iconoclast . . .'

He suddenly grinned, and added with heavy irony, 'Mind you, we are good citizens – we raised the flag this morning.'

'What about religion?' I asked quietly.

'Ah, good question! We're good Catholics – I mean, we believe *whole-heartedly* in the Catholic way, *faithfully*. It's always amazed me, because we have our own traditional practices. But now the two are melded into one. Ohhh . . . .'

Petuucche sat back in his chair. 'It's an extreme challenge for us to live today. Yet I do talk about peaceful co-existence between the invaders and us. I just hope we don't see violent revolutions. But you can't go pushing people off their land and expect them to pledge allegiance to the flag.

'As a young man I served in the armed forces of this country twice, you know. I didn't reflect on what it meant. Most Indians don't reflect on our history. In the sixties a lot of Native Americans were leaving the reservations and joining the military, to gain some knowledge of the world, to gain working skills. We served willingly. That's how it happens, you know, it's done without thinking – you fight for your oppressors! Our elders insisted we sought the white man's knowledge, without realizing what it would do to our language and rituals. By surrendering our isolation, we risk losing our identity.

Our rituals need the power of language, because our way is the oral way. Now we are losing even the language.

'Slowly we are being corrupted. Your way is commercialized, capitalistic, individualistic. Of course, we all want hot water, TV and air-conditioning. It's . . . as I say, it's a challenge for native peoples.'

'And how can you rise to that challenge?'

'We try to instil values. By ceremonies that communicate a sense of continuity, social and religious gatherings. But there aren't enough of them. One is immersed in English, on TV, in newspapers . . .'

There was a pause. I said, 'You're not making the prognosis sound very good.'

'Well, I'll tell you something. We Acoma people think we'll still be as powerful as Acomans in a thousand years as we are today. And . . . And that's all I have time for.'

CHAPTER FORTY-SEVEN

Highs and Lows

Tune in for 21 minutes four times a day, for wealth,
health and happiness.
Discover the latest skills to improve your performance
in all areas of life.

PERSONAL ACHIEVEMENT RADIO 610 was an exclusive diet of inspirational talks, alternating between business speakers and New Agers. One minute you were uplifted by New Age wisdom from hypnotherapists and Ayurvedic gurus; the next, winning your customers' confidence with a heart-felt enquiry about their family ('Put in your notebook their birthday, their wife or husband's birthday, their children's birthdays, too. They're human beings, they have their ambitions and fears just like you. Show them that you care. A customer who is your friend is more likely to want to give you his or her business').

In towns you could tune into fragments of public radio like *All Things Considered*, the patrician if low-budget voice of Washington, re-broadcast on local stations all over the US, gamely sponsored by local bookstores and European delis. That cultivated voice stuttered into silence on the edge of towns, and driving across the vast open spaces made you grateful for the lean pickings of Personal Achievement Radio. It was wisdom of a sort, better than the unironic navel-fluff fundamentalism of the Christian stations and the moronic demagogues whose spittle flecked the mikes of the talk-radio stations; and *much* better than Country.

I had to force myself to listen to Country music, and discovered that it held a terrible secret about the white agricultural communities:

that they were shallow and introverted, weak and afraid. The sentimental, grotesquely inflated Country language of the songs was delivered high in the mouth, in a metallic, nasal whine:

> *I saw an angel by your pillow*
> *I knew it from the first time that we met*
> *Give me love I'm able to trust in*
> *I'll be watching you forever, love, whatever*
> *My love will fly to you each night on angels' wings*
> *I cain't imagine any greater fear than waking up without you*
>   *near*
> *If I had to run, if I had to crawl . . .*
> *Ain't gonna play the cryin' game no more*
> *I been down that road before*
> *Stand by me like I stand by you*

The settlers had fled Europe, and wiped out the redskins, to give the world *this*?

I turned north-west, towards Hopi country.

The remote villages of the Hopi tribes cling, like Acoma, to mesa-tops, a series of spare, elongated ridges.

'We can't build a casino,' one man told me disconsolately, 'because we ain't got no highway runnin' through here. Who's gonna come here tah gamble? We Hopi bin losin' out all the way along, and we're losin' out again now.'

By way of consolation, the Hopi have their religion. Its fervent and secretive practice has been seen by some as a sort of inverted snobbery, a way of turning their back on a world that has outclassed them. Who would blame them? The Hopi are impoverished, and, like the Acomans, losing their cultural identity. At the same time, they are besieged by mystic New Agers who expect to find eternal verities among the mesas, with their TV aerials and rib-cage dogs and jacked-up cars. But the Hopi's traditionalism, poverty and isolation give them a uniqueness, and may in time come to be seen as the things that saved them.

As I sat in the afternoon light on the top of one of the mesas, an old man came out of a small, square, mud-brick house. We said nothing, but sat together for a while, staring past a battered old ice-blue flat-bed Ford at the grey-brown semi-desert.

'No rain anymore,' he said. 'I bin down to water my maize today.'

He pointed at a path just under our feet that led down past a collection of variously wooden kiosks – the houses' privies, perilously perched on the mesa's rim – to the plain.

'I go down alone now, 'cause my wife is too ill to move. Well, once upon a time it was green down there, y' know, this time o' year.' His wrinkled, slightly trembling hand indicated the entire, bleached vista. 'I don't know where the rain went. I think we did something wrong. I don't know what we Hopis did to make the gods turn their back on us. I think maybe we stopped respecting them. That's what I think happened.'

I turned north again, towards 'Ford Country', Monument Valley, the red buttes that loom a thousand feet off the desert floor, and that have become, thanks to John Ford and Hollywood, the most distinctive image of the American deserts.

I was growing tired of driving, and I knew that when I reached Monument Valley I would simply turn round and retrace my steps; but I had to see those rocks. Many times, driving through the American Southwest, I wondered why Monument Valley had become the quintessential symbol of the American desert, joining the most powerful icons of our time – the Statue of Liberty and the Taj Mahal, the Pyramids and Ayer's Rock. Countless film and video shoots have come here; and the British 'conceptual artist' Tracey Emin added value to her solipsistic oeuvre by getting herself photographed in front of one of the Valley's towering buttes. One mountain peak is much like any other, but these will always say 'Monument Valley, Arizona, USA'. To an earlier generation they conveyed the spaciousness of the West, but to a younger generation they convey the spaciousness of the Self.

Photographs usually show a reddish, flat-topped monolith sandwiched between two blocks of colour – blue and yellow. The blue is the same as the blue of the sky over a beach, the earth is sand; they suggest holiday optimism and freedom. The phallic mesa stands in an open landscape, conveying space, potential. It looks almost deliberately posed, like a sculpture (not a mountain but a Monument);

yet it is really more like a pedestal than a statue – an enormous base, inviting us to erect statues of our own gods.

This is a hostile environment, and yet we find it agreeable. Few people in the West have cause to fear the natural world, a fact that frees them to anthropomorphize Nature, to project onto it their occasional desires to escape from industrial life, to feel a little mystery and awe.

I could sing the praises of Tibesti from the rooftops, but the world's most popular tourist deserts will still be in North America. And I do not suggest that they are anything other than glorious – the sudden mountains stained red at dusk, cacti in gay spring bloom, fuming volcanic rock-pools, crystalline salt lakes, and, above all, the show-stopping number of the Grand Canyon. I will not dwell on this mammoth groove. There it is, pink and gaping, magnificently purposeless, apparently designed for the sole task of engendering awe in man.

Awe has its price. As you descend, signs remind you that you will have to climb back up again. Not everyone takes notice. You see track-suited, overweight climbers, sitting red-faced on rocks, gasping for breath. Beside them are other, inspirational signs: 'Never stop believing you can make it out.' Eventually, teams with mules come down to carry them – expensively – up.

Others pay a higher price. One of the most darkly comic things I read at the Grand Canyon was a report about people who die while having their picture taken. They stand at the canyon's edge and Pop, framing them in his viewfinder, calls out 'Just step back a bit.' And they do.

I drove through the night to the oasis of Las Vegas, wrapped up warm in the open-topped convertible, with the stereo turned up loud to keep me awake. A full moon started on the horizon as the gold-domed church of a mad desert cult, then rose and paled until it shone in the starry heights like a silver dollar, leading unwise men to Las Vegas. It was redundant, for Vegas announced itself with laser beams pencilling high into the sky, a beacon for passing UFOs.

The city materialised as blindingly bright buildings of infantile monumentality – pyramids and sphinxes, tinsel-clad industrial sheds for gambling and striptease, titanic yet fully booked hotels. In Vegas even launderettes and supermarkets have rows of one-arm bandits, where retired Americans, dressed in the gaudy checks of circus clowns and crinkled by the eternal sun, pump the machines' chromium arms like mechanical bar-maids. Las Vegas is the fastest-growing city in the US, new condos proliferating in the flat, fawn hinterland. The incomers are putting a strain on the water resources of the city, whose water larceny already empties several far-off rivers, with the result that the deserts of the Southwest are becoming ever more arid.

CHAPTER FORTY-EIGHT

Death Valley

*New Yorkers like to boast that if you can survive in
New York, you can survive anywhere. But if you can
survive anywhere, why live in New York?*

Edward Abbey

THE BREEZY T-SHIRTS on sale in the souvenir shop proclaim 'I
visited Death Valley and survived!' Just by driving through, they
suggest, you can acquire some vicarious glamour. But Death Valley –
fifteen hundred square miles of valley and mountain, consistently the
hottest place on earth – is not to be trifled with.

I was barely awake when the ranger's huge Hummer, a sort of
sideways-stretched jeep, grumbled up the track. I had pitched my tent
by an abandoned mine, well out in Death Valley's rocky wilderness,
and slept unprotected under the stars. Now I was basking naked in
early morning sunlight, reading a week-old newspaper and waiting
for the kettle to boil. Savouring the solitude. It was a perfect moment,
and perfect moments are, by definition, brief.

The ranger was wearing mirror sunglasses and a gun. He had red
hair.

'What the hell are you doing here?' he barked.

'W-well—' I stuttered. I was hastily pulling on some shorts, but my
nakedness was not, it turned out, what he had on his mind.

'Do you know how *stupid* you're being? You know, it's my job to
clean up when people like you get yourselves killed, and believe me,
it's not very nice. A few weeks ago I found a guy that had died of
dehydration. You know what somebody looks like when he dies in
the desert?'

'Well, I—'

'I'll tell you. His skin was burnt black and he was puffed up with gases like a beach ball. Lying on his back with his feet and legs in the air. You think I *enjoy* finding a sight like that?'

'Look – sir, I'm not some loony day-tripper. I'm writing a book about deserts, I've been halfway around the world in deserts. I've got food and water here for a week. And this is my hiking licence.'

'Oh.' He was taken aback. 'Well,' he said sulkily, 'no one told *me* you were here. It's my job to rescue people. How can I do my job if nobody tells me what's going on?'

I began to feel sympathetic, wishing I *had* been dying of thirst, so that he could have heroically rescued me.

Suddenly he grinned. 'Look, I'm sorry if I was a bit cranky back there, but we get a lot of nutcases out here. I have to be careful. A book about deserts, huh? You must really love 'em.'

'I do!'

'I do too! You making coffee?'

And having discovered a weirdness in common, we crouched by the bubbling saucepan and began to talk.

'When I was twenty,' he said, 'I was going to hike across the southern Sahara. I had it all planned, maps and everything. But it was too politically unstable. Have you ever been there?'

'I spent six months last year in the southern Sahara.'

'Did you go to Chad?'

The question surprised me, as many people have not even heard of Chad.

'Right up into the north,' I said, 'into the Tibesti Mountains. The heart of the Sahara.'

He looked wistful. 'Boy, I really wanted to go there. Still, you get older, your priorities change. Marriage, kids . . . I guess I won't ever hike across the Sahara now.' He stared out across many square miles of scrub, and perked up. 'Still, I do live in a desert! What do you think of the Valley?'

'I love it. From what you were saying, it lives up to its name?'

'Oh, you bet. Sometimes I just find a pair of jeans and a spinal cord – you know, after the coyotes get there. We have a lot of suicides. And homicides – it's a good place to dump a body, and we're only an hour's drive from Vegas.

'A few years back we lost a whole family. Germans – a mother and father and two young kids. Broke down in a remote corner and decided to walk out. We never even found their bones.'

'They should have stayed with the car?'

'Absolutely. Eventually we would've found them. I always tell people, if you break down in Death Valley, remember that line from *Apocalypse Now*: "Never get out of the boat!"'

Water is life. If the deserts had taught me nothing else, I would at least have learned to value, to savour, to venerate water. The previous day the mercury had stopped rising at 121°F. I stowed a light sleeping bag, sardines, oranges – and water. I was not going to repeat the mistake I had made in Pinacate, but would keep rigorously out of the sun during the midday hours. I was carrying twenty pints for a three-day hike and it weighed a ton. In addition, just in case something had happened to the car when I returned, I buried twenty pints in the earth. Desert hikes should, ideally, centre on a spring or stream. But in Death Valley water is the stuff of dreams.

There have been other dreams here too. Walking into a narrow canyon I passed a tarred shed where, almost a century ago, miners slept, dreaming of gold. They insulated their shack by gluing pages from mining journals to the walls, and fragments remain, febrile reports of new finds, claims staked, fortunes made:

> *Altogether the state has given to the world's wealth nearly a thousand million dollars. The sum almost staggers one at its mention . . . Between there and his destination was a desert that often meant death from either one of three sources – thirst, starvation or Indians. The first named was probably the most dreaded . . . Many subsidiary veins . . . stock has been offered for sale . . . Wall Street Mining and . . . Fair and Square . . . Application . . . cyanide expenses reached $345,349 . . . dry farming . . . has no water . . . the desert . . .*

No one got rich here. They burrowed some holes into the hillsides and departed after a few years, leaving behind them a litter of twisted hawsers and iron junk.

Rusty canyon walls rose around me, and pale sandstone bluffs like fortress walls. Sometimes there was a vein of tough black rock, and the water-hollowed gully grew narrow and sensuously curved. How many millennia, I wondered, had it taken for Death Valley's frugal rainfall to carve this groove in the earth?

I was hot. It was 11 a.m., time to find a rocky overhang under which to pass the next few hours.

I woke at three, feeling dopey. I drank a little water and squinted around me at the fawn rock, the white light. What was I doing here? Ah, yes. I carried on down the gully. The silence was absolute, but the gully walls threw the crunch of my boots back at me in High Fidelity, and my breathing sounded like a spouting whale. After a while, the path took an abrupt turn south. It was a place one of the Rangers had told me about, a 'secret' place, he said. For a mile here, the gully drops straight down, like a tunnel into the heart of a pyramid. 'Watch out for flash floods,' the ranger had said. 'We sometimes get sudden storms in August. If you were caught in that tunnel, you wouldn't stand a chance.'

There had been a summer storm the previous day, and a little of Death Valley's annual 1.69 inches of rain had fallen. Now I stumbled – to my amazement – on several deep hollows where water had been trapped. It was the desert traveller's dream. Suddenly, I didn't have to worry about water – I could drink my fill. And here, at the end of a long day's hike into one of the world's harshest terrains, was a warm bath, drawn and awaiting me.

CHAPTER FORTY-NINE

Red Centre

THE EARTHQUAKE HAPPENED at around six a.m.

I had watched the sun's rapid rise through ribbons of red and purple cloud, illuminating a flat landscape of grass and scrub, and I was just savouring the steam rising from an enamel mug of coffee – when the earth began to move.

I experienced an earth tremor as a child. I well remember the feeling of imbalance – the rolling pavement, the swaying telephone cables, a flowerpot falling from the balcony of an old wooden building and exploding with a shocking crack.

This morning there was first a faint, uncertain movement. I had the thought, 'I didn't know that central Australia was an earthquake zone.' Then there was another, violent shudder. Still doubting my senses, I looked through the windscreen, and saw the eastern horizon, sure enough, yawing wildly.

But then I sensed – unmistakably – that it was the jeep, not the land, that was shaking. Someone was – no, it must be two or three people – rocking it for all they were worth. I shot a panicky look through the windows. Nothing.

Uncertain and fearful, I edged towards the rear doors, and threw them open – to look down on the biggest boar I have ever seen in my life, a massive, betusked, nut-brown beast, scratching his great muscular behind on my rear bumper.

He twisted round his rosy, rubbery snout and stared at me with small, interrogative eyes. I stared back, with a mixture of fear and uncertainty. Then he lunged.

I started, banged my head on the roof, and tumbled backwards through the rear doors. At my feet, the boar was snuffling noisily. A terrified memory shot through me of a Peruvian short story I had read recently, where a man was eaten by his pig.

I heard the voice of Bill Hughes. 'Don't worry, Martin, he won't hurt you.' I sat up. Bill was standing on the porch of his tin-roofed house, drenched in dawn light, dressed in shorts and a singlet and holding a mug of tea. 'He's just enthusiastic!'

'Red Centre' is a modern marketing term, but it perfectly expresses the essence of Australia's interior, being close to the geographical centre of Terra Australis, the Great Southern Land – and being red. From 21,000 feet, the Simpson Desert was an endless terracotta plain, draped with long-veined drainage systems, like seaweed leaning in a current. The land only appeared to be flat – the striations beneath me were indications of hills, the bones of the land, row upon row of charred ribs. When the terrain did flatten, the watercourses meandered amply, their brinks defined by black shadows and their bodies dotted with black shrubs, so that they resembled bacteria magnified a billion times. Here and there, runnels radiated from a dried water-hole like veins from a swollen cornea.

Ranks of dirty fat silkworms appeared under the plane's wings, eerily lit from within. Slowly they decayed into taller, irregular tufts, each throwing an inky blob on the land: an armada in the sky, and a blight below. Straight tracks scored into the orange earth joined these black circles, resulting in a pattern of unknown meaning, a giant's board game.

I was to collect my jeep in Darwin, Australia's northernmost city. A group of watersports enthusiasts was about to swim fifty miles across the Timor Sea to Bathurst Island, inside a metal cage towed behind a boat, to protect them from sharks and salt water crocodiles. I walked down to the docks, and gazed at the sea. I had flown over eight thousand miles of ocean, from Los Angeles to Sydney. A converted railway carriage offered Fish and Chips, and I looked at the blackboard with interest: Barramundi, Blue Salmon, Jew Fish. A fat man with a red moustache was shovelling sliced potatoes into the chip frier.

'What do they taste like?' I asked him.

'What?'

'These fish – I've never tried them before. What do they taste like?'

'They all taste the bloody same – look, can't you see I'm busy, mate?'

Darwin had been rebuilt since Christmas Eve 1974, when Cyclone Tracy flattened it. To lose 80 per cent of a town to a cyclone is a misfortune, but to lose its one remaining architectural gem to 'structural unsoundness' smacks of carelessness. 'The rambling, colonial Darwin Hotel,' my guidebook maintained, 'is still going strong, its Pickled Parrot piano bar a good place to enjoy a cocktail under tropical ceiling fans.' I asked the lady in the tourist office where the Darwin Hotel was. She pointed out of the window to the adjacent plot, where a bulldozer was levelling off a quarter-acre of cinnamon-coloured soil.

'That was it,' she said, 'it was only condemned a fortnight ago, and they had it down within a week. Everyone opposed the demolition. But it was unrepairable, they say.'

'Worth a lot of money, I bet, that site?'

'Oh yes – city centre, seafront – one of the best sites in town.'

A Town like Katherine: 200 miles south of Darwin, a dusty grid of low-built, sun-baked buildings. Aboriginals sat in quiet knots under the trees in the public gardens, or wandered along the streets, matted-haired and dazed, wearing dirty nylon singlets. Open boozing among Aboriginals was causing such severe law and order problems that it was planned to create a secluded city centre area for them – a 'drinking garden'. Others claimed the way to solve the problem was to have a reduction in licensing hours – a move vigorously opposed by the town's liquor traders.

In the supermarket, adolescent whites stood behind the counters, and blushed if you addressed them directly. I was struck by the luminous whiteness of Katherine's whites. Australians were taking the skin cancer scare terribly seriously – all the way south to Uluru, I was amazed by the numbers of Australians I saw whose flesh obviously never encountered the sun without a barrier of cloth or Factor Thirty.

At the Katherine service station, the driver of a dust-caked Landcruiser was the living embodiment of Australia's macho bush culture, with a pair of the shortest possible shorts, a broad-brimmed, beat-up Akubra – and a helicopter on a trailer. His was a dying breed. I had read that the days of gigantic cattle stations that used single-seat helicopters to muster their herds had passed, now that the glossy pastures of compact New South Wales produced more beef than all the semi-desert vastnesses of the north. Australia remains the world's largest beef exporter, but US and Japanese interests have bought into the industry on a massive scale. Many small cattle stations have vanished. Some remaining owners are clinging to the belief that tourism and 'e-commerce' will come to their rescue.

The Victoria Highway runs south-west anticlockwise from twelve to eleven o'clock. It passes cattle stations' straight, determined wire fences, then imitates the broad sabre-curves of the Victoria River.

The FM stations had died out hours earlier. All I could pick up was

The only station that can add colour to your world – Triple A.

American country music had taken root in Australia, it seemed, indicative perhaps of a similar introversion to that of America's agricultural whites. I stayed tuned in long enough to hear a singer with a strong but probably fake American accent yodel,

> *I'm glad to be in this land of the free*
> *And I love my home Down Under.*
> *I guess you could say without a doubt*
> *That I'm a dinkum aussie all over.*

'Danger. Aerial shooting of feral animals in progress. Do not leave the highway.'

What, I wondered, was a feral animal. I imagined tigers, ferocious dingos . . .

'No,' the ranger told me. 'It means stock animals that have gone wild – cattle, donkeys, camel . . .' They shoot them from the air. 'We killed eight thousand last year, but only a thousand this year. The land is recovering amazingly fast.'

Carved out of unproductive farmland, the Gregory National Park is one of Australia's remoter and most hands-off national parks, where tourists are few, the roads through must be negotiated in a four-by-four, and the visitors must fend for themselves.

'Do you keep a register of who comes in and out?' I asked.

'Not any more – we found people didn't let us know when they left. We had to go in and look for them – which wasted too much time.'

'So you don't monitor people in the park at all?'

'We don't have the resources. We stress "responsible self-sufficiency".'

There are as yet few tourists to Gregory's unpretty landscape; a place of stark, flat-topped red buttes and pot-bellied, lizard-skinned boab trees, it offers solitude and tranquillity.

The park seemed to exhibit a Janus-like attitude to the continent's original inhabitants. A photographic display in the portacabin visitors' centre shows photographs of Aboriginals in chains, and concedes that the numbers of them killed by white settlers are unknown. But the park is named after Augustus Charles Gregory, who led the 1855 expedition from here south as far as the Great Sandy Desert, and a newly-placed obelisk notes that his expedition 'was a great success, opening up a large area of Northern Australia for settlement'.

Today, few people would agree that the settling of Australia was an unadulterated success. But many Australians see sympathy for the Aboriginals as the province of misguided liberals. Jim was one of them.

He had been brought up in the Queensland countryside, but had recently come up north, just to take a look, he told me. I had bought a tankful of diesel, but Jim's boss did not think it worthwhile to be polite to me.

'Don't mind him,' said Jim with a grin, offering me a cigarette. 'He's a good bloke really, it's just that he doesn't like Poms and Americans. But he's all right when he gets to know you.'

The boss muttered a few more surly words, and climbed into his Landcruiser, slamming the door. Jim patted his breast pocket for his cigarette lighter. 'You having a good journey?'

I told him about Gregory National Park's wide open spaces.

'Is that one of the parks owned by the blackfellahs? Where they charge you thirty fuckin' dollars to go in?'

'No. There's a community living in there, but it's an ordinary park.'

'"Cause you know, these days the blackfellahs are claiming every bloody corner is a sacred site or something, and the government's handing it all over to them. Then they sublet bits of it back to *us*, the poor bloody taxpayers, or else they get some money from mineral rights or whatever. And they're all sitting around on their arses on welfare as it is, spending the bloody money on booze. Don't get me wrong – I grew up in the country with the blackfellah, and he's a good worker when you give him the chance. Educate him and he's no different from us. I'll tell you the three worst things the government did for the blackfellah.' He held up three fingers, and started to count them off. '*One*, gave him the right to buy booze. Two, took him off the cattle stations – it was supposed to liberate them, but it meant they got pensions for no work, and nobody who doesn't do a good day's work is ever going to have any self-respect, are they? And three, all this financial assistance. Now they're never going to have to do a day's work. There's blackfellahs haven't worked for two generations. Petrol sniffing, sexually transmitted diseases, Christ knows what. The only way to solve their problem is to *integrate* them – get them in the cities, give them jobs and self-respect. Work.'

Jim lit his second *Escort*, short, thin weeds that came in packets of thirty-five on which the enormous black-edged warning SMOKING KILLS loomed larger than the brand name. He grinned self-consciously. 'Keep trying to give 'em up. I spend forty-seven fuckin' bucks a week on these things.'

'You have to admit that the Aboriginals' situation would be a lot better if the white man hadn't taken away all their land in the first place,' I suggested.

Jim admitted nothing. 'Sooner or later, somebody was going to get here, mate,' he said. 'That's history. Who would they rather have had here – the Dutch? The French? The Germans? The place is run like a bloody holiday camp compared with what some of them would have done! I'd have more respect for the blackfellahs if they'd banded together and formed militias, and put up a decent fight. Maybe they would have chucked us back into the ocean.'

'Jim, these people were Stone Age. They were never going to be able to resist Europeans armed with guns.'

'The American Indians did.'

'Well, look what happened to them. We almost wiped them out.'

'It's history! The strongest tribe bloody wins, and the blackfellahs were fighting amongst themselves long before we got here, make no mistake! One tribe pushing another one off some piece of land. Even now, some of them are pissed off because they're stuck with a piece of useless land and the next parcels's got minerals in it, and they're saying, "That land used to belong to us before the so-and-so tribe took it off us".'

I might have retorted that the British settlers whom he admired so much interpreted the concept of *terra nullius*, uninhabited wasteland, as meaning that the aboriginals did not in fact exist, and hastened to actualize this legal condition by exterminating them; but I did not. Instead, I accepted the second coffee that Jim offered me free-of-charge, and we parted friends, having agreed on the less controversial proposition that all politicians are bastards.

CHAPTER FIFTY

Diesel Rock

I TURNED ONTO THE TANAMI TRACK with Jim's words ringing in my ears: 'It's a bastard. Corrugations all the way.' The Tanami is a thousand-odd kilometres of rough and (after rain) occasionally impassable dirt track, curving inward from the Kimberlies to Alice Springs and the Red Centre.

The rainy season was due, but had not arrived yet. Driving south-east towards Alice, I was travelling through land explored by Augustus Gregory a century and a half earlier. After a couple of hours, I saw the sign for the Billiluna Aboriginal Community, close to the southernmost point reached by the Gregory expedition – and Lake Gregory, the explorer's compensation for failing to discover that dream of Australia's early explorers, a great inland sea. (The remains of freshwater dolphins and flamingoes have been found in central Australia: the explorers arrived a few million years too late.)

At around eight-thirty p.m. I saw the lights of Billiluna. The availability of diesel on the track ahead was uncertain, and I decided to try and buy some.

Occasional light-bulbs revealed knots of people sitting outside small houses with open doors. There were dozens of dogs. I stopped and asked the way. An Aboriginal man pointed. 'Over there, see that white house?'

I could see a pinprick of light.

'You see two Toyotas – you find it dead easy mate.'

A wire-fenced compound contained several tiny cabins. A single bulb shed a thin, orange light, and the air was filled with the whine of generators, and the smell of boiled meat. Two young children and two puppies played in the dust.

I knocked on the door of the first cabin. It was opened by a chubby, smiling woman with sandy-coloured skin.

'Sorry to disturb you so late,' I said.

'No worries,' she replied, turning towards the bed that nearly filled the room. Sitting up against two upended pillows, the sheets just over his waist, was a pink-skinned man with a white beard and long hair combed back over his ears. He had a vigorous, leonine look. He was not sick, simply in bed.

'Whaddya want, mate?'

'Can I buy any diesel tonight?'

He shook his head. 'Eight o'clock in the morning.'

'Thanks,' I said.

He returned his eyes to the television.

Dawn unveiled Billiluna as a decayed assortment of prefabricated huts, almost all of them dented or torn. Litter lay everywhere – upended junk cars, disused fridges, scraps of clothing, polythene bottles. Few people had yet risen. A few Aboriginal people trudged through the dust, their dust-caked clothes hanging off them. They raised their heads and stared at me.

The Tanami Track cuts a straight orange line across a landscape that is almost wholly flat, though spotted with shrubs and dwarfed trees, and anthills resembling a million melted candles. The Tanami Desert does not contain mountains of sand like, for example, the Simpson Desert. But bleak and barren it is, and mostly useless for cattle-rearing, which may explain why a huge proportion of it is officially Aboriginal Land.

In the first decade of the 20th century, Tanami experienced a short-lived gold rush, the prospectors' haste hampered by the inaccessibility of the land. They hurried south-east from Halls Creek and north-west from Alice Springs, following Aboriginal trails between water holes, in some cases dying as they struggled to navigate their ways across the unmarked desert. More died in the mining camps, from fever and scurvy, suicide and murder. Little gold was found. In a locally printed history of the mines I read: 'The outback prospector is a natural philosopher, he regards even the eating of a single orange conveyed from far-off Alice Springs as the height of effeminacy . . .'

A second rush occurred in the 1930s, with similarly disappointing results; but the latest Tanami Mine, opened in 1987, has yielded over four tonnes of gold. Large boards warn that it is Off-Limits.

Soon after the mine, a sign pointed left to Rabbit Flat, one of the remotest roadhouses in Australia. I left the Tanami and drove a couple of miles along a track lined with discarded drinks cans. The roadhouse was a pleasant shock, an oasis of closely-planted shady trees and bougainvillaea. Sprinklers whispered 'hush-hush', and a dozen or so Aboriginal people sat on the lush grass lawn. One of the group, a pretty girl of seventeen or eighteen, greeted me in good, though slurred English. 'G'day mate, where y' from?' We exchanged a few words, and I walked on past an oil drum half-full of beer cans. A couple of dozen cans that had not made it into the bin lay on the grass.

I went through a shady room with chairs and tables to a cubicle off the back. The narrow counter was covered by a deep, prison-like grill, through which it was just possible to glimpse some tins and cigarettes. There was a bell with a piece of wire attatched, and I rang it. A woman in her fifties appeared on the other side of the steel grill, and looked vaguely in my direction.

'Hello,' I said.

'Do you want something?'

'What cold drinks do you have?'

'All of them.'

'Well, I mean—'

'Coke,' she said tersely, 'Sprite, Fanta. Ginger beer—'

'I'll have a ginger beer, please.'

She turned to the invisible refrigerator, then pushed a can through the grill. 'Two dollars fifty.' She spoke with a strong French accent.

'*Vous êtes francaise?*' I asked.

'Long time ago,' she said, and disappeared.

People around here must ask a lot of irritating questions, because the walls were covered with signs explaining in pedantic detail why the roadhouse's prices were so high – why fuel, for example, was two-thirds more than the city price. Reasonable in the middle of the desert, you might say, but at the next Aboriginal community, fuel cost 25 per cent less than at Rabbit.

What Rabbit did have, and what some of the Aboriginals outside had driven 200 kilometres to obtain, was alcohol. A foursome of men called me over, and I sat down beside them. They were a motley crowd, a fat young man in an American sports shirt, with his hair tied back in a pony-tail, a couple of older men with black matted mops of hair and ragged beards, and a white-haired fellow lying on one elbow like an Arab. It was he who asked me, 'Where you from?'

We exchanged the usual pleasantries, and the old man told me he had trained as a teacher in Adelaide. 'I'm an educated man, not like this mob,' he said. The others tittered. 'I'm like you mate, I'm as good as you.'

'I'm sure you are.' I sipped my ginger beer.

'Well, aren't you going to buy *us* a drink?'

I hesitated. 'Why?'

'*Why?* 'Cause you a rich man and us we got poor bloody nothing.'

I played for time. 'What sort of drink?'

'A beer.'

The others chimed in, 'Beer, yeah, beer.'

'At ten a.m.?'

'Why not?' The old man regarded me with unwavering, humourless eyes.

I gave in. 'OK,' I said. 'Who's going to get them?'

The young fat guy clambered to his feet. I gave him a twenty, and he trotted away, returning with four beers; he kept the change.

When the other Aboriginals saw the beer, they wandered over. 'You gonna buy me a drink?' asked the pretty girl. Her voice had taken on a wheedling note.

'Shopping's finished,' I said.

'Buy me a drink,' she insisted.

'What sort of drink?'

'*Export*,' she said with irritation, 'same as them!'

'No.'

'No, no, no, *no*,' she mimicked, flouncing away.

'Look mate, I gotta problem,' the old man said. 'I got a sick family member in Alice – real sick – and I need some money. Naturally, I wouldn't ask if we weren't friends, but you seem like a nice fellah, I feel like we're friends. Now if you could give me *some* money, I could give you a piece of paper, and when you get to Alice with that paper they'll give you a painting . . .'

'No,' I said.

He rolled onto his other elbow, turning his back on me. 'You English the worst people in the whole fuckin' world.'

No one doubts what the future is for Alice Springs. *Tourism is everyone's business!* said a sign. 'Last year, tourism was up by 16 per cent,' enthused a radio announcer, 'and optimistic predictions say it could be up next year by as much as fifty per cent.'

Surrounded by barren hills, Alice is a very Westernized sort of oasis, where the pool of green that hovers ahead of you in the desert quickly hardens into the golden arches of Macdonald's and the oversized logos of Pizza Hut and K-Mart.

By nightfall I was sitting in a 'Mediterranean' café, chatting to a man called Mario, who moves around Australia inflicting arts events on the inhabitants of small towns. A Sydneysider, he viewed Alice Springs with a caustic urban eye.

'Alice is a backwater,' he said. 'Take away the tourists and you take away everything. You're left with two shops selling aboriginal paintings and three cut-price supermarkets. It's still basically a supply post. Makes Darwin feel sophisticated, for Christ's sake.'

Mario was drunk, and almost yelling, but it did not matter. No one sitting around us was from Alice. His companions were a theatre director from Adelaide and an Irish doctor. On either side of us sat groups attending a conference on futurology at a resort on the edge of town. Our waitresses were English and Canadian, cheerfully and illegally employed.

Connor, the doctor, smiled indulgently. 'Would you allow me to maintain that Alice is essentially a good-natured town? I like it. It's unpretentious, it's surprised to find itself turning from a backwater into a sort of Australian Santa Fe. It doesn't take itself overseriously.'

Jo said, 'Has anybody else noticed that the floors are covered with cockroaches?'

'You know what's offensive about Alice Springs?' Mario yelled. 'Its very inoffensiveness! Behind the picket fences and the bland, wilful mediocrity, and the well-watered shrubs and the swimming pools (although they endlessly piously tell *us* not to waste water), there are Aboriginals sleeping in the riverbed! It's not so much that they are an underclass – they are more like visitors from another universe. Like half of Australia, this place rests on a constantly reiterated heritage of

317

bloody bush stations, homesteaders, flying doctors. Unlike the suburbs of Perth, however, this place is full of Aboriginals, and it still manages to marginalize them! The radio phone-ins are all British pensioners reminiscing about petrol irons and the days when a half-pound steak cost one-an'-sixpence! And the relentless wholesomeness of it all cannot disguise the fact that it was born out of cruelty and spilled blood.'

'I don't know what your problem is,' Jo intervened. 'You make a good living out of bringing Bulgarian dance troupes into the bush.'

'Because somebody has to tell you people what you did – are doing – to these people. My dad was a bloody barber! The worst thing he ever did to an Aboriginal was cut his hair. Your lot went around cutting their *throats*!'

'I don't think there's exactly a shortage of people reminding us about the lot of the Aboriginal,' said Connor dryly. 'Did your father really ever cut an Aboriginal's hair?'

'Did he protest?' asked Jo. 'Man the barricades with his hairdryer?'

Mario glared at her, and Connor laughed. 'Look,' he said, 'I'll give you a professional opinion. I may be the only person around this table who has actually lived in the bush. I worked with Aboriginal people for five years, and I . . . And I . . . I don't know the bloody answer.'

'This place used to be a cattle farm,' said David, a mild-mannered, grey-templed Aboriginal. 'Now it belong to us. I reckon it's pretty good. It let us go back to the old life.'

'What is the old life?' I asked him.

'Like, before, I had to work. I worked for the old fellah before he die, work for him as a stock man. It was pretty hard work, pretty hot! I be out walking from early in the morning. Now I can take it easy, maybe I do some cooking here, make some billy tea, or if I like I take a swim in this waterhole! I show the tourist how we live, our traditional ways. Japanese tourist, German tourist. We had one, he said he's an American Indian. He very interested in us. I reckon it's all right.'

The good days David was saluting had turned him from an employee into a living exemplar of Aboriginal heritage. Neither he nor any of the other people from his community that I spoke to objected to this. They were once more the masters of the land on which, for a century, they had been forced to serve others. David was gregarious, and German and Japanese tourists were grateful for the glimpses he gave them of spear hunting and fire-lighting, traditions from which clearly he himself felt remote. But it was a good life. And the government, he told me, looked after the road, the rubbish, mowed the grass, did all the gravel.

This Aboriginal-run cultural tour was really a sort of job creation scheme, according to Bill Hughes, the man whose pig had given me such a start. An old Scot, Bill had tried his hands at most things – stock-rearing, buffalo-hunting, croc-shooting – living in a wide variety of habitations throughout central Australia, including a house made from straw. 'But while I was away,' he said, 'the horse ate the bathroom.'

He was now employed by an Aboriginal-run quango to oversee the smooth running of this very Australian exercise in cultural tourism.

'Martin, a massive smokescreen prevents the public from knowing how much is spent on the Aboriginals. I'm not at liberty to tell you how much *this* set-up is out of pocket, but it's a fuckin' fortune.'

'And do you find that objectionable?'

'Not in the least. We took this land away from them. Now it behoves us to give them some sort of restoration, doesn't it? Something like this strikes me as an OK halfway house. But we've been giving them everything for a long time now. Sooner or later, for their own good, they have to learn that it can't go on. But Christ knows, the kids growing up on this community are still not ready to blend into modern Australian life – I don't know how far down the line that is.'

I told him about the old man who had talked me into standing him and his friends a round.

'Well, I wouldn't have done it, Martin. And nor will you when you've been here a few weeks – you'd bloody soon run out of money. Grog's a problem, your blackfellah makes a bad drunk, that's for sure. And grog affects everyone. Take David there – he's a good man, he has a real dignity. But put half a bottle of rum in him and he's not worth knowing. He attacked his wife a few months ago, y'know – head wound, could've killed her.'

I found it hard to reconcile this image with the grave, benign figure I had met.

'The truth is, Martin, they're no different from us – no, that's not true. They have a different problem. The land is sacred to them, it relates them in the universe. It's that break with the universe that causes a loss of the sense of belonging – especially up around Alice, and all the Aboriginals who got displaced from their land when they first put the telegraph through. They're the ones you see wandering around like lost souls. They *are* lost souls. Their sacred sites are down underneath all those houses and supermarkets. When the blackfellah is healed – when the break with his land is mended – grog won't be a problem any more.'

Dawn at the sunrise viewing point: perhaps three dozen vehicles parked, a hundred and fifty expectant faces, and the sound of diesel engines. If I had been shocked by the roads through the Arches National Park in America, I found the road the Australian Government has built around Ayers Rock/Uluru even worse. Heavily pushed by Australian Tourism as the ultimate in awe, serenity and silence, it is closer at dawn to a race track, with late arrivals roaring around the ring-road and screeching to a halt in the car park.

Through the open door of a coach, a Japanese pop song played. Photographers fiddled with their tripod-mounted cameras. Three Australian girls, trendily thin and barefoot, stood smoking cigarettes.

Everyone waited for the sun. They had all travelled thousands of miles for a glimpse of the orange monolith bathed in a light that turned it to the colour of tinned tomato soup. Alas, the sky, source of the awaited light, was overcast.

At last the great orb appeared on the horizon, a sore shade of red, and for a few moments Uluru began to glow like an ember left from some immense conflagration. People said, 'Oooh,' and camera shutters clattered. Then the sun was swallowed by a veil of grey, and the rock grew dim again. There were moans of disappointment. People stared east for signs that the sun might break through once more. Then they began to leave.

For two or three minutes there was the stuttering of starter motors and the roar of engines as tourists raced away. Ten minutes later, when the clouds parted again, only four vehicles, and seven people, remained.

I walked around the rock, and found I had it almost to myself. The monolith has a stunning, silencing presence, its mottled copper skin at times sinuous, at times seeming to sheath stressed muscle. There were marks on its flanks like clouds of smoke, serpents frozen mid-strike and spear-gashes, which have been incorporated into the folklore of the Aboriginal people in subtle ways that betray long observation of the rock. Here and there are paintings and engravings, and vast and dark, imploding sacred caverns inaccessible (on pain of death) to any but tribal initiates.

I went on a guided walk led by an old Aboriginal known as Jacob. He told us stories and pointed with his club at the rock, and the line of tourists making their way up its shallowest slopes, clinging to chains, silhouetted against the morning sky. 'Look at the ants,' he said. 'You can climb too if you want. But maybe first you could think about what I have told you, and the sacred significance of the rock.'

'Why are you against people climbing Uluru?' I asked.

'Every year there are deaths,' he replied. 'Do you think we want that – to know that people's children or parents are falling to their deaths or collapsing from heart attacks on our rock?'

'I understand that,' I said, 'but does it make you feel bad in some way that they climb? Would you like the government to make them stop?'

'*You* ask the government to make them stop, if you think they should,' he said angrily. 'They're far more likely to listen to you than me!'

I persisted. 'I still don't understand exactly why you don't want people to climb it.'

'Questions, questions! All you people know how to do is ask questions. But you don't know how to listen!' And he turned his back on me, and shuffled stiffly away.

On my way back to Alice Springs, I camped for one night on the edge of a large Aboriginal community. Outsiders are discouraged from visiting the 'communities', partly because the Aboriginals do not welcome them, and partly because white Australia is ashamed of what they would see. It was dark when I arrived, and I walked tentatively towards the edge of the village.

A religious meeting was in progress. A single lightbulb hung in front of a windowless portacabin, and a man paced up and down in the semi-gloom, a large Bible open in his hands, chanting into a microphone.

'Matthew tells us, He is the Salt, Praise the Lord, Amen. Halleluiah. You ever tried meat without salt? It's no good, is it!'

There was good-natured if thin laughter from the dozen people sitting before him on a selection of packing crates and garden furniture. I approached hesitantly, and a woman waved for me to be seated on a broken plastic chair.

'It's no good without salt, and life's like that, it's no good without the salt of Jesus Christ – Amen, Halleluiah.'

'Amen,' mumbled the gathered dozen. 'Amen, Halleluiah.'

There were blankets spread out in the dust in front of us, and four children splayed out asleep. Over a dozen dogs slept or prowled or scrapped, with people occasionally shooing them away or throwing a stone.

'And we gotta be the salt in the lives of others, Halleluiah! We gotta bring the Lord Jesus Christ into their lives, Amen, Halleluiah! We gotta go and take the word of our Lord Jesus Christ out into the world – Halleluiah!'

'Halleluiah!'

'Amen!'

'Praise the Lord!'

'Now I don't like long talks, so I'm not gonna go on now, brothers and sisters . . .'

Three women had walked forward, and taken their places before a second microphone, and now they began quietly to sing.

> *There is only one way to Jesus.*
> *And it will always be the same.*
> *There's one thing that you have to remember –*
> *Just believe when you call on his name.*

They repeated this verse twenty, thirty times, louder all the time, and we moved forward and stood in a row, swaying together. Some of the women raised their hands above their heads and waved them enthusiastically. Others stared at the dust, and mumbled.

Meanwhile, the preacher was moving along the row. As he

reached each man or woman he put a hand on their forehead and another on their shoulder, loudly exhorting them for three or four minutes to receive the Lord into their lives. We swayed and sang. I felt a sense of heat, of expansion, in my chest. When the preacher reached certain women he placed both hands on their foreheads, and his voice became higher and higher, while two more women gathered behind them like fielders, until those being prayed for suddenly collapsed into their sisters' arms, and were laid gently in the dust, where they remained, arms akimbo and perfectly immobile.

At last, the preacher reached me. I felt warm, and calm, and curiously happy.

'Breathe your Holy Spirit on him, Lord! Oh, breathe your Holy Spirit on him. Wherever he goes, Lord, surround his camp with your guardian angels! And wherever he goes tomorrow, go with him, Lord!'

PART IX

From the Gobi
to the Empty Quarter

CHAPTER FIFTY-ONE

Le désert, c'est l'avenir

THE HILLS NORTH OF BEIJING were shaggy-wooded belljar domes. The railway track wound among them, flinging out occasional views of the Great Wall zig-zagging its own ancient trajectory. The train climbed towards drier terrain, shadowing the Yellow River, north through brown hills contrasting harshly with the flood plain of glinting paddy. Peasants in blue overalls and straw hats crouched among the rice shoots. The Great Wall came and went, occasionally shadowing the track. We passed still-inhabited ancient forts, magic villages where trees burst through shattered roofs and children played among mazes of crumbling wall. Numerous hilltops were capped with stone towers thirty feet high, fire beacons that, seven hundred years ago, brought news of invasion to the capital in twenty-four hours – it took several months to reach it by road. The journey ended in a vast desert, which for the Chinese was the end of civilization.

The Gobi was the natural barrier that protected ancient China to the north, a baked wilderness of dust and gravel from which, nevertheless, invading Mongols came to overwhelm the Chinese emperors. The Great Wall – or, as the Chinese call it, the Ming Wall – had been built to keep out these murderous horse-born raiders, who forsook their steppes and thrust south across the desert. But what they crossed was not called the Gobi – that is a Western name. *Gobi* means gravel, and much of that arid band running almost from the Pacific to the mountains of Central Asia is gravel, interrupted by sandy deserts with distinct geologies, populations and names – the Ordos, the Alashan, the Taklamakan.

I planned to travel by train for about twenty-eight hours, north towards the Mongolian border, then west. The railway line follows the Yellow River as it drives a curving green corridor 500 miles into the Gobi. I would alight in Yinchuan, and attempt to go by road across that central part of the Gobi known as the Alashan Desert.

The train left Beijing at eleven-thirty that morning. Families struggled aboard with suitcases and picnics, while farmers with sunworn faces stowed city plunder – stereos, small articles of furniture, and one delicate bonsai tree. Everyone chattered and ate, hawkers bellowed, orchestral versions of 'Greensleeves' and 'The Girl From Ipanema' screeched from the tannoys, conductors officiously checked our tickets. I felt part of an enormous and unpredictable enterprise about whose outcome everyone felt optimistic but not entirely certain. The conductors re-checked our tickets. They were edgy, they made you nervous. Would they uncover some flaw in this great enterprise? But at last the train pulled away. On the platform, uniformed officials stood to attention and saluted our departure.

The three Chinese men sharing my sleeping compartment were already tackling a meal of fried fish, pork knuckles, chickens' claws, pickled eggs and rice. Gorging, they ignored me, their chins glistening with dribbles and their shirts dotted with grease. They washed down the food with beer and vodka, then jasmine tea brewed in jam jars filled from a giant thermos on the floor. Then they fell asleep.

The two other people in the compartment were Antoine and Jimmy, a Frenchman and a Nepali. We had exchanged slightly stiff greetings in English. Clad in black, tall and stooped, with a large nose and a Monkish thatch of greyish hair, Antoine quietly sat reading a book of Taoist meditation. He had the wasted, malarial look that comes from decades in the tropics. Rope veins stood out on his sun-burned hands. He chain-smoked cigarettes, and occasionally took a chaste sip from a large bottle of beer. His friend, Jimmy, wore a Dodgers baseball cap and a T-shirt with cut-off sleeves that revealed his powerful biceps. He smiled a lot, but kept silent, smoking, playing with his Zippo lighter and staring out of the window.

At one o'clock a turquoise-capped attendant imposed a midday curfew, moving through the compartment shooting the dividing curtains and shushing at the noisy. The piped music stopped halfway through an easy-listening version of Boney M's 'Rah, Rah Rasputin'. Silence fell.

I studied my maps and watched the landscape grow dry. After a while, Antoine put his book down and offered a few considered words. We exchanged cards of identity. He said he bought and sold antiques and Jimmy was his assistant. They travelled. Now they were headed for a remote mountain region in western China, where Antoine implied rather than stated that interesting antiques could be acquired cheaply. I suspected that he was up to something slightly dubious; but maybe he just liked to play at being mysterious.

I asked Antoine if he had a home and he gave a wistful smile. He had lived in a number of places, he said – India, the Philippines, Thailand, but he always moved on. '*Mais j'aime bien le Ceylon,*' he added, enigmatically.

I told him a little about my own journey, and suddenly he became animated. Reaching inside his bag, he pulled out a book on global warming. 'It seems,' he said, 'that the deserts are going to expand. The Sahara will absorb most of the Sahel countries. There will be no grassland left in Central Asia. The same will happen in the American prairies, Argentina . . . There will be deserts everywhere. Even southern Europe will be a desert.'

Ever since the terrible Sahelian droughts of the 1970s and 1980s, respectable scientists have been predicting the juggernaut march of the deserts. Scorched earth! Death! Apocalypse! More recently, satellite photographs have shown that, with the restoration of average rainfall, 'desertified' areas have a high capacity for recovery. More than a few scientists have been left with egg on their faces, and many now maintain that the scientific data we have available are far too limited to allow us yet to draw firm conclusions about the patterns of natural vegetation in deserts. In other words, *natural* desertification – despite its politicization by arid countries in search of a piece of the environmental cake usually distributed to rainforest countries – is a mystery, and perhaps a myth. But will rainfall *increase* over the next 100 years? Certainly the deserts seem unlikely to grow cooler. So *man-made* desertification, a product of the poor management of fragile, arid environments, and almost certainly of global warming, is a reality, though certain scientists, including some backed by money from industrial interests, claim that even this is a myth. My own conclusion is that human beings are manifestly degrading arid and semi-arid environments into irrecoverable deserts. But I could not emphathize with Antoine's apocalyptic relish.

Antoine's was a sort of negative romanticism – *if humanity is set on destroying nature, then let it and everything else be swept away!*

'Well,' I said facetiously, 'given that scientists have only just about made up their minds that global warming is definitely happening, how can they predict that in a few years there are going to be giant cacti all the way up the Champs-Elysées?'

'This man', said Antoine conclusively, 'is one of France's foremost climatologists. There is no doubt about this.' The Academy had spoken: *le désert, c'est l'avenir.*

The train was moving now through an increasingly bare landscape, drawing close to the Gobi and the doomed grasslands of Mongolia. Antoine's long nose drooped over his black Tibetan shirt; Jimmy stared past the curtains at the landscape's unending rush; I dozed.

At three, the canned music kicked in: 'Good King Wenceslas' performed by a Chinese orchestra, with cymbals. The eaters woke up and began to eat. But after meat, eggs and pancakes, they showed signs of exhaustion. One man, tubbier than the other two, kept pressing extra helpings on his friends, but they groaned and massaged their drum-taut bellies. They contented themselves with a large water melon, while finishing off the bottle of vodka. Finally they re-filled their jam jars with green tea and lifted their eyes from the table to contemplate the foreigners.

The tubby one tried to communicate, but our poor attempts at Mandarin frustrated him, and his attention wandered to my maps and books. He gazed wide-eyed at black-and-white pictures of Central Asia, asking me where each one was. I pulled from my rucksack a picture book with colour photographs of China, and the three Chinese fell on it with a new hunger, turning the pages reverentially, gasping in wonder at each fort, waterfall and panda, arguing over the location and asking me to translate the captions to settle their debates. Their enthusiasm was touching. I already realized how few well-illustrated books or magazines were available in China. The news-stands sold cheaply printed journals with muddy photographs on coarse paper. I wondered to what extent the whole culture of books had perished with the Cultural Revolution. These men had not grown up with access to libraries full of *National Geographics* and the *Encyclopaedia Britannica*, or with natural-history films roaring and squawking from television screens. They treated my book like a treasure.

At dusk we reached the town of Datong. Scraps of red flag fluttered on impromptu flagpoles, brickworks bled smoke into the orange sky; a giant factory was a six-storey plateau of grey-pink brick. As we pulled into the station, our ever-hungry passengers were disgorged onto a platform crowded with food vendors' trolleys. It was dark when the train eased away. As the trolley lights swayed past I was reminded off a far-off scene – the ribbons of bulbs along the Embankment of the Thames, soaring and dipping from pillar to pillar like the flight of a bird.

Next morning I rose at first light and stood in the corridor with a soldier who was smoking away his insomnia. As the sun flared on the jagged horizon it flung its first beams over a landscape of cracked hills. I found my way to the restaurant car. It was not yet open, but a female cook, tightly bound in catering white, smiled and passed me a free glass of jasmine tea. In the night the train had swung through ninety degrees and was now heading west through Inner Mongolia, the sun on its trail. Peering into the yolky light I saw some sand dunes. I saw three camels. It was the Gobi.

CHAPTER FIFTY-TWO

Desert is Too Hot

THE GOBI IS A DESERT 2,500 miles long, of 500,000 square miles in area. Its central tract is called the Alashan. A horseshoe of roads curves up from the Qilian Mountains in the south towards Mongolia, bracketing the Alashan. A short stretch of road parallel to the Mongolian border connects the heel of this shoe, making a ring around the desert. I wanted to start my Gobi journey by following these remote paths to a small settlement shown on my maps as Ejin Qi.

That morning I had left Yinchuan, a blank industrial town on the banks of the Yellow River. The counter clerks at the bus station insisted that there was no bus across the desert. I found this implausible. After an hour of mutual incomprehension, in which my patient attempts to communicate were met with escalating scorn and anger, I was approached by an English-language student who asked on my behalf and received the same tart reply. Puzzled, he took me out to the busmen's cafeteria, where we heard a different story. Yes, there was a minibus to the desert town of Suhait, and from there you could get another one to Ejin Qi. And the Suhait bus left in five minutes.

Most of the passengers were young Mongolian miners. As the bus veered through Yinchuan they drained beer bottles and threw them out of the windows – obviously an established local practice, for the driver kept swerving to avoid puddles of glittering jade glass. At the bus station a goatee-bearded, Mao-jacketed ancient had been selling beers from a galvanized bucket: three yuan each, cheaper than water. The miners, returning to darkness under the Alashan after a day on the town, relished one last drink. With their sooty faces, jackets sprouting tufts of padding and trouser knees torn and drooping, they spilled onto the bus, slapping each other, spitting nut shells and urging beer on the other passengers. The conductor roared ferocious threats to get their fares out of them, but kept slyly grinning at me like a roguish uncle. I had settled a little uneasily in the middle of this troupe, a foreigner with odd clothes and a strange hat and a lot of *things* – bags, cameras, notebooks, and now a tall bottle of beer . . .

By the time we reached the outskirts of Yinchuan, the miners were succumbing to the combination of alcohol and heat, their heads lolling one after another. As I looked out of my window, I felt something like an insect crawling on my arm. I looked down and saw the hand of a miner, thick fingers and black nails, caressing the blond hairs on my forearm with the tenderness of a lover. He looked up at me and held my gaze, then pointed at his own, hairless arm. He grinned a mouthful of broken teeth, and we both began to laugh.

The air became sulphurous. Black factories spouted plumes of vivid smoke – burnt umber, burnt sienna. One industrial monolith framed a curtain of yellow flame two storeys high.

The barrier of the Helan Hills appeared ahead. We passed through a crumbling strip of Great Wall into the badlands beyond, climbing hills a pale terracotta in colour and cracked like mis-baked pots. The road was broken, and a new tarmac one was being laid. Engineers stood in purposeful groups around theodolites, and road gangs shovelled gravel in the full heat of the sun. As our bus approached, two Mongolian workers lifted their spades high, raising a cloud of amber dust that drifted across our path. As the bus drove into it the conductor yelled at the passengers to close their windows, and angrily cursed the workmen, who leaned on each other as they crumpled up with laughter.

The bus was frequently diverted over stretches of rocky ground, negotiated in first gear. With no draught passing through the windows the atmosphere grew molten, and as the bus wallowed, the passengers lolled sweatily against each other. At length we regained a road, and emerged from the foothills of the Helan Hills onto a brilliant, featureless plain. A haze hovered on the western horizon between a cloudless sky and a golden strand like the distant edge of a sea. It was so like a seashore that my mind painted in the turquoise water that surely lay, deep and cool, just beyond the haze. But it was the shore of a sand-sea – the Alashan Desert.

The headgear of a colliery grew out of the flat landscape. The miners stumbled into the bright sunshine. The conductor winked, cupped his flies and nudged in the direction of the doorway. I gratefully joined the miners to empty my bladder onto the fine brown sand. They jerked up their zips and turned towards the colliery. I watched them in their shabby suits, hair stiff with grime, shambling away. In the desert to the west, a dozen camels were shimmering blurs in the heat-haze.

An hour later, the Helan Hills threw a blunt ridge down into the desert, and I saw some buildings on its steep slopes. It was the oasis of Alxa Zuoqi. The town was breeze-block grey against the brown hillside, with a few strips of vegetable cultivation on the western slopes. We passed coiled ropes and buckets piled in front of a hardware store, and pool tables set up on the pavements outside tea shops. The faces here were no longer pale and round like those of the Han Chinese, but long, high-cheeked, bronze. This was the 'autonomous region' of Inner Mongolia, on the very edge of Han China.

'Stop,' said the conductor.

He meant that the bus was not going any further. Confused, I asked him about Suhait, which my map showed as a few hours further on.

'*Ming-tian*,' he said – tomorrow – and gestured for me to follow him.

I climbed down and shouldered my rucksack. A policeman passing on a bicycle turned and stared with such profound astonishment that he wobbled dangerously and nearly fell off. I hastened after the conductor.

The door into the bus station was the kind of heavy heat-impervious plastic flap they use in abattoirs. In a gloomy village hall to the left, a roller-skating derby was in progress. The youth of Alxa Zuoqi revolved on the lino, while a girl strained her voice in agonizing karaoke. The bus conductor handed me to the ticket clerks.

An eager crowd gathered. Their faces were leathery, their teeth strangers to dentistry. They were dressed in cheap, torn, darned and redarned clothing. I pointed at my map and asked one of the clerks for a ticket to Ejin Qi. Unfortunately the map was in English, and he could recognize neither the places on it nor my pronunciation of them. Humorously he scratched his head, amidst banter from the assembly. We all laughed. Then a tiny dumpling of a girl pushed her way into the circle. She wore a short, boxy jacket in yellow plastic and thick spectacles. Though less than five feet high, she looked about sixteen. She inclined her head backwards and peered up at the foreign colossus.

'Please,' she said, 'I speak English. I can help you?'

I explained where I wanted to go, and her face slowly creased with confusion. She pulled an exercise book and fountain pen from her schoolbag. 'My speaking English not good,' she said. 'Please write.'

I wrote, and she held my block capitals up to her lenses as though deciphering an archaic script.

'Go Ejin Qi?'

'Yes.'

She spoke to the clerk, and I was led into the office behind the counters, followed by fifteen chattering people. While the ticket was being issued, my tiny translator transcribed questions from the crowd in her spidery royal-blue script – what was my name, which country was I from, what was I doing here.

I asked if there was a hotel in town. Many hands pushed me after a man in a red motorcycle helmet, and I followed him out to a parked motorcycle. With my knapsack half on, hands clutching books, maps and hat, I crouched towards him and we sped away. The hotel was nearby, a squat concrete building with red flags and a militaristic gatepost. My translator had already arrived, and was standing holding the handlebars of her bicycle as though on parade – polite, formal, calm.

No room looked out onto the dramatic prospect of the Alashan: the tall windows afforded views only of a concrete wall. I expressed deep gratitude to my translator and she, wreathed in smiles, giving little bows, left me in a room of heroic furniture.

Like everything else in this officious hotel, the furniture was unnecessarily large. Party members no doubt took it as evidence of their importance, but I found it oppressive, a totalitarian three-piece suite. The bath and sink were by any standards huge, yet there was no running water. A bulbous cistern crouched unplumbed over a blackened toilet bowl.

Ten minutes later my translator was back, but grim-faced, and accompanied by a policewoman. The officer, neat in her green uniform with its berry-brown holster, perched on the arm of one of the giant armchairs, like a pixie in a children's tale. She obviously did not want to arrest a foreigner, one of the first to visit this remote outpost for years; but it was a closed area, she pointed out testily. Without a permit, I had no right to be there.

My translator's attitude had undergone a sinister change, from Helpful Student to Junior Interrogator. 'Policeman think maybe you spy,' she said, with a smirk. I returned what I hoped was a disarming smile.

'Where are you going?' the policewoman asked in translation.

'I'm going to Ejin Qi,' I stated firmly.

They conferred. 'But it is across . . . *shamoa*,' said the schoolgirl, unable to find the word in English.

'Desert,' I said.

'Yes; it is across desert. That is forbidden. That is frontier Mongolia, not foreigner allow go here.'

I pulled out a letter furnished in London by a friend at the School of Oriental and African Studies, claiming my serious academic intent in visiting deserts. I opened geographical textbooks with colour photographs of desert vistas, and grinned with a half-witted enthusiasm sure to dispel any suspicion. All this was received with frowns; they wanted a permit.

They conferred once more; the translator took up her fountain pen and wrote, 'You must pay fine in order to suffer Our Policeman of the search room.'

Paranoia possessed me. How did this schoolgirl know phrases like *spy* and *fine* and *search room*? I had heard rumours of the English language textbook that is said to contain an entire chapter on the people's arrest of a captured foreign airman. Anyhow, if the room were searched and my recording equipment, GPS, compasses and flight maps covered with mysterious arrows and hieroglyphs were subjected to scrutiny, things might take a nasty turn. I decided to try and change the subject.

'Can *you* – she – give me a permit?'

'No, you must pay fine and go back.'

'How much is the fine?'

'Five Yuan.'

I almost laughed with relief: it was less than a pound. I tried to look miserable.

336

'It is small fine,' said the schoolgirl primly, 'You should not coming here then there is no fine.'

'No,' I said, 'but I came a long way to visit the desert.'

'You are very unhappy?' she asked, archly.

'Yes.'

Her expression softened. 'This policeman is very nice. She only make very small fine.'

'I know. She is very kind. But I wanted to cross the Alashan.'

She looked at me through her curved Mongolian eyes, so narrow I could not see her pupils. 'Why you go to desert?' she demanded. 'Desert is too hot.'

The policewoman stood with an air of finality and solemnly brushed the creases out of her jacket. The schoolgirl announced, 'You take first bus tomorrow seven o'clock.'

'Yes,' I said, 'yes.'

The door closed, and I leaned back on a gargantuan sofa.

I did not seem to be under house arrest. I left the room and walked down the corridor, half-expecting one of the tense floor-maids to call out, *He's trying to escape!* I passed a red guard of enamel potties squatting by each door, the reception with its shiny vinyl banquettes like the back seats of fifties Buicks, the concrete gatehouse where a guard was levering chopsticks into a mound of rice. I dived into an alleyway that seemed to lead west, towards the desert. It twisted downhill for half a mile, between walls of dried mud, and occasional open doors that offered glimpses of shady interiors. Two wide-eyed boys in school uniform sidled past me, giggling. Behind a low wall a girl-child tossed cabbage leaves to a scampering piglet. An old lady sat in cool evening sunlight. Clad entirely in black velvet, with a quilted cap and golden tassel, she swung infantile feet that had been, seventy years earlier, deliberately deformed.

The alley debouched on the town dump, a pyramid of yellowed vegetable leaves and broken flip-flops. The hillside on which Alxa Zuoqi was built fell away in a long ridge, with a vista into the pale infinity of the Alashan Desert. Lower down it became a graveyard, scores of unmarked mounds of sand in rings of stones. A wind blew from the west, and pennants of tattered plastic blown from the garbage dump fluttered noisily. It would soon be dusk, and a burnt-out sun was slipping to earth. I squinted at the desert, and watched a

337

truck driving along its edge, trailing a cornucopia of dust.

It took me an hour to walk down to the fringe of the Alashan. At first I kept looking anxiously at the peasants hoeing their narrow strip of fields, imagining they were all spying on me. Horizontal sunbeams threw my shadow like a black bridal train towards Alxa Zuoqi. I crunched over rocks and gravel, and the sun had almost set when I reached sand, pale and fine, and knelt and let it run through my fingers. The evening wind blew grains into my eyes and mouth.

CHAPTER FIFTY-THREE

The Desert Blooms

THE 2,000 MILES OF DESERT ahead would take me from Mongol lands into the territories of the Turkic peoples known as Uiygurs. Before the revolution of 1949, few Han Chinese were found outside the boundaries of Imperial China. From the 1950s onwards, the goal was to tame these frontier territories, exploit them economically, and overwhelm their original populations with Han Chinese. Railway tracks were driven west, and with them went settlers, some fired up with revolutionary fervour, many unwilling, all charged with extending civilization beyond the pale. The large obstacle that stood in the way of the Chinese was the desert itself. The railway line had to cross long stretches of dunes where drifting sands made it impossible to erect any permanent structure. The engineers of the Shapotou Desert Centre were instructed to find a way to stabilize the desert so that the foundations of a railway track could be laid reliably, and drifting sand be prevented from burying the line.

Back in Yinchuan I found a seat in a clean, carpeted carriage, where the seats had bright white antimacassars, printed – this being born-again capitalist China – with adverts for motor cars. Many of the passengers were unready for such gentility. One man, waking from a doze, cleared his throat, spat on the carpet and closed his eyes again. It was a warm, lazy afternoon. The track ran through fields beside the Yellow River, flanked in the east and west by the constant presence of the desert.

Zhongwei is a small town with little of renown apart from the desert research institute. As I stepped out of the railway station into bright sunlight, a cycle rickshaw driver ran up to me and said, 'Shapotou?'

I looked at his beaming face. I had read that the research centre was about twenty kilometres away, and that the road was bad. Yet this man not only knew my destination, he gave the impression that he took people there all the time. I peered into the distance, and tried to gauge twenty kilometres. I looked up at the sun. I looked back at his smiling face.

'Yes!' he urged, 'Go Shapotou!'

Wasn't it, I asked, very far?

'No-no-no!' His hand described a small arc, suggesting that it was close by.

'How much?' I asked sceptically, '*Duo-shao qian?*'

'*Wu yuan.*'

That clinched it. Five yuan was the cost of a two-mile taxi ride, so the place must be nearby after all. As I climbed aboard the driver was beside himself with joy, and set off at a tremendous pace. We passed another rickshaw going in the opposite direction and he jubilantly called out our destination. The other man braked and stared at us open-mouthed. It did not seem a good sign.

We had barely gone half a mile when my driver turned onto a dirt track and pulled up at a cycle workshop beside a swollen, muddy stream. 'Fi' minute', he said, then jumped on a borrowed bike and disappeared. Experience told me that when taxi drivers disappear before a fare, it usually means they are warning someone that they are going to be gone a long time. Another bad omen.

The cycle mechanic, bleary-eyed and white-goateed like a Confucian sage, lifted the bike up on one side, inadvertently emptying the contents of my bag onto the ground. Exposed films rolled towards the churning waters.

The mechanic expertly levered the inner tube out of the tyre. It was so completely covered in pink patches that it had become a rubber mosaic. My driver was taking advantage of a doubtless wealthy foreign passenger to effect a long-overdue repair. In fifteen minutes he was back, and we embarked for Shapotou.

A long road stretched between paddy fields. The road surface was a smooth, shiny black, and my qualms about the journey quietened.

My driver wore no hat against the sun, and I watched the sweat trickle down his neck. A stain grew on his back, first a crack between his shoulder blades, then a broad dark shadow. His calf muscles bulged as his feet, in their red nylon socks and black cotton pumps, steadily drove the pedals, heaving the tricycle forward. It was made from thick metal piping with heavy frictive wheels, and had no gears. Half an hour passed. Pointing at the pale hills in the distance, he turned to me and grinned, 'Shapotou!' I felt horror. It *was* a twenty-kilometre journey. How could I sit still while a man drove himself to exhaustion for my benefit? It felt like a nightmare of Victorian exploitation.

The road became a mass of potholes, and he had to strain to keep moving. I stared anxiously ahead, hoping against hope that the road would improve, or Shapotou miraculously appear. He began to grunt with every effort to force down a pedal. We reached an area of factories, where smoke darkened the sky and lorries thundered by.

'Are you all right?' I asked him.

'Yes,' he called back.

'Would you like some water?'

'No,' he said, turning to look at me with a face utterly drained and focussed on his task.

I drank from my water bottle, and the wind caught drops and splashed them against his muscled calf, where I watched them dribble through the dust on the bronze skin.

When at last we entered a drab township, I made him stop, and he almost fell off his saddle. I went into the market square to look at the mean collection of shrivelled vegetables. When I returned, Aoni – that was his name – handed me an ice cream. He was once more wreathed in smiles, and enjoying a lolly of his own. It was over an hour since we had left the station.

We carried on, Aoni with renewed spring in his calves, me sinking deeper into a pit of disbelief and guilt. The road grew hilly, and Aoni had to get off and push the bike. Despite his protests, I climbed down and walked beside him. We turned a corner and saw something that made our hearts sink: a tailback of trucks blocking the road. One carriage was being rebuilt, and the other was jammed by a truck, its spinning wheels spouting sand into the air. Another truck's load of dark syrup was leaking, attracting a swarm of the biggest wasps I have ever seen. Into this confusion Aoni and I jointly propelled the tricycle. He pleaded with the road workers to let us cross their construction work, and we forced the bike over earth banks, across planks stretched over drying cement and up ramps, finally leaving the stranded lorries behind us. We had surmounted an obstacle together; we were joined now in this, our aim to get the rickshaw to Shapotou. At the top of the hill, Aoni splashed his face in the water of an irrigation channel, and we both sat, laughing and sharing a cigarette. Then he told me to get in my seat, and broke the spell. We were master and servant once more.

But after half a mile the road climbed, and Aoni clutched his chest in exhaustion. I ordered him off the bike, and into my seat, and I took his place.

The road ran alongside the railway. Beyond it a plain of orange sand stretched towards distant charcoal hills. I strained to keep the gas-pipe bike frame moving forward. After fifteen minutes Aoni ordered me off the saddle. His pride would not let him be driven. Then, thank God, we saw Shapotou. Aoni grinned, jubilant once more. The journey had taken us two and a half hours.

We coasted down a gentle slope to a concrete viewing platform overlooking a giant sand dune that plummeted to the broad Yellow River. A collection of people emerged from tea-shops and stalls to meet us. Aoni, gasping, began to boast about his achievement. One man, a tout who worked for the guesthouse by the research station, quickly appraised the situation. He spoke good English, and with a shrewd grin asked me, 'How much you pay him?'

'I agreed five,' I said.

The tout, whose name was Tong Jian Hua, communicated this to Aoni, who began to scream loudly at the crowd. They started to debate among themselves what might be a reasonable fee.

'But of course I'll pay him more,' I yelled irritably over the uproar. Tong Jian translated this to Aoni, who was engaged in a Brechtian dramatization of his duping and exploitation at the hands of the foreign devil – how long the journey had taken, how it had almost killed him, how late he would return to his anxious family. He turned to Tong Jian, and uttered a few hysterical syllables.

'He say he want a hundred,' yelled Jian Hua. 'How much you pay?'

'Fifty!' I yelled back.

When Aoni heard this he clutched his chest and began to scream about how the journey had permanently damaged his heart.

'He told me five yuan,' I said, hotly. 'I ended up cycling *him*. And now he wants twenty times as much.' Jian Hua grinned malignly.

Aoni wailed and the crowd's hubbub grew to a crescendo. 'OK,' I told them, 'eighty.'

Aoni stuck out his lower lip. Tears glittered in his tragic eyes.

'All right, dammit, a hundred.'

About £10.

Aoni snatched the note, and stomped back to his cycle with a show of wounded pride. He cycled away without even glancing at me. I turned to Jian Hua. 'So how much do *you* think I should have given him?' I demanded.

He grinned. 'I don't know. Maybe thirty, forty!'

The next morning I stood outside my room in the guesthouse and looked down at the Yellow River. It was a quarter of a mile wide and I could not guess how deep, a boiling confusion of muddy water. Each year this torrent disgorges 3 billion pounds of silt into the ocean. At Shapotou it sweeps in a broad brown curve past the 300-foot sand dunes that mark the southernmost edge of the Alashan. Here the Chinese government located the research station that has been its chief weapon against the arid lands that make up a third of the country.

The station has been built on shallow dunes at the river's edge – a walled compound with a canteen, a billiard hall and orderly rows of huts. From one of these emerged Associate Professor Li Yu-Jun, a compact man of fifty wearing a cotton zip-top and straw hat that made him look like a Sunday gardener. But the professor's garden is unique, a little Eden where plants blossom in the sand.

He led me to a sort of patio where imported plants from all over the world were growing. 'Jojoba bush from America', said the professor, running his fingers through the leaves of a small tree. He flung out his arm to take in a hundred neatly labelled pot-plants with spines and berries and flowers, 'Acacia, *Ephedra przewalski, Hanloxylon ammudendra*. This one likes salty conditions. The jasmine is from the Mongolian desert – it has pink flowers that smell very sweet. That is *Prosopis alba*, from India, and this is a special asparagus from Japan – it is medicinal.'

The aim had been to see if any of these plants might flourish in the Chinese desert. Perhaps the Gobi could be greened with beneficial plants that harvested water or offered medicinal benefits or attracted useful insect life. I asked how successful the project had been. Li Yu-

Jun smiled. 'Not very successful. These imported plants will grow here in controlled conditions, but rarely in the wild. Their demands are very specific – soil type, climate, altitude.' He gestured for us to proceed. The successes, I gathered, lay ahead.

We left the tree-shaded plant pots and climbed towards the dunes. It was 10 a.m. and the sun was already strong. The ground levelled and we saw a shimmering rectangle like a swimming pool – a small paddy field in the midst of sand. 'Many people cannot believe this,' said Li Yu-Jun, 'that we are growing rice in the desert.'

He led me on, pausing here and there to point out a flowering desert shrub. Water ran brightly through the irrigation channels that criss-crossed the land. We passed dense orchards and open acres of corn, pausing at last before a vineyard. The unexpectedness of the rows of fat vine trunks stretching their branches over arched trellises made me gasp.

'All this land was desert before,' said Li Yu-Jun with quiet pride.

Something troubled me. I knew that much of the earth's desert consists of cultivable soil, which lacks only water. Topsoil can even be imported. To show that plants can be made to grow in the desert seemed to me to have more an emotional impact than a scientific one, to be merely a demonstration of irrigation. To grow plants beside a river was not enough. It merely raised the question of whether one could somehow bring water to vast tracts of desert.

If I felt sceptical about the long-term viability of irrigation, there was one achievement of Shapotou which seemed unquestionable. The dunes above the gardens looked as though someone has turned a 3-D computer map into reality, with a grid of graph-paper lines etched on them. Coming closer, I saw that the grid was made from ridges of thick, reed-like grass driven into the sand. Li Yu-Jun explained that this technique of weaving grass three feet into the sand had stabilized the dunes, preventing them from growing, or – more importantly – from moving laterally. In this way, the 500-mile stretch of railway which I had just travelled had been assured. 'Nowhere else on earth,' said the professor, 'has a railway been built over sand dunes.'

To follow the Gobi west I had to pass through Lanzhou, the major regional city, and I reached it late the following night. My hotel was the sort that has paper strips across the toilet bowl proclaiming its hygiene, and a fur of hair, dust and nail-clippings under the bed. The lobby teemed with the life of New China – heavy-drinking businessmen sweatily cutting deals. I was intrigued by the sight of two beautiful female police officers with superbly sculpted hair and high-heeled shoes, arrogantly parading uniforms that might have been tailored by Lagerfeld.

The next morning a sandstorm hit Lanzhou. I was walking down the city's main street when the warning came – a sudden high wind. With the abruptness of a damburst, a wall of sand surged down the street. Heavy noticeboards propped against buildings took to the air like leaves, and a blue neon shop sign exploded, showering me with glass. Twenty bicycles parked in parallel performed a choreographed domino-flip, like some insane fairground attraction. Heads down, as though scattering in an air raid, we pedestrians ran for shelter, bracing ourselves against the wind, squinting through our fingers into the sand. I reached the security of a shop doorway and kept running, only when I was well inside turning to view the chaos in the street. A boy stumbled past me, his eyes as pink as wounds, seeping water. Last year, a few miles from here, six people had died in a sandstorm. The sand is thick enough to choke you. They had jumped into a canal to escape it – and drowned.

Lanzhou did not interest me. My guide book talked of tourist sights, but I impatiently stalked the streets around my hotel, staring into windows full of agricultural machinery. Getting a train for the next stage of the journey was proving difficult. The ticket office told me there were no seats available for two days, and I spent several hours exploring the availability of buses. Then I was approached by a tout, a wiry, well-dressed young Moslem who scorned the alleged ticket shortage. 'Of course I can get you a ticket,' he said haughtily, 'that is my job.'

He sent an assistant off with my cash, and sat with me in a pavement café, under the disapproving eyes of pale Han waitresses. 'They do not like me,' he said, 'because I earn much more money than they do. But they are stupid. The only people in China with money are Party people or business people. Why should I work for 500 yuan a month? *And*,' he insisted, between aggressive drags on his cigarette, 'they are all Han Chinese. The Han hate the Moslems. They knocked down all our mosques – for twenty years they wouldn't let us pray. Even now we can't get a good education, or jobs in government service.' He sneered, stabbing out his cigarette in the plastic ashtray. 'I hate them, too. I'm not stupid – I went to university in Pakistan. All I want to do is get out of this fucking country.'

I hoped he had sold me a genuine ticket. With two hours to spare before my train left, I went for dinner to Huimin Market, a street of extravagant food stalls where anything conceivably edible is steamed, fried, baked, boiled or grilled under your nose. I sat in the light of a paraffin lamp and ate steamed snails while a drunken urchin of about sixteen reeled around me, clutching a beer bottle and demanding, 'Dollar, dollar!'

CHAPTER FIFTY-FOUR

Heaven's Gate

THE TRAIN SAT MOTIONLESS. 'You like the deserts?' asked Zhon
Qin Reng, as we stared through the bars on the window at the
haze beyond. 'Well, *I* don't. I *dislike* them. A lot of people in China
were killed by deserts. Families were broken up and people sent to
labour camps in them. People died out here doing hard labour –
teachers, doctors, lawyers . . .' He subsided onto his seat. 'Deserts . . .'
he said. He blew a long plume of blue cigarette smoke out of the
window and watched it dissipate. Then he brightened. 'I know a
poem you might like:

> *In the desert*
> *the smoke rises straight up*
> *because there is no wind.'*

He sat back on his couchette bunk, smiling. 'Twenty years ago,' he
said, 'that poem would have been banned by the Cultural
Revolution.'

There are three classes of train in China: Express, Fast Train and
Slow Train. My cunning tout had sold me a ticket for a Slow at the
cost of an Express. Now I sat in an immobile train, stifling in the heat.
Unreasonably humble, it stopped at every halt, giving precedence to
everyone – even to goods trains. A nine-hour journey was going to
take eighteen.

I had left behind the Yellow River and was once more travelling
west, along the southern edge of the Gobi, past the foothills of the
Qilian Shan mountains. This train was bound for Urumqi, the capital
of Xinjiang province, a Turkic city that lies on the border between
Central Asia and China. I intended to get off long before Urumqi, at

Jiayuguan, the town that has traditionally marked the point where China ends and the western deserts begin.

Dust wafted through the window. Zhon Qin Reng and I stared at our respective books. He was slight, nervy, chain-smoking, a neurosurgeon at a provincial hospital in one of the string of small oases watered by ice-melt from the Qilian Shan. He had just finished his annual holiday, a fortnight's sight-seeing in southern China. He had travelled without his wife. 'Our holidays come at different times of the year,' he explained.

Late into the previous night and all this morning, we had been gabbling like men starved of conversation – which we were. For me, it was the first chance for a long conversation with an educated Chinese who would talk openly. For his part, Qin Reng was fascinated to talk in depth with a foreigner.

He revealed a mind split between anger with the Chinese authorities and stubborn patriotism. He had not heard of Tibet (it took some time to establish that I was referring to the Autonomous Region of Xizang Zizhiqu), and my assertion that many of its people owed allegiance to some ancient religious figure he had never heard of – the Dalai Lama – was greeted as absurd. He would jauntily accuse me of belonging to the country that had tried to addict the Chinese to opium, or taunt me with his certainty that China would be the world's dominant economy in the twenty-first century. But then he would mention some repressive aspect of Chinese life, and sit back with his arms crossed, staring moodily out of the window. Slowly he became more frank about his political views.

'The Cultural Revolution was a disaster. I was poor during that time, *hungry!*' His father, a teacher, had been imprisoned for three years for some mildly dissenting remark uttered in class. His mother had to take in washing to provide an income.

'What about Tiananmen Square?' I asked.

'Another disaster.'

'How many do you think died there?'

'Maybe thirty?'

I snorted. 'More like three thousand!'

'Maybe . . .' he said.

I passed across a book on modern Chinese history, open at a

description of the massacre. For half an hour he sat, quietly reading it. At last, with a little gasp of frustration, he stood up. 'You want some food?' he said.

It was too early for lunch, and when we reached the dining car we interrupted a tableau of feminine relaxation. The four female restaurant workers were lounging around a table, smoking and gossiping. One had loosened her cooking smock, revealing a satin slip, and when she saw us she quickly pulled the smock tight around her. Another was leaning back on the rear legs of her chair and had her skirt up around her thighs and her feet on the window sill. One arm was crossed over her belly, cupping an elbow, and she held a cigarette to her lips. I noticed the irritation with which she put down her cigarette, dropped her legs and, holding her hem, wriggled her rear to pull the skirt back to her knees.

Qin Reng showed indifference to the inconvenience we were causing, and sauntered to a seat halfway down the compartment.

'Aren't we too early?' I whispered.

He looked at his watch. 'They will serve us,' he said. Reng had a certain authority about him. It seemed to emanate partly from his professional status, and partly from a combative spirit. He did not suffer fools.

'What do you want to eat?' he asked.

'What's good?' I returned, hopelessly. Breakfast had been a stomach-turning pile of boiled rice with a few fragments of onion in rancid meat-oil, my worst meal in China.

His brow furrowed slightly, as if this were an inappropriate question. He resolved it neatly. 'It is all good!' he said.

I looked around the carriage. It was lined in weary sky-blue melamine, with pink art deco lamps, sea-green ceiling fans and bottle-green curtains. Each table featured a vase with a pink lace flower like a psychedelic butterfly. The air was stifling, heat pumping from the kitchen stoves and oozing around the drawn curtains.

Qin Reng ordered beef and pork. I ordered an omelette.

'They tried to explain everything in terms of class warfare,' Reng said, returning to his theme, 'but you cannot. You cannot only blame the rich. Everyone needs to be able to work. After all, a businessman takes risks, and he provides other people with work. He can lose all his money, too.' As an apologia for the market economy, this seemed to me a little thin. But Qin Reng was building to a climax. 'No, you can't just blame certain groups, even greedy people. It did not work. It was too . . . too . . .' He reached for my Chinese-English dictionary. 'Too *imprecise*. Frankly it was a shameful episode. It led to much cruelty.'

I asked about Mao, and Qin Reng told me that as a schoolboy he would be interrogated about his feelings for the great man. 'It was very dangerous. You could not be even slightly uncertain. You must—' he searched for the word— '*worship* him!'

I asked if he had heard about the book by Mao's physician, Zhisui Li, an exposé that had been banned in China.

'No,' he grinned, 'I would *very much* like to read it! It is time the people were told the truth about that man.'

As a schoolboy, he had had to memorize and chant passages from Mao's *Little Red Book*. Now Qin Reng began to shake an imaginary book over his head, chanting the words in a strained, high-pitched voice, so that the kitchen staff and a policeman who had wandered in turned and stared. Qin Reng sang on almost maniacally, then broke off and laughed loudly. 'Stupid,' he said, 'so stupid.'

Towards dusk, as Qin Reng gathered his things together, he asked if I wanted to get down from the train with him and stay at his home. I hesitated. I was enjoying his company enormously. It would be fascinating to see how he lived, to meet the wife and daughter he had told me about, maybe to be able to visit the hospital where he worked. But something made me suspect that the offer was made out of politeness. It was a feeling that haunted me through many of my dealings with the Chinese. In a culture of caution and undemonstrativeness, with the addition of recent political repression and paranoia, many of them were reluctant to be seen with a foreigner. Speaking his language, laughing with him, exposed them to the suspicion of their compatriots. So they kept quiet. I found it difficult to imagine the perky Qin Reng experiencing such doubts. Yet some instinct told me otherwise.

And so I declined. We parted warmly, but without even exchanging addresses. As the train pulled away from the platform, and Qin Reng's bird-like figure busied towards the ticket barrier, I already intensely regretted not accepting his offer. This is the melancholy of travel. The too-brief encounters, the never-to-be explorations of a person, a place. I held my stomach. Lunch with its sickening oil had left me feeling nauseous and bloated.

By late evening, when the train reached Jiayuguan, I was sick. I tumbled into a taxi and gave the driver the name of a hotel. Scenting a fast buck, he tried to overcharge me by about five times. Feeling faint, I asked the hotel receptionist to arbitrate, and when she agreed with me, the driver became so enraged I thought he was going to hit me. By now I was perspiring and feeling faint. I filled in the multiple check-in forms, mumbled that I was ill and asked not to be disturbed. They gave me the last room in the house, opposite a gurgling and stinking public toilet. It was all I could do to fall into bed. For the next twenty hours I lay helplessly in a fever.

At seven the following evening I awoke, exhausted but clear-headed. I washed, and tottered outside to explore the town.

Somewhere in China, workers labour tirelessly to produce blue glass panels of about a metre square. Every town has a rash of new buildings with blue glass faces. In Jiayuguan it was the preening new post office opposite my hotel. Traditional buildings in desert regions are cool by design – their walls and their ceilings tall, an elaborate punctuation of vents and shafts inhaling the soothing breeze. Modern glass buildings are wrong in hot climates, they magnify sunlight and have to be air-conditioned to stay cool, increasing electricity consumption and pollution. Yet in every desert town you see the glitter of glass – often government offices – as architects play with the shiny material and the civil servants look with pride at their crystal boxes.

Halfway down the main road I came to the market, a narrow side street crowded with shops and stalls. Office workers heading home threaded bicycles through the throng, and many of the peasants had also come by bike, displaying their spring onions or apricots or chillies in straw panniers on handle-bars.

The butchers slaughtered in the shade of brown awnings in the middle of the market. Blood ran along fractured concrete gutters, or dried in the dust. Galvanized buckets slopped with shiny purple organs and entrails, carcasses dangled on hooks. Never was meat so fresh. Cuts were sliced off before your eyes, customers prodding the dark red flesh to test its tenderness, or urging the butcher – sawing with a knife thinned by sharpening to half its original width – to avoid the creamy lumps of congealed fat. Every morsel of meat was stripped from the bones, until all that remained was a reddish flock. One butcher had slaughtered a calf that morning. Nothing remained but a near-complete skeleton hanging neatly on a hook like an exhibit in a medical school. Lying unsold on the ground beneath the bones was the head of the calf, brown-furred, its long-lashed eye-lids closed as if in meditation.

As I walked past the fast-food stalls by the market entrance, some boys called me over. Sipping a beer, I watched them produce the fast food in question. An old man sat by the hanging carcass of a goat, scraping the flesh and fat from the bones with brain-surgical precision. Next, one of the boys, his son, sliced the meat with equal care into gobbets of identical size, carefully separating flesh from fat. A second son fanned out the skewers and balanced them on a brazier. Customers sat on benches and gossiped, and enjoyed the deft twists the chef gave his kebabs, the scent of roasting goat, the sight of the red flesh growing crisp and brown, the fat turning yellow and oozing oil which hissed as it dripped on the glowing charcoal.

I found the primitive sensuality seductive. Here there was an honesty, the animals not raised in concentration camps and butchered in shame and secret, then gift-wrapped in plastic for the supermarket shelves.

Soon, the first brother was trying to persuade me to eat some meat raw. It was the best part, he told me, as he popped tufts of flesh onto his tongue and rolled his eyes like a gourmand.

Few towns could less deserve the title Heaven's Gate. Jiayuguan is an industrial conurbation, bleak, isolated, bound by factories and ring-roads and wired-off plots where breeze-block houses are rising. What makes it famous is the Ming dynasty fort that lies on its outskirts, just far enough from the urban sprawl for you to lose yourself in its delirious ambience. For ancient China, the Great Wall was the threshold between the civilization and plenty of the Celestial Kingdom, and the arid barbarian chaos of the imperial backyard. This fort – they called it the Impregnable Pass Beneath Heaven's Gate – was traditionally the last outpost on the Great Wall. It stands on a ridge overlooking a desert vacancy, from which winds gust sand against its massive ramparts. There are inner and outer walls, both thirty feet tall and thirty feet thick, and over the gates gorgeous towers rise in three tiers. They lean on crimson-painted columns, each the heart of a great tree. Their roofs are lapped by watery green tiles, their upturned eaves capped at each corner by a belching dragon.

From this fabulous fort the Ming Wall snakes over a couple of hillocks then dives straight into the desert. But it never went much further than here. The silk caravans that wound past these ramparts went forth into otherness. Although somewhere in the west, unimaginable distances away, lay Rome, no Chinese emperor ever saw the Mediterranean, and no Roman emissary was presented at court in Xian. The obstacles to such a journey were immense, and it took a thousand years from the Romans hearing of a land of silk to the day when Kublai Khan is said to have first laid eyes on Europeans – Niccolo and Maffeo Polo. For a millennium, the extended thread of the Silk Road bound together two civilizations that knew almost nothing of each other. It was not so much a highway as a fabulous well into whose darkness you lowered a gondola laden with riches, later to haul it up and claim the marvels magically exchanged. The route had innumerable intermediary traders who bought and sold each on his own patch. Goods changed hands often, and rarely completed the entire inter-continental journey. The great exception was silk, for which Romans developed such a mania that the Empire nearly bankrupted itself. In AD 14 the Senate anxiously imposed a quota on imports of the material. By the fifteenth century, the closely-guarded secret of silk manufacture had been betrayed to the Europeans, and Lyon was the centre of a flourishing industry. The Silk Road went into decline.

Jiayuguan's fort has a west-facing gate which looks straight into the desert. A dank, reverberant tunnel leads through the outer rampart, its curving brick walls until recently covered with messages and poems left by political exiles and criminals as they went out into oblivion. The outer world was a place not only of hardship, but of separation from the metaphysical harmony of the Celestial Kingdom. Central Asia, India, Egypt and Europe were but circles of hell. I watched sand slip through the cracks around the vast wooden door. The wind from the west penetrated a corridor of urine stains and crushed drinks cans.

CHAPTER FIFTY-FIVE

Sand City

A DAY'S DRIVE FROM JIAYUGUAN took me to the last staging post on the Silk Road before a thousand miles of sand. 'It is a very bad road,' the hotel receptionist had told me. 'Take food.' The minibus pulled out of Jiayuguan at seven in the morning under a barrelling cobalt sky. The crumbling road played tag with the railway track as both wove across a desert landscape of exposed shale, and low dunes fringed with scrubby grass. Dust-devils, miniature tornadoes like whirling bolts of amber silk, danced on the sands. For long stretches the bus clawed across broken fired earth, trundling past road gangs. All over western China I saw urgent road-building, the heavy lorry extending the railroad's influence to the remotest corners of the region. China's west is much poorer than the Pacific-facing east, and constantly agitates for subsidy. The titanic Three Gorges Dam project on the Yangtse, the most grandiose dam scheme ever undertaken in the world, is being justified by the claim that it will open a navigable sea-lane deep into the west, along which investment will flow. That remains moot, but in any case, its influence will be felt only in Han China; the desert provinces of the far west will have to wait for their own boats to come in – although, as I was to discover, they have not entirely escaped the eye of central government.

The first harbinger of the future appeared on the left side of the road a few miles from the small oasis of Qiaowan: a serene stretch of turquoise. I peered at my map, looking for a lake, but it was a new dam, yet to appear on maps. The Shule river, which the road and rail track shadow, has carved a meandering and wildly verdant gorge into the arid plain. Now the Shule is being drained for agriculture. The oasis towns are spreading, as heavy machines level soil and dig trenches, and the waters are tamed and siphoned into heroic irrigation schemes. Why let the river bury itself in the Taklamakan Desert, when its higher reaches can become gardens?

The oases around the Gobi have a shady laziness they share with the small towns of the Mediterranean. The bus bowls along dusty streets of small mud-brick or white-painted houses, where creepers climb and trees cast pools of ink-black shade. The paddy fields are outnumbered by grain and tomatoes, melons, apricot orchards and vineyards. Cows and goats amble freely. Human movement is slow, by foot, bicycle or donkey-cart. I was slumped in my seat as we entered the outskirts of an oasis, musing that I might be in Spain or Crete, when my attention was caught by a barren hill looming over the outskirts. With a start, I sat up and stared. The hill was a gigantic sand dune, perhaps 400 feet high. This was Dunhuang, Marco Polo's 'sand city'. Here the Silk Road bifurcated into northern and southern legs, ringing the terrible Lop and Taklamakan deserts which constitute the western half of the Gobi.

Twelve miles from Dunhuang is the site that made it a legend of the Silk Route, the Mogao or 'Thousand Buddha' caves. Buddhism had started in India in the fifth century BC, and quickly spread north and east, ultimately becoming the most common religion in Asia. From the first century AD, Buddhist pilgrims joined merchants on the Silk Road. They travelled south between desert and mountain, where shifting sands and murderous tribal raiders presented equally deadly threats. Finally they crossed lofty passes through the Karakorum or Pamir mountains, to reach India. Their goal was Bodh Gaya, the north Indian town where, cross-legged under a Bodi tree, Gautama Buddha attained enlightenment. For these brave pilgrims and entrepreneurs, Dunhuang was a logical place for worship. They might solicit the Buddha's beneficence before embarking on their perilous desert crossing, or show gratitude many months later for their safe return to China. Dunhuang swelled into a spiritual metropolis, where many thousands of monks and laymen venerated the divine.

Foreigners pay heavily for access to China's historic attractions. At Dunhuang, the price includes the hire of an English-language guide, and for a little more money a very necessary torch is thrown in. I tried eight of these to find the one with the least weak batteries. My guide was a sweet creature, who stuttered in agonizing English that she had just qualified in guiding, and this was her first morning at work.

Virtually nothing she said was comprehensible, and the few sentences I grasped were shallow compared with the guide book I was carrying. But she had the keys, and so we processed, with no historic or aesthetic logic, from one cave to the next. The site is a blunt sandstone honeycomb of hundreds of caves. Once they were open to the desert winds, re-discovered in the nineteenth century full of a fine sand which had protected the richly decorated walls for half a millennium. Now the caves have wooden walls and doors which block out both wind and light. I ran the torch beam over ceilings where numberless Buddhas and disciples, winged *apsaras* and creatures of the forest were arrayed in illustration of Buddhist legends and Gautama's many lives. The earliest paintings are Indian, later acquiring the Grecian lines of Gandaran art – the Buddhist civilization intriguingly influenced by Alexander the Great – and still later the features and aesthetic become elaborately Chinese.

Over the centuries monks hewed the caves into caverns up to sixty feet high. The tallest are columns of air enclosing monumental Buddhas, feet like trucks, hands like oak crowns, faces soft-curved spheres on which the narrow eyes and cupid lips curl like esoteric symbols. The painted, gold-leafed colossi have an air of magic, of chthonian monster-gods that demand propitiation. Three balconies face them for a hierarchy of worshippers, the richest donors allowed closest to the blessed face. As in gothic Europe, wealthy donors had their images painted into the murals. But before it became decadent, the art of Dunhuang was delicate, allusive, suggestive of inwardness. Life-size painted Buddhas sit in meditative postures, eloquent of detachment and compassion, their features suggested by exquisite brush-strokes in which their quiet ecstasies quiver like the shadows of clouds.

Lying on the ground before one statue was a poor scrap of red cloth and some coins, a sign of living worship.

I turned to my guide. 'Do people still come here to worship? I mean, Buddhists?'

'No. Nobody come here for prayer. This is . . . museum.'

The Dunhuang caves and their dusty Buddhas exude the poignancy of faded glory – a feeling always more acute when it is a religious building that has decayed. Shrines are expressions of the highest human ambitions, employing mysterious proportions and geometries in an attempt to approach the ineffable. When they fall, it is as though their gods have fallen with them.

But only a materialist feels pain at the decline of a religious structure; a shrine, or the material image of a deity, is merely His or Her or Its temporary residence. This indifference to form is crucial to all Indian religions, whose artists never sign works of sacred art. The quality of artistry is not pre-eminent, the vital thing is not what it gives us, but what we bring to it. The analytical European eye sees a work of art, the product of an identifiable individual artist – and when art is damaged, it should be preserved or restored. For a thousand years, monks at Dunhuang revered another reality, one without attachments to form or time.

I had to hire a car to take me from Dunhuang to Liuyuan, the nearest railhead on the Lanzhou-Urumqi line. The driver was a friendly woman who spoke English well. We drove north across the desert, with dust-dervishes whirling around us.

There are few cars in private hands in China, but this was her own. She told me that her husband, who ran a travel agency, had bought it – second-hand – two years earlier. They were members of the enterprise culture of modern China. Sitting on the dashboard was a sun-faded picture of the husband with two giggling toddlers.

'When', I asked mechanically, 'was this taken?'

'This year,' she said. 'They are both boy – two and four.'

I had assumed the photograph was a few years old. I felt her looking at me closely, and saw a slight smile playing on her lips. She asked, 'How old you think I am?'

I blushed. She had the neat, compact features of an ethnic Han, another daughter of that army of forced or voluntary migrants whose job was to settle far from home and civilize the wild west. I had taken in the fine lines around her eyes, and I thought she looked in her mid-thirties. I lied: 'Twenty-six?'

'No,' she laughed, 'twenty-eight.'

'You look younger,' I said.

'No, I look *old*.' She pointed through the windscreen, up at the sun. 'It is very bad to live in the desert. Too hot, sun always burning. It make the skin very dry and dark. It make us – especially lady – old very quickly.' She gave a wistful smile. 'I have many family in south

China. But we have to live here, we cannot go back to live in other place. We will always live in the sun.'

Hanging on the rear-view mirror was a small golden medallion with a scarlet circle in the centre. I lifted my hand to examine it. There was a golden figure against the red background, seated on a lotus and holding out the palm of his hand.

'Are you a Buddhist?' I asked.

She smiled. 'Yes, I believe in Lord Buddha – Dunhuang has famous Buddhist caves, you don't see them?'

'Yes, that's why I came here. But you're the first actual Buddhist I've met. At the caves they told me that no one around here practises Buddhism any more. They said no one worships at the caves.'

'Oh no, of course there are Buddhist – many Buddhist. Buddha protect us. We go to caves, it is very important shrine for us. People worship Buddha in Dunhuang for hundreds of years – thousands of years.'

'And you're allowed to practise your religion?'

'Why not?'

'Because wasn't it rather difficult a few years ago? Didn't the government try to ban religion?'

'Ban?'

'Er, stop it. Tell people it was not permitted.'

'Not permitted, yes. That is before. Government say religion is bad thing for people. But people . . .' she gave a sweet smile. 'Government cannot tell people what to feel in heart.'

She took the medallion, which hung on a red ribbon, from the mirror, and gave it to me. 'Keep this, please. Buddha will protect you.'

I thanked her, and slipped the ribbon around my passport so that the gilt disc lay inside it.

The journey lasted nearly three hours. At last we reached Liuyuan, a pragmatic assemblage of ugly buildings.

'Liuyuan means "red town",' she told me. 'They change name at time of communist revolution.'

'And . . .' I wasn't quite sure how to ask. 'Was the revolution a good thing?'

'What thing?'

'Well, I mean . . . is everyone happy about that? I mean . . . Is everyone a communist?'

'Oh yes,' said this Buddhist entrepreneur, with no discernible irony, 'everyone is communist.'

CHAPTER FIFTY-SIX

Silk Cut

IN THE NIGHT the train entered the giant desert province of Xinjiang Uiygur Zizhiqu, the Uiygur Autonomous Region, an enormous area of western China and the homeland of the Turkic-speaking Uiygurs. Islamic for many centuries, and falling on the natural frontier between China and Central Asia, Xinjiang has been variously overrun by the Mongols and the Chinese, though its predominant population of Uiygur peasants and nomads remained under the rule of local warlords until the Chinese communist party brought freedom and democracy in 1949. But there are many in Xinjiang who claim that the province is run by and for the Han Chinese, who are now as a matter of policy swelling their numbers with the aim of out-numbering the Uiygurs. The Han, who run the police and civil service, and control business and tourism, are not loved here. From time to time, especially in the state capital of Urumqi, Chinese settlers are murdered. But this is not much talked about by either group. Officially, comradely love prevails.

The Uiygur land is, like the rest of the Gobi, a buffer between the fertile south and the countries to the north and west. Here the Gobi blossoms into a gigantic sand-sea, the largest of any desert in the world, gripped between the Tien Shan mountains to the north and the Kunlun Shan to the south. The name of this oven-hot waste, Taklamakan, means 'You go in but you don't come out'. The Silk Road divided and passed north and south of the Taklamakan, running the notorious twin gauntlets of treacherous sands and murderous tribesmen.

At dawn I reached Turpan, an island of green in a sea of roseate sand. From here you can travel west to Kashgar along the northern leg of the Silk Road, or, theoretically, go south for a few hundred kilometres to pick up the southern leg. But no one had been able to tell me anything about the road south, and the receptionist at my hotel was equally vacant. 'You go Kashgar,' she said.

'But I want to go south,' I said.

'No, no is possible. No bus. Tourist go Kashgar.' She smiled enthusiastically. 'Kashgar very good!'

I thanked her, and looked up at the map on the wall behind her. It showed something unthinkable: a road, drawn on with a felt-tip pen, passing north – south through the middle of the Taklamakan. 'What', I asked, 'is *that*?'

She shrugged. 'It is new road.'

'Can I take it?'

'No. It is new. No bus, no car. I don't know if finish.'

I went in search of someone who might know if it was finished. But even at a nearby travel agency, no one had heard of the road. Beginning to doubt its existence, I spent the morning wandering around Turpan. For many centuries the main industry here has been vine-growing. Vineyards thrust into the suburbs, and in the town centre vines have been trained across huge frames to turn streets into tunnels of grapes.

Turpan had a recognizable humanity absent from the other Chinese towns I had visited, a *joie de vivre*, a sense of identity and a pride in being different. There were tours of the vineyards with their subterranean irrigation tunnels, and even an annual grape festival. But when I wondered how much of Turpan's liveliness has to do with that traditional grape derivative – wine – I was firmly told *nothing*.

At midday, people began to settle at the dozens of open-air restaurants that line the shady pavements, eating a spicy cuisine that reminded me of the proximity of India. I saw no wine, but few tables were not crowded with beer bottles.

I found a second travel agent who claimed to know all about the new road. He told me that it was certainly open – even if construction was not yet quite complete – and that I could definitely cross the desert by it. As for why anyone should have built a short cut across 300 miles of sand dunes to a string of remote oases, the reason was a reservoir of liquid wealth lying *beneath* the sand: oil – a resource the Chinese government is eager to exploit. Every book I had read said there was no mineral wealth of any kind in Xinjiang.

Its poverty had meant that the outside world had left the Taklamakan alone since the decline of the Silk Road. Clearly that was about to change.

It was impossible to hire a car in Turpan without a driver. As I signed a punitive six-day hire agreement, I couldn't help wondering if the agent's confidence in the new road was influenced by a desire to get his hands on my money. An unworthy thought, perhaps, but as it turned out, the driver shared it. I watched his face as the agent and I clinched the deal. He was not happy.

When he turned up at seven o'clock the next morning, his anxiety about the long trip had congealed into a superstitious pessimism. Accustomed to driving tourists around the handful of archaeological sites on the fringes of Turpan, here he was being asked to go away for six days, driving his precious car on a road he knew nothing about, into a desert generations before him had had every reason to fear. He demanded 40 per cent extra as danger money, and said he wanted to bring along a mechanic, also at my expense. We had an acrimonious row, which gave him the excuse to pull out altogether.

It took several hours to find another driver, this time a madcap Tibetan whose attitude was so cavalier I had to force him to find some spare water for the radiator. This was at dawn the next morning, after a day-long drive along the northern rim of the desert. We had reached an oil-refining complex where huge clouds of billowing orange flame challenged the early morning sun. The only shop was an old caravan, from the back of which a grumpy young woman emerged in her night-dress, rubbing the sleep from her eyes, to sell us a crate of bottled water. Then we drove south, into the Taklamakan.

The first dunes were low, but gradually they grew into monstrous humps hundreds of feet high. Hot, dry, desert air rushed in through the open windows, drying our lips to parchment and making us crave water. Occasionally a road sign pointed to an oil-well invisible in the dunes, and oil-tankers thundered by. This might well be the best road in China, I thought, a shiny black-top uninterrupted by junctions or even road-side restaurants. Certainly my driver saw it that way. I wanted to savour the trip, while he wanted to complete it in the shortest time possible, roaring along the snaking road at insane speeds. He only slowed down after we passed an accident – an oil company jeep had just hit a sand-drift on the road and skidded off, capsizing.

Luckily for the driver and passenger, who were still sitting, dazed, by the vehicle, it had buried itself in soft sand.

I felt a romantic guilt at driving across the desert. This was one of the world's most pristine natural phenomena, a sand-sea that had earned the respect of explorers from Marco Polo to Sir Aurel Stein, the Briton who did much to uncover the Buddhist culture of this desert a century ago. Indeed, many of Stein's colleagues perished during their excavations. Now this fearsome desert had been humbled by a strip of tarmac. But as we drove, I began to feel a grudging respect for the men who had built the road. It is a great feat of engineering, and represents the second major achievement of the desert scientists at Shapotou: hundreds of acres of shifting sand have been stabilized with vast patchworks of dried grass. Even so, men and bulldozers labour constantly to keep the invading drifts from closing the road. We passed the camps where they live – rows of huts huge distances from any shop or home or sign of human life.

My driver's dare-devilry had its limits. At one o'clock we stopped for lunch among some of the highest dunes. I announced that I was going for a walk, and asked if he wanted to join me. His expression plainly showed that he thought I was mad. I waded a mile or so into the desert, until he was invisible and his anxious hooting inaudible. At the top of a tall dune I tumbled into the soft sand and stared west, across a solid ocean of pale gold. The desert had an untouched, monochromatic beauty. Somewhere out there, I knew, was the point where the sand abruptly ended and three great mountain chains collided – the meeting point of Central Asia, The Indian subcontinent and China. I felt an uncomfortable, complex emotion, alone in this spot where I could be certain no human being had been before. The desert enfolded me in itself like an indifferent parent.

What vision of an unpopulated utopia drives us into places our fellow human beings have not yet sullied? A revulsion at civilization was the prime motive of the desert traveller Wilfred Thesiger. Now there are no unexplored places left, he says, and an essential experience has died, leaving us the poorer. I cannot agree with Thesiger that exploration has ended. It is etched into our DNA to clamber over the world, and when the world becomes too familiar, to turn inside ourselves. Why did the Christian Desert Fathers build their monasteries in the Sinai? Perhaps, like Aldous Huxley, they found in the desert's vast sameness the integration of all things, a vision of unity. Perhaps in its monochromatic unsensory numbness they found a release from the burden of their flesh. Pierre Loti, French exoticist, found on a long camel trip across the Sinai that the lack of stimulus liberated his imagination, freed his spirit. The first Desert Father, St Anthony, became celebrated for the temptations he underwent – he was wracked by the fiendish contents of his own mind. The pursuit of the self is the ultimate encounter, the confrontation with what irreducibly *is*.

Towards nightfall we reached the southern fringes of the Taklamakan, driving through marshy land where one of the ice-melt rivers out of the Kunlun Mountains is extinguished in the sands of the desert. The water-logged landscape with its low-growing shrubs had precisely the feel of seashore mud-flats, and a chill went down my spine as, despite myself, I craned my neck for a glimpse of the sea that must be near at hand. The paradox was that I was in one of the places on earth that is farthest from any seashore.

The finished section of the road ended, and we drove twenty miles over a dirt track that had my driver cursing and complaining about his axles. 'Road very bad for car!' he bellowed, inexpertly charging the urban saloon car across huge crusty corrugations, where it threatened to shake itself to death.

'If you're so bloody worried, why don't you slow down a bit?' I demanded. He flashed me an angry look, and kept up his speed. In darkness we reached the southern Silk Road.

It was a morning of wind and racing clouds and an intense, clean light. Sand dunes floated alongside the road like basking whales. We passed two men on a donkey-cart. They wore skull-caps and wispy beards, and narrowed their already narrow eyes against the sand that gauzed the road.

In mid-morning the driver announced with grim satisfaction, 'Axle break.' I gave attention to the motion of the car. There was a distinct rhythmic shudder from the rear. 'I tell you it break, this bad road,' he said.

'I tell *you* go slow,' I retorted, '*your* problem, not *my* problem.'

In the mirror I saw his eyes narrow angrily. 'We go repair car. You must pay!'

'I pay nothing! I pay very much for this journey – too much! *You* pay for car!'

At the edge of the next oasis we found a tiny auto-repair shop, and discovered to our mutual relief that not the axle but the rear sub-frame was fractured – a much easier repair. A grubby young mechanic without shoes or protective clothing set to work with a welding torch.

The small marketplace nearby sold only the minimal household necessities – oil lamps, padlocks, needles and thread, enamelled bowls, flashlights, aluminium spoons – all of Chinese manufacture. A more abundant sight were the vegetable stalls, with onions, chillies, melons and tomatoes grown plump under the desert sun.

After losing half a day on repairs, we were back on the road. The sun slid over the constant high wall of the Kunlun Mountains, their peaks snow-white over the desert dun. In oases the road was lined with avenues of poplars that threw a delicious shade. Such Mediterranean touches, marks of an earlier generation's forethought, were rare touches of humanity in a country that often seems inhumanely bureaucratic; I felt grateful for it.

At dawn the next morning we pulled up at a roadside bakery. The baker was a cheery man in a white singlet with fat, muscled arms and a handlebar moustache. He sighed a ritualistic '*Al-lah*' as he peeled each pastry from his open-air oven. I peered into the oven's mouth, and saw its concave surface lined with golden globes, like a many-breasted statue of Artemis. I bought several of the six-inch rolls, and thought I had been overcharged for mere bread; then I bit into the crusts, and grey meat in creamy mutton fat oozed down my chin.

For a mile's length, an irrigation ditch carrying snow-melt water from the Kunlun slopes had burst a bank, covering the road with a film of brown water. A single man was labouring with a shovel to mend the gash. These towns were even less touched by modernity than I had anticipated. The main mode of transport was the donkey-cart.

I marvelled at the faces I was beginning to see. I was used to the features of the Han Chinese, the Mongolians and the Tibetans, but now there were the broad flat faces of southern Russia, and the kind of chubby Caucasian features you could find in Greece or Armenia, or, more darkly complexioned, in Pakistan.

Most of the native Uiygur men wore traditional dress – wrapped-around frock coats, calf-length leather boots and tall fur hats. The women were clad in baggy floral pyjamas and draped with scarves. Gold glinted when they grinned. Just as the nineteenth-century Indian civil servant wore the British suit, the Turkic labourers of Xinjiang were kitted out in blue cotton collarless suits styled on those worn by Mao Zedong. International video and satellite culture had yet to make its mark on rural Xinjiang, and the T-shirt was unseen. There was television, however.

In a small town where we stopped for lunch, I sat under a rough wooden awning among cheerful Uiygurs during a broadcast of a concert. The female singer wore a tight turquoise sequinned dress of a type seen at European cocktail parties in the 1960s, and sang an Uiygur folk song. The TV editor had inserted frequent cutaways, obviously filmed out of sequence, of a nervous and unentertained audience wearing the dullest western-style dress. Spookily, they were all Han Chinese.

The Uiygurs around me were cheerfully eating their mutton stews and conversing in the jolly but slightly shy way of country people when they come to town. A jeep pulled up, and a Han Chinese couple climbed out. A hush fell over the restaurant. The man was presumably some big-shot local bureaucrat – he was, as they say in Africa, 'as fat as a politician'. His turtleneck velveteen top was emblazoned with the Playboy logo (in Beijing I had seen the Chicago bunny on T-shirts, bags and running shoes). His even more corpulent wife wore a tight black nylon dress with transparent sleeves. Her puffy white-powdered face had lines of frustration etched into it, and her mouth was smeared with lurid lipstick.

They rattled off their orders with expressions of contempt, and looked at no one. I tried to read some clue as to what it was that justified these scowls. Had they had an argument? No, because they were speaking to one another. But they did not want to know where they were, whom they were among. Was it fear, in a climate of occasional revolt and assassination, disguised as aloofness? Anger at being exiled to a bureaucratic post on the very rim of the empire? Perhaps even a sort of self-disgust at their manifest under-achievement, expressed as hatred of the people they had been sent to rule?

When their – enormous – portions of food arrived, they gorged. Satisfied, they threw down money and drove off. The atmosphere relaxed at once. On the box, another lady in a sequinned dress sang.

At the centre of another small oasis I saw a statue of an Uiygur warmly grasping the hand of his Mao-suited Chinese brother. It made me angry. Any notion of Uiygur nationhood seems hopeless. Here, as everywhere in their empire, the Chinese are pouring in Han migrants, relieving the overpopulation in the south, and outnumbering a potentially separatist ethnic population. The mountain rivers that drain into the Taklamakan, making these isolated oases, also water alpine grasslands that have provided grazing to Uiygur herdsmen for millennia. Under a new scheme, the Chinese plan to dam these rivers to make gigantic new oases for paddy cultivation. They will be populated with Han settlers.

In our lifetime, we are likely to see the systematic destruction of another Central Asian nation, an act of cultural genocide that parallels what is happening in that land no one in China has heard of, Tibet. The Uiygurs, deprived of their economic base as farmers or nomadic herders, kept out of jobs in the civil service and the new oil industry, will drift into southern China as cheap labour. They are the archaic remnants of a primitive and anarchic nomadic tradition, and they are ethnically non-Chinese: they will have to go.

As we drove through the dusk, I wondered how long it would be before the oil industry and settlers, television and tourism, would drag Xinjiang into the modern age. But perhaps, one day, the oil will run out and it will become prohibitively expensive to keep the road across the Taklamakan open. Then the tireless desert will reclaim it, as it has swallowed whole cities in the past.

CHAPTER FIFTY-SEVEN

Kashgar

A FEW MILES BEFORE KASHGAR we came to the town of Yangi Hissar. There is little to distinguish it from the other settlements around the Taklamakan; but in 1902 a man experienced a revelation here.

The Armenian George Gurdjieff had been a cog in the Great Game machinations of the British and the Russians, vying for influence over the arid Central Asian buffers between the two empires; the turn of the nineteenth century found him in the employ of the Tsar of all the Russias, as a spy. In 1902 he was involved in some sort of skirmish, and 'punctured by a stray bullet'. Friends brought him north, out of the mountains, and down to the oasis of Yangi Hissar.

Gurdjieff was a mystic; over the years he had gathered elements of the psycho-religious teachings of many traditions, from Cabbalism to Tantra, and come to believe that the Christian doctrine of free will is meaningless. Psychologically, human beings were a competing mass of desires and fears, their main preoccupations being food, sex, vanity, pride, jealousy, revenge – and a host of other unelevated appetites. He had committed his life to trying to attain some permanent state of self-awareness that would enable him to rise above this automatism. So far, he had failed.

After six weeks of convalescence Gurdjieff was strong enough to continue on his way and was awaiting the arrival of a man with camels. His last night in Yangi Hissar found him depressed and reflecting late into the night on his past mistakes. So, he thought, he was well again; he must resume his normal life, with its familiar patterns of loneliness, disappointment, satiety and a sense of inner emptiness.

At dawn he walked to a spring on the edge of the desert, undressed, and began to pour the chill water over himself. Suddenly he experienced a revelation. To remain permanently self-aware, he would have to set up some sort of permanent inner tension. He would have to make a kind of psychological sacrifice . . .

My driver did not want to stop at Yangi Hissar. We were just a few miles from where he would leave me and return to Turpan along the better-frequented northern Silk Road. But I made him pull up beside a stream, and I walked down to the water. The air was hot and crystalline, and the water chill.

I imagined Gurdjieff bathing in the ice-melt waters at dawn, the sleeping camels grunting occasionally nearby. A man sitting on the edge of the desert, trying to dominate his internal world like some god ordering the destiny of a planet.

Ever since we had crossed the southern Taklamakan, the peaks of the Kunlun Mountains had been a visible presence in the south, a great arid range that divides the high plateau of Central Asia from Tibet. Some way before Kashgar, the peaks of the Tien, or Celestial Mountains, had appeared in the north. The two chains curve in like the pinching claws of a crab, to meet at Kashgar. Within a few hundred square miles the Kunlun, the Tien and four other mountain chains – the Himalayas, Karakorum, Hindu Kush and Pamirs – collide. That geological pile-up has made these mountains the highest land on earth.

Kashgar has been a great city in the past, at the crowded confluence not only of these mountains but of numerous countries and cultures. Nearby cluster the frontiers of India, Pakistan, Afghanistan, Tajikistan and Kyrgyzstan. A giddy sequence of nations, races and religions – Indo-Aryan, Turkic, Chinese, Hindu, Moslem, Buddhist – took turns in controlling the northernmost extensions of the lush Indus valleys and the onward trading route into China. Through this fissure came Alexander's men, contributing to the Indo-Greek fusion of Gandaran Buddhism. Archaeologists have found the remains of red-haired people, possibly Celtic soldiers of the Roman armies. Arab armies first reached Kashgar in the eighth century, and during the eleventh century Islam became established here. For centuries under the Mughals, the imperial language was Persian, and Persian is said to be spoken to this day in pockets of the Karakorum. In the late 1800s,

Kashgar became the centre of the Great Game between the British and Russian empires.

Today, Kashgar has lost all strategic influence and commercial importance, but remains the principal market for all the farmers and herders of western Xinjiang. Even in its decline, Kashgar was bigger than I expected. Almost obliterated as a Central Asian city by communist concrete blocks, it still boasts a vast mosque, Id Kah, first built in 1442, and reinvigorated since the re-legalization of religious worship. There is also a colossus of fat flesh-into-bronze Mao, straddling the city like Big Brother. It is said that recent attempts were made to remove him, but demolition proved so difficult and dangerous that it was decided to leave the bloated dictator in place.

That night I walked to the street markets where fish (from nearby lakes) and chicken claws sizzled in batter. The air was full of the sound of merriment, as stalls served home-made beer that was cheaper than bottled water. People drank as though it were water, raising their beer mugs aloft, yelling jokes and quaffing away the hot dry night. I had already concluded that Chinese beer was the worst I had ever drunk, but I thought this watery brew might be less toxic than usual. I should have known better. What the hell did they put in it, anyway? (According to a brewer friend of mine, the answer is overgenerous dosages of a toxic clarifying agent, polyvinylpolypyrrolidone.)

The next morning I was sitting in a café, nursing a hangover and a mug of ersatz coffee, when a neurasthenic student-type took the seat next to me. He offered me a cigarette, and held his own cigarette in a strange, cupped-handed fashion, peering shiftily around him. It was obvious that he wanted something, and I let him talk. It didn't take him long.

'Do you have any jeans to sell me?'

'I don't wear jeans. This is a desert, I'd boil.'

'I see. Of course. And, perhaps you have T-shirts?'

'I never wear T-shirts. Especially ones like yours, with writing all over them.'

'I see.'

My mocking attitude was unkind. After all, he was only doing what people always have done in Kashgar: traded in the commodities that passed through their city.

He left, and as I ordered another coffee I noticed that the waitress was reading a fat book. I thought it might be some 'back-packers' bible', but ironically it was *the* Bible, translated into Mandarin. She explained that she could not go to church today – Sunday – because she had to work in the café.

'You speak excellent English. Where did you learn?'

'From my teacher, Mr Johnstone.' She gave a shy, infatuated grin, and pulled out a photograph of a buck-toothed, red-headed man in a hand-knitted jumper, grinning for the camera with his Chinese wife. 'He is a real English gentleman.'

'I am not allowed to be a Christian in China,' she went on in a hushed voice. 'The government bans students from religious practice. If they found out, I would be expelled from university, and I would not have any other chance of being educated.'

'The government can monitor whether any student in the country has religious beliefs?'

'Since Tiananmen Square, they have taken much more of an interest in students. So there are spies among us. Communist party fanatics.'

'Fanatics? There are still people who *believe* in communism?'

'Some of them are just trying to get advancement. That is how the system works here.'

'So how did you manage to study Christianity?'

'We said I had to go to Mr Johnstone's house for special English lessons. But he is an evangelist, we held secret services in his house. I was baptized last year.' She smiled again, the memory making her eyes brim with emotion.

'What is it that appeals to you about Christianity?'

'I like the prayer. It makes me very calm.'

'Yet becoming calm is not one of the principal aims of Christianity, is it? Calm is more of a Buddhist ambition, wouldn't you say? Have you looked at more oriental forms of spirituality – Lao-tzu, the Buddha?'

'Oh yes,' she said dismissively. 'I have read Lao-tzu and Buddhism. I used to be influenced by them. But then Christianity took over.'

'It's the other way around in the West. Christianity is declining and the influence of oriental religion is spreading tremendously.'

'Yes, I know.' She smiled complacently. 'Maybe it will be different in China.'

'Why did you reject the oriental religions?'

'Oh!' She laughed, and took a deep breath. 'I found them not, not . . . active enough. Confucianism, you know, is still very much deep in Chinese people's thought, their way of seeing the world. It is not really a religion, it is a sort of social code. It is not individualistic, it is collective. We can say that Christianity is far more individualistic, because the individual must face his responsibility. Personal responsibility. The Kingdom of God must be made on earth. I like this.'

'Liberation theology?'

'I have heard of it, of course, but no. That is politics. It is wrong to use Christianity to attack the state. I am interested in religion, not politics. Mr Johnstone – I mean, we – we preach Christian love, Christian acceptance. We believe that people will be won over by love.'

'You dislike Chinese society because it isn't individualistic enough, and you stress Christian action; but you say Christianity is essentially about acceptance.'

She smiled brightly. I could not decide whether it was an all-accepting Christian smile, or the kind of tight smile the Chinese use on foreigners who are being rude enough to disagree with them. Disagreement is very unconfucian.

'How do you feel about the view that evangelical Christianity is a type of cultural imperialism?' I asked. 'It preaches certitude, something that people find very comforting. A bit like communism, I suppose, in a way.'

She looked displeased. 'I do not understand your question.'

Sunday is the day of the weekly market. After the Buddhist caves of Dunhuang, this is the most celebrated spectacle in north-western China. By nine, the streets of the eastern quarter of Kashgar were already jammed with thousands of people, all heading one way. Hundreds of donkey-drawn carts were pushing their way through them, piled-up with people and produce.

The market takes place in a network of small lanes that have somehow escaped the communists' bulldozing zeal. Like all markets, it is divided into sections. Whether people need a galvanized bucket or a hairpin, they know exactly where to go to. There are lanes of zips and cotton reels, lipsticks, combs and cream plastic folding fans with photographs of Japanese models; there are richly painted tin trunks, water pistols and artificial flowers; grass brooms a yard wide; carrier-bags made from rice sacks; Russian binoculars; wooden pitchforks (here, far from any steel mill, metal is costly); fibre fanbelts for tractors and sawmills; bridles and harnesses, lynx-fur hats and hand-made clasp knives. I saw a huge pile of women's shoes, heaped unpaired like the possessions of the dead, and poor women picking them over.

Shopping is a hot and hungry business. Women sat beside metal pails of watery yoghurt, chilled by floating icebergs. A podgy vendor propelled a barrow of doughnuts made from knotted ten-inch twists of dough. Boys stood behind tin carts with glasses of watery tea poised on top of them. A woman clad in layers of diaphanous scarves ladled from a mountain of cooked noodles.

Many of the lanes are roofed with coloured sheets, making cool corridors of blue or yellow or red light. As I wandered under each filter, the white paper of the notebook I was scribbling in changed colour. Columns of burning sunlight pierced holes in the cotton.

The street of men's suitings was black-awned. There were corduroy frock-coats, and even — extraordinary cross-fertilization — fur Mao jackets. Suits dangled like hanged men, and I saw a poor-looking peasant woman bargaining to buy one. The stallholder grunted disinterestedly, but when she turned and walked away, he hastily called her back.

Men were wearing fur-trimmed felt hats, or square, four-pointed skull-caps with a single tassel. Behind one stallholder I glimpsed a workshop where these were being made, on four-cornered wooden lasts. They were often perched on shaved heads, and there was an exuberant spectrum of facial hair, from pointed goatees to square Quakerish beards framing clean-shaven chins.

I came to a street of open-air barbers' shops, where two dozen barbers were vigorously soaping and scraping heads and jaws. A couple of men with freshly shaven heads and neat goatees were crouching on their hams. They caught my eye with their narrow, twinkling eyes, and enquired whether I was 'Musselman' or 'Kefir'.

'Buddhist,' I replied, for the hell of it.

'Ah, *Buddhist!*' they grinned, bringing their hands together as in prayer.

On the edge of the market was the section where chickens and goats, donkeys and camels were bought and sold. I watched a negotiation for a donkey, a tall man counting out 850 yen and trying to push it into the hand of the vendor, who bucked with apparent anger. A third man intervened – the negotiator. He had been standing to one side rolling a cigarette, a big grin on his sun-lined red face with its single remaining tooth located centrally beneath a handlebar moustache. Under his frock-coat he wore a Russian shirt with an embroidered collar, and, like a symbol of office, a large blue scarf tied low around his waist and knotted over his groin.

He reasoned with them, first individually, then jointly, and, concluding his rhetorical spiel, seized their hands and tried to bring them together. But the deal was not clinched, and the two men broke away with false acrimony, the old vendor putting on a look of disgust. The young would-be buyer stormed off – but did not go far. A deal was close.

The negotiator took the old man to one side and whispered intensely, man-to-man. Then he took the purchaser aside, and, with an arm around his shoulder, reasoned in a fatherly way. I noticed a slight physical change in the tall young man, an almost imperceptible release of tension around the shoulders. The negotiator led him back towards the older man and brought their hands together again. Reluctantly, they shook. The negotiator gave a belly-roar of laughter, knelt and slapped the earth. Youth handed Age 1,100 yen in grubby, pulpy notes. The donkey had a new owner.

All across the market, similar deals were being struck, and butchery was also going on. As I turned to leave the market, a man ran past me holding a sheep's stomach in his hands, white and bulging and slopping like three plastic bags full of water.

CHAPTER FIFTY-EIGHT

Karakorum Highway

I WAS STANDING in the early-morning cool waiting for the bus to leave Kashgar, and wondering why there was a delay. I realized there was an argument going on behind the bus.

Passengers were asked to heave their baggage onto a large weighing machine to check for excessive weight. Mostly this meant Pakistani merchants, returning home with bulky bales of silk wrapped in tricolour nylon, but for ritualistic or bureaucratic reasons everything was weighed. My heavily stuffed rucksack was deemed overweight, and some trifling sum levied. So, it seemed, was the bag of another foreigner, who was now making all the fuss.

He was a small, sallow man in his early thirties, with sleek dark hair and a neat moustache. He was unleashing a torrent of abuse on the official who had wrongly weighed his bag and falsely levied a charge – which made him (the little man informed him), like most Chinese, a fucking crook.

'Excuse me, where are you from?' I asked.

He turned and grinned, assuming that I was offering Western solidarity. 'Belgium,' he said.

'Well if you must be abusive, please do it in your own language, not mine.'

His face darkened. 'This man is a fucking liar!'

'The bus is going up a vertical mountain pass, it's not unreasonable if they want to know its weight. The charge is minute.'

'My bag is not overweight! He is corrupt!'

I looked at the Uiygur official. His impassive determination made me suspect that he just might be honest.

'Even if you're right, you don't help your case by abusing him.'

The Belgian began to scream. 'You don't know what I have been through! I had my bag stolen in Beijing! I had my camera stolen when I was asleep on a train. Everywhere in China they are overcharging, trying to rip us off.'

'They're poor.'

'I have been to *other* poor countries! In South America the people are as poor as the Chinese, but they are happy, friendly! The Chinese are greedy, unhelpful, the worst people I have ever met.'

'And how do you think *you're* behaving? Anyway, make your bloody mind up, you're holding us up.'

I turned and walked away. Even as I sat in my seat, and my anger began to subside, I realized how hypocritical I had been. Every traveller knows the moment when the frustrations of travel finally boil over into incoherent rage. In truth, I too had found China frustrating. But in my determination to be unpatronizing and understanding, I had taken it all out on the Belgian.

Maps suggest that the broad band of desert from the Atlantic to north-eastern China is interrupted by the Himalayas. But much of this mountain range is dry – most of Tibet, for example, consists of arid high plains, technically a desert. To continue my journey west, I wanted to travel from the Taklamakan Desert to the Thar and Baluchistan Deserts of the Indian subcontinent. The route I would have to take cuts through the Karakorum range of the Himalayas – the Karakorum Highway.

The world's highest strip of tarmac, it runs 800 miles from Kashgar to Rawalpindi and Islamabad, via the 14,000 foot-high Khunjerab Pass. It took twenty years to build, ten more years to pave, and cost the lives of hundreds of workers, many of them criminals and political exiles. It connected and cemented accord between China and Pakistan, and infuriated the neighbour and long-standing enemy of both countries, India.

The bus climbed south into the eastern Pamir Mountains, parallel with the border of Tajikistan, near the westernmost point of China. The road was cut into blood-red sandstone and grey granite cliffs, in which veins of white rock forked like lightning. Grey rivers plunged by. The road attained a plain dotted with reddish sand dunes, and climbed another pass, before descending to a treeless valley ringed with snowy peaks. There was a broad lake, with clouds of yellow dust drifting across it like sulphurous steam. By the shore I saw a small settlement of yurts, with grazing sheep nearby. Men in fur hats herded a couple of dozen shaggy-coated cows.

The white slopes came closer, and the temperature in the unheated bus fell. For once, the silk long-johns I had carried through a dozen deserts proved useful. Even the grazing camels were wrapped up against the cold in woollen overcoats. Himalayan marmots, bundles of brown fur with erect furry tails attached, tumbled down the rocky hillsides like playful schoolboys. Sometimes they stopped, and their pert, triangular heads watched us pass.

The road wriggled and climbed. We crossed some fertile river valleys, but more barren, boulder-strewn plains. We saw almost no other vehicles, and few human beings, though I remember one girl of sixteen or so, rosy-cheeked and wrapped up in furs, who moved her sheep out of the path of the bus and turned to wave as we went by.

Apart from the engineering achievement, I found it hard to see the real purpose of what the Chinese call the 'China-Pak Big Road'. Sea transport killed off the Silk Road long ago, and scant traffic passes by this route; Kashgar has been telescoped closer to the capital of Pakistan, but it remains an enormous distance from any centre of population or industry in China. As a road, the Karakorum Highway is pointless; but as a piece of Cold War propaganda, and a means of enraging the Indians, it has succeeded brilliantly.

Of course, I am being cynical; in twenty years' time, when Kashgar is full of Western fast-food franchises, the Karakorum Highway has been upgraded so that Pakistanis think nothing of driving across for holidays, and the pass is blocked by traffic on summer weekends, it will be obvious to everyone why it was built: because human beings have a compulsion to make the world smaller.

After a day of slow ascent, we passed a night in Tashkurghan, a bleak little town magnificently enfolded in mountains. At dusk I walked to an old fort on the edge of town, and looked out across the wide, spongy green flood meadows. Sunset came in with a roar, monstrous orange and violet clouds churning down the valley.

Late into the night, squads of Chinese soldiers in T-shirts, combat trousers and heavy boots stomped up and down the main street, chanting Chinese marching songs.

At its apogee, the Karakorum Highway was not overshadowed by jagged white peaks. The surrounding summits, though snowy, seemed little higher than Scottish mountains, giving the very misleading impression that we were at sea-level. Although the Karakorum Highway is on the roof of the world, where plate tectonics are forcing the mountains higher at the rate of two inches a year, the highland greens and greys and purples are for the most part not ravishingly scenic. The Hunza Valley further south numbers among the loveliest mountain landscapes in the world, but the Khunjerab Pass has a testosterone toughness. As the bus looped down towards Pakistan, I stepped into the open doorway and, hanging onto a pole, leaned out, so that the rocky river valley seemed to be rushing up towards me. There were purple rocks shining darkly as though wet, with stains of ochre like trickles of paint stealing across the face of a fresh watercolour. Knuckle-white streams clawed by.

As we descended further, the landscape became more dramatic. There were steep sandstone canyons with towered buttes almost indistinguishable from man-made castles, hectic confusions of angles like vorticist canvasses. The road darted under cliffs that leaned forward as though frozen in mid-collapse. It looped around granite protrusions like snakes' heads, slid through carved-out clefts and tunnelled into the sheer rock.

Much of the Karakorum is not a mountain road in a sense that would be meaningful to a Swiss road-engineer. There are no expensive concrete underpinnings, no cliff-faces secured with steel ropes and bolts, no slicks of hardened tar. The road is essentially a bigger version of the donkey trail it replaced — provisional, and in some places as unstable as the slip-face of a sand dune. The friable sandstone cliffs are in constant revolt, hurling themselves in the paths of vehicles. The Karakorum Highway is not so much a road as an unreliable truce.

Gangs of workers labour constantly to keep the road open. The man next to me had journeyed from Pakistan to Kashgar a week earlier, when a landslide cut the road off, as it often does. 'The entire hillside slipped,' he told me. 'I'd never seen a natural object that large in motion, and I never expect to again. It just took away the road behind us, and if we'd come along thirty seconds later, we would have been dead. It threw a mass of rocks into the air, quite big ones, and they hailed down on us – it was like being under mortar fire. The roof was loaded with cloth bales that absorbed the impact, but a rock the size of a fist came through the window next to my face. It bounced off the back of a seat and hit a woman on the other side of the aisle in the face, and she started to bleed everywhere. I tell you, this is not a safe place.'

Our bus came to a halt behind a queue of seven or eight vehicles. We climbed down and went to take a look. Some of them had been waiting two days, we were told. This was the second landslide since my neighbour had been through. Now bulldozers were clearing away the last dust and rocks. Half the road's width had tumbled into the river, boiling greyly thirty feet below, and mechanical diggers had had to construct a new road. Massive rocks dislodged from the cliffside had left craters in the remaining road surface, which the bulldozers had filled with rubble.

The whole hillside just ahead of us was quivering, like a living thing whose massive flank might stir at any moment. High up, rocks would occasionally break loose and roll a few feet down the hill, throwing up gouts of dust, their rumble and the higher-pitched rattle of a score of smaller rocks amplified in the cone of the canyon. When rocks came all the way down, they threatened to bring the whole hillside with them.

We watched as the driver of a caterpillar truck smoothed out the last bumps in the newly carved road. Each time he shovelled a mound of shale and dust into the river, another miniature landslide would deposit a fresh heap somewhere along the fifty-foot embankment. Every few seconds his head would jerk over his shoulder and he would squint at the slope above him to see whether it was about to slide. With infinite patience and delicacy he would push the new deposits over the edge, trying not to disturb the equilibrium of the hillside. It was perilous work. I was beginning to see how so many hundreds of lives might have been lost in the construction of this highway.

At last the road was clear. The drivers of the halted vehicles looked at each other. Would the hillside remain stable long enough to let them pass? A car or bus could only advance across the potholed surface in first or second gear. If a landslip started, it was unlikely that a vehicle would be able to accelerate to safety. High on the hillside, a massive, finger-shaped boulder was tilted forward at a crazy angle. Small stones kept rolling away from its base, sending up geysers of dust. Obviously there would be another landslide before long.

It was decided that we would advance, but very slowly, and one at a time. I, along with several others, decided to cross the space on foot. I knew I could run faster than the bus could drive.

Several times, pools of dust oozed onto the road, but it remained passable. Our bus was the last to cross. I stood, watching it inch forward, the hillside above it trembling. As we drove away, the caterpillar trucks manoeuvred into position to keep working on the road. I did not envy them the job.

At last we crossed the border into Pakistan, and were greeted by two tall and handsome guards in knee-length *shalwar* shirts and smiles full of big white teeth. These strapping men were yet another part of the ethnic jigsaw of these mountains, racially quite different from anyone I had seen in Xinjiang.

I climbed out of the bus and smiled at one of them. He grinned back, and said, as though he had read my mind, 'There is a *chai* shop here, behind this building.'

I thanked him, and went to find some *chai*. After a month of drinking green jasmine tea, at last: black tea, served with milk. It felt like home.

CHAPTER FIFTY-NINE

Pakistan

THE HUNZA RIVER is born close to the Karakorum Highway and flows south into the Indus, which continues south, bisecting Pakistan, to greet the Arabian Sea a little below Karachi. Above the river, the Highway is scored into the edge of a cliff with no crash barriers, the slopes falling vertically hundreds of giddying feet. The road was made more perilous by our driver's decision to have a race with another bus.

As is usual in these situations, none of the passengers protested. Our laden buses raced to overtake each other on straights, and careered into corners on the wrong sides of the road. The passengers grinned lop-sidedly at each other and shrugged, as though to say, 'Ineluctable forces are at work. Mere humans can do nothing to resist.' We were sentient sheep, grinning all the way to the slaughterhouse. But then again, I have yelled myself hoarse at drivers trying to make them slow down: it never works.

'A large bus went over the edge recently,' the man in the seat next to me confided, thrown against me as the bus hurtled round a corner. He grinned fatalistically. 'Everybody died.'

'I'm beginning to wish I'd flown.'

'Oh, one or two years back a plane crashed in the river too. Everyone was killed. The airlines don't have a very good record.' Still the mad, fate-accepting grin. Evidently a statistician, he went on, 'Of course, there are far more accidents with buses than planes, but then again, there are far fewer flights. Proportionally, it is difficult to know which is safest.'

Eventually the drivers got bored with racing, and the odds on our staying alive lengthened dramatically. Southward down the Indus Valley we went. Night came on, the statistician got out, and I sat beside the open window, enjoying the mountain breeze.

A German couple in their late twenties were seated next to me, Brigitte and Peter, two typical low-rent Western back-packers, all crumpled T-shirts and filthy fingernails. They came from Berlin, where they had both worked as taxi-drivers. In their own vocabulary, they were 'marginal', and their marginalized perspective saw society as a heaving amoral ant-heap in which the pursuit of any objective good was impossible; politically, therefore, one was radical, yet inert. They had both dropped out of middle-class families too early to have acquired much education, and they were travelling, as young Westerners often do, through a world of which they knew nothing.

Brigitte told me hair-raising stories of her life as a Berlin cab-driver – by night, for God's sake, as it paid better – violent drunks, masturbating businessmen, customers having sex in the cab, theft, violence, near-rape. They had been on the road for five months, and now, having almost run out of funds, were heading for Delhi, where a bank transfer awaited them. Meanwhile, they had barely enough cash to buy food. 'I see,' I said harshly, thinking she might be after a hand-out – it would not be the first time freeloading hippies had tried to touch me for cash.

I dozed. Occasionally I opened my eyes to see the headlights raking the sides of a cliff. In the darkness one felt safer: the great vertiginous drops were invisible, so somehow unreal.

In the middle of the night, everyone was woken by screaming. A man in the middle of the row in front of us had turned and squeezed Brigitte's breast. A noisy row ensued with much yelling from Brigitte, while Peter, a wispy specimen with a vestigial beard, sat looking uncomfortable and saying nothing.

'You wouldn't do that to a Pakistani woman,' I remonstrated with the breast-squeezer; 'you'd get your throat cut!' He looked sulky, and replied that Western women were different from Pakistani women.

Some time after dawn, the bus broke down, but after three hours the driver managed, to everyone's surprise, to repair it. We crawled on towards Rawalpindi, reaching it in the full heat of a hot day. I wanted to leave the city as soon as possible, and invited the Germans to share my taxi. We drove straight to the bus station and climbed on the first available coach to Lahore.

The 'luxury video coach' had nylon velour seats humidly thick with sweat and dust. Brigitte and Peter slumped together, ate some samosas in greasy brown paper, then slept. I found myself sitting next to an architect, who began a lecture on the state of Pakistani architecture. I mentioned my admiration for the traditional architecture of the subcontinent, with its tall, cool rooms, and he told me about some buildings he had designed for the Pakistani military, using traditional Islamic techniques to keep the interiors cool. 'With tall ceilings, ventilation shafts and courtyards you can control the temperature of a building naturally. There's absolutely nothing you could teach the Mughals about architecture. Today, we are less intelligent. I managed to persuade the military to let me use some traditional techniques, because with soldiers, at least, you can appeal to rationality. Not so with the stupid civil servants or businessmen who want to make some sort of vulgar statement. We end up with south-facing, glass-fronted buildings. The air-conditioning bills are enormous for a poor country – an *impoverished* country. Very evidently a *corrupt* country.'

The last comment was a reference to the three times the coach had been stopped by policemen who tried to extort bribes from two female passengers who were bringing textiles from Iran.

'Can you imagine the cynicism of these cops?' the architect asked. 'These women have travelled thousands of kilometres to buy fabrics and bring them to a marketplace where they can make some profit. And all these bloody bastards of policemen want to do is exploit them for showing some enterprise. How can this country ever evolve?'

I nodded sympathetically. Then the architect tired of bashing his own country and turned on the West.

'Why are there so many murders in America and England?' he demanded.

I protested. 'But there are at least ten a day in Karachi!'

'That's different; it's tribal warfare. But in the West you have these mad people in the streets. Serial killers . . .'

'Well, from time to time, I suppose . . .'

'But why?'

'Well, I . . .'

'At Nottingham University my teacher told me that there are two reasons; one, all the murders are committed by blacks; two, there are not enough police.'

'Then it sounds as though your teacher was an idio—'

'He said there is a spiritual vacuum in the West. I agree with that. Look at Saudi Arabia, it's a country where religion is respected. And they know how to punish criminals: *severely*. So there is no crime. In the West, you have Amnesty International telling everybody they should abolish capital punishment. How can you hope to contain crime if you have no way to punish the wicked?'

'I see. So the answer is a pure Islamic state? Hard line *sharia* law, rigid application of the rules?'

'Yes. It is the only way.'

'Keep women veiled, out of politics and so on? Punish blasphemy? Cut the hands off thieves?'

'The Holy Koran is very clear about these things. The West is corrupt because it has turned away from the word of God. We in Pakistan must not make this mistake.'

'I see. And why is Islamic Pakistan a hundred times more corrupt than the West?'

'We are a young country. And we have been corrupted by you, your colonialism that introduced cronyism here, and by your Western media, your capitalism which introduces greed and irreligion.'

'I see, I see.'

The Germans were becoming a problem.

When we arrived in Lahore, they told me they had virtually no money to spend on a hotel, and what was more, they were terrified of even looking for one. Their guidebook informed them that the cheap hotels of Lahore were the most crime-ridden on the subcontinent; it even offered a nightmare anecdote about some criminals who cut a hole in someone's hotel-room wall to steal their bags. The streetwise, hardened Berlin taxi-drivers were freaked out.

'Look,' I told them, 'I'm planning to stay at Faletti's, it's an old colonial joint. I'll take a double room and you can both share it with me. Free of charge.'

They stared at me.

'It's safe; trust me.'

At Faletti's I duly took a room with two double beds. The place had gone to seed, but the room was huge, with polished black mahogany panels. The adjoining bathroom had a cavernous bath, and dirt-caked Brigitte whooped with joy when she saw it. Steam billowed through the doorway as she filled the bath. For half an hour, there was much splashing and laughter.

I was sitting on my bed reading when Brigitte came out of the bathroom. Naked, she walked across the room, enthusing about the bath. Clothed she had seemed of waif-like thinness, but though her torso was slender, she had big hips and strong legs. Her whole body was rosy from the heat and exuded cleanliness. Where I had felt a sort of disgust at the filthy clothes, broken nails and tangled hair, I now experienced a mild erotic frisson.

She went to her rucksack and rummaged through it. Peter came out with a towel wrapped around him, and blinked at the sight of his woman naked before another man.

To my amazement, Brigitte found a clean T-shirt in the bottom of her rucksack and pulled it on, though nothing else. She bounced joyfully on their bed, her large pubic bush dancing before me.

'Have fun, you two,' I said. 'I'm going to get a drink. I'll be back in an hour.'

Faletti's bar had closed. Next-door was a Holiday Inn, a modern building with all the charm of a cigarette packet, busy with designer label-clad businessmen. Nevertheless, the bar had closed at 8 p.m. I walked back to Faletti's and grumbled bitterly at the receptionist. He leaned forward, and with a nervous look around, whispered, 'Sir, I can get you whisky.'

'How about a cold beer?'

'No, sir. Only whisky.'

'Do you have ice?'

'I can get you ice, sir.'

'Please do. And where can I drink it?'

'I'll bring it to your room, sir.'

'No you won't. Do you have any Coca-Cola?'

'Yes, sir.'

'Well, stick it in a Coke. I'll be reading out on the veranda.'

As we set out the next morning for the Indian border, Peter was feverish and half-conscious. It was a complicated drive, involving a taxi and two crowded buses. Then it took two hours to complete the customs formalities. At Wagah, just inside India, there was a government-run guesthouse, and I went to check it out. Like most new Indian government establishments, it was galloping towards premature decrepitude, but inexpensive. Peter needed somewhere reasonably clean and reliable. I installed him in a room, and he fell on the bed and went to sleep at once. Outside with Brigitte, I asked her how much money they had. It was more than I had expected, and more than enough for a few nights here.

'I'm carrying on now,' I said.

She looked at me wide-eyed. She did not want me to desert them. I did not care: she would just have to look after them both.

The first public bus out of Wagah was badly overcrowded, and I climbed up on the roof, which was occupied by a group of itinerant musicians complete with drums, and a porter from Amritsar railway station. 'It is illegal to travel up here,' he told me, 'so when we go through towns we have to keep our heads down in case the police see us.'

'What will they do if they catch us?'

'They will want money.'

It began to rain. The day had been long and hot, and the cool rain soothed us.

CHAPTER SIXTY

The Great Indian Desert

RIDGES OF SAFFRON SAND stood out on the gently undulating landscape, as did camels and peacocks. There were so many peacocks – birds I had only known as living ornaments – that Rajasthan took on the character of some enormous and enchanted garden. An occasional flame of mango, vermilion or violet flickered in the heat haze – a sari. A woman used her crooked arm to support a pot on her head, the pot's brass throwing off golden light; golden her forearm too, a column of forty or fifty bangles. Men in red turbans; vultures gnawing at a bloody road casualty; buildings made of red stone, and stone used for gateposts, fenceposts. Here were the leaning walls of an abandoned house, like a miniature megalithic stone circle. The sun grew red.

Penny slept beside me, her head on my shoulder. She had flown to Delhi to meet me, and we had spent several recuperative days – we were both exhausted – by a swimming pool. The heat in Rajasthan was fierce, and acted on Penny like a sedative. I, more accustomed to heat, sat with a notebook on my sweaty knee, my perspiring palm rendering the paper impervious to ink.

Sand drifted across the road. It was no one's responsibility to clear it away, but someone had marked it for night-time drivers with a row of white stones. Our driver, Raja, giving his moustache a vigorous twist, launched the Ambassador at it, the car slowing, slewing, bumping gently into the soft drifts, before it regripped the road and surged on.

It was dark when we turned onto the road that led to the old palace – very dark indeed. We pulled into a courtyard whose walls were only dimly visible. I walked through a tall sandstone arch where a single candle flickered. Soon the manager appeared, all solicitude and explanations – 'Sir, the local electricity is down.' He began to issue instructions. Wicker chairs were brought, and *nimbu* – lime – sodas. After ten minutes, there was the distant rumble of a generator and around us lightbulbs began to glow, revealing the pink façades of the courtyard and a garden of tall trees. It was more than the obscure, former Raja's hunting lodge a friend had told me about – and the prices were much higher. 'We're the new owners, sir,' the manager told me, 'we've only just taken the place over. We're going to turn it into a palace hotel.'

'We?'

He named a fast-growing Indian hotel chain.

'But you've more than *doubled* the prices.'

'Wait till you see the rooms, sir. I don't think you'll complain.'

He was right, of course. As the incipient palace hotel was empty, we were given the best suite. It was the guest room of a Victorian country house, updated in the 1920s, when the owners returned from a Mediterranean tour, with gold-framed watercolours of Beaulieu and Napoli. Two gothic windows that formerly faced the courtyard had been turned into wardrobes. On the far side of the room, an arch framed an entire wall of windows, where huge insect screens were black with scores of light-infatuated geckoes and grasshoppers. Under the windows, two art deco armchairs faced each other, with broad blue and cream stripes like the deckchairs on a steamer. The white marble floors were covered with a thick wool rug in wine red and turquoise, while the walls were papered with a warm brown floral design. The cast-iron English fireplace had a carved pink Rajasthani stone surround. It was surmounted by the room's only other touch of India, a statue of Lord Shiva, dancing as Nataraja in a halo of flames. The bed was enormous.

The next morning we rose early, and walked past the broken swimming pool and overgrown tennis courts to the quiet, misty lake. The lakeside buildings had been vandalized and the old walls were crumbling, but this forest around a lake in the heart of the Rajasthani desert was once a source of prolific wildlife, and the scene of prodigious slaughter.

A room in the lodge contained lists of the hundreds of tiger and many thousands of fowl shot here – mallard, pintail, gadwall, teal, pochard, wigeon, red-crested pochard, tufted pochard, white-eyed

pochard, shoveller . . . . There were framed photographs of
Anglicized Indians in Sam Brownes and Norfolk jackets, or Nehru
jackets with front pockets stitched on to hold cartridges, standing with
their broken-barrelled rifles on the corpses of giant tiger. Endless
paintings and lithographs showed scenes of fox-hunting and stag-
hunting; one depicted an elephant crushing a tiger, the dying beast
clawing the elephant's trunk; another lovingly evoked the
convulsions of a grouse at the instant of being caught in a hail of
shotgun pellets. There were photographs of His Majesty King George
VI, and of cloaked viceroys – Harcourt Butler, 1928, Harding of
Penshurst, 1920. Other photographs froze the family in the 1960s, the
last time they inhabited this lodge, in shiny narrow-lapelled suits,
Jackie Kennedy cocktail dresses and wraparound sunglasses.

One of the hotel staff, an old man, a former retainer, had a genteel,
subservient family memory of the Maharajah. 'He was a great man,
sir, one of the great rulers of India. He helped the poor very much.
He went to Britain and he brought back the railway, which
connected us with outside world. He brought telephones also. He
made this place what it is today. Now we are governed by corrupt
government ministers and officials, sir. We would very much like to
be governed by a man like the Maharajah again.'

We travelled west to Jaisalmer, the westernmost city in Rajasthan,
established almost a thousand years ago on the trade route between
Europe, the Middle East and Iran, and Indian civilization on the
Ganges. We were close to the frontier with Pakistan, closed since
Partition. A train formerly crossed the desert, connecting Jodhpur,
Jaipur and New Delhi to Karachi in Pakistan. These days, Jaisalmer is
a terminus. It is a fine, fortified city, its ramparts and buildings hewn
from the soft, golden local sandstone. The city's *havelis* are renowned,
eighteenth- and nineteenth-century mansions of nobles and wealthy
traders, their exquisitely carved façades with oriel windows rising over
the narrow streets in increasingly jutting tiers until their uppermost
bays almost touch, the buildings nestled so close that illicit lovers might
easily open the shutters and kiss. Inside there are warrens of cool rooms
smelling of damp and bat droppings, and scores of elaborate *jali*
shutters, the light-porous lattices that offered wives and concubines
discreet glimpses of the street and the shadowed inner courtyards.

The Thar (or Great Indian) Desert is not as intensely a desert as the Sahara or the Taklamakan. It does not reward the spectator with titanic dunes – even at its barest, the Thar allows glimpses of green, a patch of hardy millet, amidst the brown. It is the world's most populated and most cultivated desert, where even in what seems like profoundest wilderness one looks up and sees a turbaned head bobbing along on the other side of a hill, or a camel, or even a bicycle. But it is a desert nonetheless, picking up, after the interruption of the Himalayas, the Gobi's arid theme, which then plays almost uninterrupted until West Africa meets the Atlantic. And the Thar has experienced the whole gamut of horrors that permanently haunt the earth's dry lands – drought, crop failure and famine.

The climax of our journey was a camel safari through the dunes south-west of Jaisalmer. At night, our guides baked us bread, fisting lumps of dough and burying them in the sand under the fire. It was a technique I had seen in the Sahara; the bread rolls emerged caked with grit, carbonized and heavy, but good; good dipped in the fresh curries cooked by four or five of our entourage, slicing and chopping together, while Raja, our driver, comfortably playing the urban sophisticate, made them laugh with anecdotes from a life on the road.

Someone – I think it must have been Raja, a cheerfully self-confessed alcoholic – had brought a pint or two of 'desi whisky', or hooch, and long after midnight our four camel-handlers, one guesthouse-owner and one car-driver would lean contentedly together, wrapped in shawls, and sing old songs. One of them, Salam Singh, insisted that they sing us the 'Makana' song, the many-versed lament of a desert shepherdess.

'"Makana my love," she is singing, "I remember how we met at the well where I had gone to collect water, and you had brought your goats, and we would make excuses to stay out in the desert, and sit under our tree. But now my mother and father are pressuring me to marry another man – but Makana, I want to marry only you. Makana, I am in love with you!"'

'This is the most popular, the most traditional song of the desert. Every desert man, every boy taking his sheep out into the desert to feed, they are always singing the Makana song . . .'

There is music everywhere in Rajasthan, the music of troupes in elegant costume when one visits a palace, the plangent flute of a beggar-musician when one pauses at an ill-frequented ruined castle or temple deep in the desert. There is a highly developed visual tradition too: the mirrored, embroidered saris in gorgeous colours that defy the desert's uniform shade, the patterns carved into the havelis' soft sandstone. Every Rajasthani village has its singers; indeed, music plays a vital role in the life of this bleached and bleak desert state, a greater role, it seemed to me, than any other state in India, and any other desert I had visited.

In Jodhpur, we met Kamal Khotari, a musicologist who has spent his life collecting the folk music of Rajasthan. 'Desert Rajasthan is musically very rich,' he told us. 'Music plays a fundamental role in life, in the distinctiveness, the sense of identity, of the various different social groups in this region. A number of musical instruments indigenous to Rajasthan are found nowhere else in the world – they're not just rudimentary sound-making objects, either, but highly sophisticated musical instruments.'

The question in my mind was Why? I had visited oases in some parts of the world where isolation has clearly inhibited the development of decorative and artistic traditions.

'Well, the Indian desert is of course heavily populated, hot and terrible, but not so vast that people are completely isolated from one another. It *is* a social place, with trips to attend weddings and religious ceremonies, and with hundreds of groups of travelling musicians. And one thing is sure: people have time, here. They are not caught in the drudgery of running for the office at eight o'clock, and running back again at five; they have long nights in the winters, and that is the growth period of the cultural year in Rajasthan. Life otherwise is tortuous and hard, survival is difficult, and people use music to make their lives aesthetically valuable.'

We also met Professor S. M. Mahnot, a Rajasthani and a renowned environmentalist, who agreed with Kamal Kothari that the climate was the crucial factor in Rajasthani life.

'You have to understand that desert people are lethargic.'

We laughed. 'Lethargic?'

'Well, obviously. In the summer months, when the temperature goes up to fifty degrees in the desert, you cannot work – your body physically does not permit you to work. By and large, the crops are grown after the rains, which takes about three to four months; so you're only busy for, let's say, five months of the year. The rest of the year you have no work. An enormous amount of time is available for other activities — social, cultural, religious activities. This gives the desert a cultural uniqueness, despite the terrible lethargy!'

Penny was going to return to London, but I would carry on west, to end my desert circumnavigation in the Middle East. But to reach the next desert, the desert of Baluchistan, I had to make an enormous loop north.

CHAPTER SIXTY-ONE

Civilization

Wandering seemed no more than the happiness
of an anxious man

Albert Camus

IT WAS IT MELANCHOLY to be travelling alone again, and tiresome to be retracing my steps overland via Delhi and Amritsar to Lahore. On the Indian side of the frontier, I went into a tea-shop. A boy of sixteen brought me tea, and as two Moslem women went past, heads covered, he muttered something I didn't understand. 'What?' I asked.

He made a ring with his thumb and forefinger and inserted his other forefinger, parting his lips in an unpleasant grin. 'Good fucking.'

'Shut up,' I said, taking back my penknife, which he had picked up and had been using to clip his filthy nails.

I had waited days in Delhi for an Iranian visa. Halfway to Amritsar, I had realized that my Indian visa was a day out of date. Would the customs man notice?

Of course he did. I affected surprise and horror. He remained low-key, leaning towards me and murmuring, 'One hundred dollars.'

I tried to appeal to his good nature.

'One hundred dollars,' he said, 'or I will send you back to Delhi.'

Sotto voce, we bargained. Finally, he told me to go and wait in the men's toilets. He sidled in after me and I handed over forty dollars – I had got off lightly.

Approaching Lahore, the road followed a canal, the grassy banks lined with feathery willows. Near-naked small boys swam in the silken water. It was an image from somebody's paradise. I am a camera, capturing the lean form of a boy frozen in a swallow dive, or in the midst of a coronet of spray, hands outstretched, exalting in the fusion of water and sunlight.

Lahore: spreading trees, classical palaces in paradise gardens, new cars with grinning passengers happy to be insulated from the slower, pedestrian mass.

I planned to take the next available train south, to Karachi, then on to Moenjo Daro, the great ruined city of Indus Valley civilization. After a lengthy enquiry I was told that I might, just *might*, get a seat on a train the next day.

Suddenly I felt irritated – tired and lonely and fed up. Lahore, a leafy and civilized city, was devolving into a hell-hole. I wanted out. I walked into a travel agency and asked when the next flight for Karachi left. In ninety minutes, they said. I bought the ticket, hailed a taxi and raced to the airport.

The flight purser had a resonant and beautiful voice. 'Ladies and Gentlemen,' he said, 'let us begin our flight with a prayer to our Holy Prophet, Peace Be Upon Him.'

As I came out of Arrivals, a boy attached himself to me, gently lifting away my rucksack. I tried to shake him off, but he deflected my rebuffs with a quiet, dignified smile.

I planned to fly on to Moenjo Daro, 200 miles north of Karachi. The one daily flight left at dawn. The PIA ticket-counter here resembled a wasp's nest, with thirty jostling people. I plunged in.

Suddenly I realized I had forgotten about my rucksack and jerked my head around: the boy met my eyes with a reassuring and slightly reproachful smile that said, 'How could you doubt me?'

I reached the front of the scrum, to be told I would have to come back an hour before the flight.

'Where can I spend the night?' I asked my porter. 'I need to get up in four hours.'

He led me, as one might a sick child, to floor seven, negotiating a grumpy guard with the soothing touch of a young wife. Under a sign saying IN-FLIGHT FOOD COMPANY, a door led to a canteen.

'You can sleep on the floor here, sir,' he said. 'I will go and talk to them.'

'What's your name?' I asked.

'Saleem, sir.'

He returned with a pot of tea and poured me a cup.

'So how much do I owe you, Saleem?'

His face became sublime. 'As you like it, sir.'

I had observed his trousers, badly worn but pressed that morning; the skin on his delicate hands and forearms, marked by some sort of disorder; and his dignified, angelic smile. I handed him a note. He folded the bill, smiled again gravely, then almost skipped across the carpet and out of the door.

It was an average night in Karachi. About eight people were shot or stabbed or incendiarized, according to the next day's papers, but I slept soundly on the floor of the staff canteen.

At five-thirty I woke, drank some tea, scraped off some bristles in the canteen toilet, and rejoined the PIA scrum. My ticket could still not be confirmed.

'So,' I asked, not unhysterically, 'I spent the night on a restaurant floor for nothing?'

'You must try to check in,' said the ticketing clerk.

I followed his advice. The check-in clerk laughed in my face. 'But this is an *unconfirmed* ticket!'

'Yes. How much room is there on the plane?'

'Please wait.' He walked away. I waited. The plane was due to fly in twenty minutes. He returned.

'Give me your ticket, sir.'

'Are there any seats?'

'Yes, sir.'

'Thank God.'

'Sir, you are our guest in Pakistan.'

'But if there are seats available, why couldn't they confirm my ticket just now?'

He eyed me superciliously. 'Sir, we don't *have* to let you onto this plane. We're doing so to make you happy. Have a nice stay in Pakistan.'

Chastened, I walked to the departure gate. The plane I climbed into, a couple of minutes later, was half-empty.

The ruined city of Moenjo Daro is said to be the cradle of civilization on the Indian sub-continent, the most important of the scores of Indus Valley sites that extended for a thousand miles from the Arabian

Sea to the northernmost tributaries of the Indus. At its height, Indus Valley civilization covered almost a million square miles, as much as the Mesopotamian or Egyptian civilizations. Moenjo Daro was born about 4,000 years ago, and was at its peak from 2300 to 1700 BC. Archaeologists believe forty or fifty thousand people may have lived here.

It was supposed to be dangerous to visit Moenjo Daro. My guidebook talked of murderous dacoits lurking in the woods, and troops of stern soldiers who frog-marched visitors around the site. But, as the government bus drove me from the airport, I was struck by a sense of absolute rural tranquillity. I was able to wander around the Moenjo Daro unguarded, indeed alone.

The heat was astounding – a humid heat, because the broad Indus was only a couple of miles away. I perspired heavily. The river used to flow past Moenjo Daro's walls, and over the centuries it rose, slowly threatening the town. No one knows how Moenjo Daro fell, but some theorize that it was due to a catastrophic flood of the Indus. (Others believe that the Moenjo Darans, apparently a Central Asian people, were overwhelmed by Aryan invaders.) What is certain is that today the Indus threatens the town anew. The water-table is rising, eating away at Moenjo Daro's brick foundations. UNESCO has worked with the Pakistani government to preserve the site, and gangs of men labour in the sun, digging new damp-courses, while pumps chug all day long, draining off the seeping river water.

The well-ordered, neat-bricked, two- and three-storey homes of prosperous merchants reminded me of Roman buildings with their regular red bricks under the stucco and marble. Finely worked sculpture and jewellery have been found here, in copper, silver and gold, testifying to the high quality of life. On the eastern flank of the town there are granaries, and a sloping-bottomed swimming bath very similar to what we would build today, though Moenjo Daro's bath is assumed to have had a ritual significance. The length of the city's decline is demonstrated by the fact that its most prominent building, a Buddhist stupa, was erected in the second century AD – two millennia after its zenith.

For fear of accelerating the water damage, archaeological digs have not been conducted for decades. But it is known that the rich, agricultural Indus Valley civilization had trade contacts with Mesopotamia, Central Asia, India and China. Moenjo Daran religion included elements of mother-goddess worship and the cult of the bull, which have clear parallels with Hinduism. Two centuries after the peak of Moenjo Daro, Vedic culture, the foundation of Hinduism, was established in what is now India, on the land between the Yamuna and the Ganges.

Two great fascinations of Moenjo Daro remain to this day. One is that its script, on which archaeologists have toiled since 1875 – most recently with the aid of computers – has never been deciphered. The second is that no weapons have ever been found in any of the Indus Valley sites, or any evidence of a warrior class. Is it possible that a specifically non-violent religion was the unifying force behind this long-lived and successful civilization?

My hotel room had a filthy bathroom and a broken shower. I had a boy bring me a bucket of water to wash off the sweat, then I went out to lie under a tree on the coarse, sun-burned grass. At a standpipe some boys were stripping the scales from a basket-full of large, freshly caught Indus fish.

It was a quiet night, with nothing under a starry sky but the clicking of cicadas.

I reached the rather dismal town of Larkana early the next morning, to find that I would have to wait all day for a train. Larkana is proud of being the birthplace of Prime Minister Zulfikar Ali Bhutto, even though his political career ended in 1979 with him dangling at the end of a judicial noose. I installed my bags at the utilitarian Asia Hotel, and explored. It did not take long, for Larkana's attractions are limited. I wondered how, and where, I was going to spend the day.

Just off the marketplace I was surprised to stumble upon a doorway leading to what looked like the squat dome of a small Hindu temple. It was a reminder that Hinduism and Islam mingled across the subcontinent, and that not all of Pakistan's Hindus had fled during Partition in 1947. Nevertheless, it felt strange to go from the streets of an Islamic town – Islam being fiercely antithetic to any form of idolatry – into a building whose centrepiece was a graphic life-size statue of the self-decapitating, blood-pluming Chinnamasta, who unites sex and death, standing on top of copulating Rati and Kama – Lord of Desire – and using the energy generated by their intercourse to nourish her disciples with her own blood.

Yet it also felt oddly soothing. I crouched on a cement ledge in the tree-shaded courtyard. An indifferent cat was stretched out near me, and a man lay sleeping too, his singletted belly slowly rising and falling. From time to time throughout the day, worshippers came in to kneel and pray before Kali and her fellow deities, sometimes leaving behind them petals or a scattering of rice. No one objected to my presence, I was accepted; Hinduism is that sort of religion.

At 8 p.m., I boarded the night train to Quetta, sharing a wooden bench with five other people. It was a night of punishing heat; we sat packed together, sweating and peering past the barred windows at the nothingness beyond. The train passed through the neighbouring towns of Sibi and Dadhar, renowned in Pakistan for the overwhelming heat of their summers. A Persian proverb asks, 'Oh Allah, having created places as hot as Sibi and Dadhar, why bother to conceive of Hell?'

Outside Dadhar is Mehrgar, the earliest site of settled agriculture in southern Asia. Evidence has been found here of human settlement 10,500 years ago, when Britain was still connected to continental Europe by land, and the ice age had yet to retreat.

When I came to, soon after dawn, the train was climbing through the Bolan Pass, one of the oldest trade routes in the world. It passed dusty hills cross-hatched with goat trails, and dome-shaped rocks pocked with caves, where, as clothes lines and hanging washing proved, people lived.

CHAPTER SIXTY-TWO

Hell is Other People

PLANTED AT 1,000 FEET, where the Toba and Kakar mountains begin their ascent towards Afghanistan, Quetta has long provided a retreat from the heat of Sibi, Dadhar and the plains. Destroyed in 1935 by an earthquake that killed 40,000 people, it has been rebuilt in uninspiring concrete. The town's flavour comes from its tree-lined streets, and the presence of tall, imposing men in beards and sheepskin coats, some of them carrying guns. I had been in town a day before I realized that I had yet to see a single woman – even in the marketplace, it was men who did all the shopping. The cinema was showing a Hong Kong movie called *The Policewoman and the Whore*.

Rarely, in the course of a long journey, had I deliberately travelled with a companion (with the exception of Penny). In a café in Quetta, writing up my notes and waiting for the twice-weekly train that crosses the Baluchistan Desert to the border with Iran, I had the misfortune to fall into conversation with a thin, manchu-bearded Englishman called Barry. He had driven a Land Rover from England to Nepal, and was now returning home. His route would take him through Iran, and he expressed an interest in seeing some of the desert places that I wanted to see. We agreed that if I helped Barry service his Land Rover, he would drive me through the Iranian desert.

There had seemed to be something inherently interesting about someone with the nous to set off on an adventure across continents in an old vehicle he maintained himself. And we decided to share a hotel room to save money. But after two nights of Barry's non-conversation, I had lost the will to live.

The vehicle was parked in the hotel's open courtyard, and all day long it baked in the sunlight while I basted underneath it. Over three days we executed a by-the-book 10,000-mile service that involved me lying under the vehicle while oil ran down the insides of my arms and gathered in my armpits. The ex-GPO Land Rover contained, like an earthworm, the means of its own regeneration. Its insides were a mobile workshop, with every conceivable spare split pin and sprocket-thingy taxonomically arranged. Most of them needed to be inserted into arcane, dark and unpleasantly sticky recesses of the vehicle's nether parts.

One morning, a sprightly person with a French accent poked his nose under the Land Rover's bumpers and enquired whether we were going to Iran. Yes, we were. Well, he, Olivier Noël, and a friend were driving a Hindustan Ambassador from Delhi to Paris. Would we by any chance be interested in crossing the desert in tandem?

The Ambassador is the car that symbolizes Indian industry in its state-socialist phase, all Five Year Plans and abject inefficiency. This ghastly yet much-loved former Morris Oxford dates from the 1950s, and is still being chucked together at a factory in Calcutta. The car's technical performance, Olivier told us, was disappointing. Little did we know that the rattling Ambassador would effortlessly outperform our Land Rover.

A few miles from Quetta, the Landy served notice that it had no intention of travelling at any more than forty miles per hour. By mid-morning, it had revised the figure down to thirty. When the road was bad, this did not matter. But when there was tarmac in an approximation of good repair, the Ambassador disappeared over the horizon. It could not, however, surpass the Land Rover's ability to retain heat within its cabin while diffusing a nauseating odour of petrol.

Olivier was Belgian, a refined, sensitive person with impeccable manners. His French companion, also called Olivier, owner and principal driver of a zombie slice of British automotive history, was a stocky and freckled aristocratic type with a gung-ho enthusiasm for life. They both wore immaculate designer leisurewear, contrasting with my cotton Indian *kurta* and Barry's filthy never-removed T-shirt.

Two incidents have stuck in my mind from that snail's-eye view of the largely featureless Baluchistan Desert. Halfway across, the good-natured Olivier Noël suggested that the passengers might swap cars for a while. I remember his bright-eyed expression as he climbed into my seat; and three hours later, the look of shocked sobriety as he staggered out again. 'That man has . . . no conversation,' he managed to utter. 'No,' I said grimly, climbing back into the Land Rover.

The second incident was French Olivier's tea-making. He seized a plastic bottle of water that had become superheated in the sun, popped in two tea bags, shook it for a minute into a hazelnut-brown scum and announced, '*Voilà*, eh naice keurp eurve teeee.'

This unprovoked stirring of ancient race hatreds might have caused an incident – had it not been for our unity as rational beings in the face of the thing that was Barry. 'I was travelling with my girlfriend,' Barry told us in his dismal monotone. 'We'd planned this trip of a lifetime together for years. But we'd only been on the road for three weeks when she met a German motorcyclist in Istanbul and went off with him. She told me I was too boring.'

The Oliviers and I stared at the table, terrified that if we met each other's eyes we would laugh out loud. 'How terrible,' one of them managed to force out at last, through clenched jaws.

On the evening of our second trans-Baluchistan day we reached the vilest town I saw in this entire journey, Taftan, a collection of hovels on the edge of the desert, existing only to fleece hapless travellers and import contraband petrol from Iran. We ended up camping in the police station compound.

It took an entire day to clear the Pakistani and Iranian customs, a day of hallucinatory frustration. The unfortunate Olivier Noël was found to have used up his Iranian visa on a night in transit in Tehran, and obliged to return to Quetta. The rest of us were allowed into Iran, but forced to spend the entire night locked in the compound with fifty Russian truck-drivers.

The next morning French Olivier invited me to join him for the drive to Istanbul. I would have loved to do so; but I had my own plans.

CHAPTER SIXTY-THREE

Iran

AT DAWN WE ESCAPED from the the Iranian border post. Barry and I arranged to rendezvous with Noel in the next town, Zahedan, and set off. Noel disappeared ahead of us. The landscape was softer, more green-flecked, than Pakistan – we had left behind the yellow nothingness of the Baluchistan Desert.

The road was different, too, macadamized and cats'-eyed. There was orderliness and relative prosperity here, and I felt a sense of relief. I was about to learn how eccentrically Iran occupies a space between the Orient and the Occident. Europe was still many miles to the west.

I was obsessed by how to be rid of Barry, with whom I had made this insane pact to travel through Iran. Yet in the event, separation proved astonishingly easy – as he was equally eager to be rid of me.

'We don't really want to do the same things, do we, Barry?' I ventured.

'No,' he said.

And that was that.

Zahedan might have been some light-industrial Italian town, with dull new buildings, dry, clean streets, and a scattering of shady trees. Unlike Italy, however, every woman was covered top to toe with a black *chador*.

Noel did not turn up at the appointed meeting place; after we had waited half an hour, Barry drove to a curious government campsite with permanent glass-fibre tents, and asked me if it was OK to drop me there. It would have cost him nothing to take me to the bus station, but I was so eager to escape that I couldn't give a damn. Poor Barry's worst sin was to be boring, but it took a sort of malignant genius to raise boredom to such a pitch.

I was sitting in a bus, staring past the words painted on the window:

You are in my heart
Remember God

It was a pleasantly bulbous children's-book bus, with newly changed antimacassars on the seat-backs, and a spray of flowers in a vase at the driver's wrist. There were only three seats across its width, therefore enormous leg-room – this was First Class bus travel. It fairly glid on the smooth Iranian highway.

We passed acres of grey warehouses plastered with signs saying 'No Photography'. Everywhere I looked there were soldiers – which I attributed to the proximity of the sensitive Afghani border. It seemed ironic that the world's two most fundamentalist Islamic governments had so signally failed to hit it off. The soldiers put me in mind of the eight-year-long Iran-Iraq War, that had sucked to their deaths hundreds of thousands of young men, including many under-age village boys, lured by promises of rewards in the afterlife. Two years after this futile and inconclusive carnage, Iran and Iraq had been allies again, united against the common enemy, the great Satan, America.

The bus halted, and a boy climbed aboard with a neatly arranged tray of travellers' snacks. 'Nuts,' he murmured, 'chewing gum . . .' One man bought a small cellophane bag of shelled peanuts, and the boy silently withdrew. The contrast with the raucous Indian subcontinent was startling. In my ignorance, I had imagined neighbouring Pakistan and Iran to be similar countries, not having grasped the extent to which Iran was a developed country. This made its impoverishment through fanatic self-isolation and trade sanctions even more ironic. The Islamic government has also discouraged tourism from non-Moslem countries. Satanic influences must be kept at bay.

The bus windows were all wide open, the drawn black curtains billowing like spinnakers. Another adolescent, one of the driver's assistants, moved along the aisle, taking down plastic cups that were clipped over each seat and filling them with iced water. I thought of the last coach I had taken in Pakistan, a 'luxury bus' with filth-sodden seats and a blaring video about rape and death.

We climbed into strident sandstone hills with eruptions of purple, red and green. Iran has fertile hilly regions, but most of the country is arid. In the south-east there is the continuation of the Great Indian and Baluchistan Deserts, while the heart of Iran consists of two deserts, the Dasht-e Kavir and the Dasht-e Lut. In the north-east they run into Turkmenistan, the Karakum Desert and the arid lands of Central Asia.

Lest I should be starting to enjoy the journey, the realities of Iranian life intervened. The bus stopped at a road block, and surly policemen came aboard to inspect us. Half an hour later it was stopped again, and this time everyone had to climb down while the bus was searched. At the third road block, we had to unpack our luggage so that it too could be examined – though the women were allowed to stay on board. I imagined them sitting there, contraband stuffed under their enveloping black *chadors* (though at later road blocks even they were searched). The halts continued. By the time we had been on the road for three and a half hours, I estimated that we had spent an hour and a half being searched. On smooth, near-empty roads, we were averaging maybe forty miles an hour. In the course of the journey, the bus was stopped eighteen times.

I wondered why the government subjected the ordinary people of Iran to these indignities. Could it be that like many a previous revolution, the Islamic one had replaced a repressive system with something far worse?

I noticed, halfway down the bus, a young man of extraordinary beauty, with a Grecian cast of features, blond hair, and golden skin. As we climbed down for the bus to be frisked yet again, we found ourselves urinating against the same wall. When we joined the queue, he offered me a cigarette.

'Are you English?' he asked.

'Yes, I am,' I said. I wondered how he had guessed.

'Welcome to Iran.' He grinned sardonically. 'I'm sorry this journey is so boring.'

He introduced himself as Alireza. I complimented him on his English, and he told me that he had studied it at university. Now, he said, he 'bought and sold'. We shuffled towards the knot of policemen who were rifling through the open suitcases.

'What the hell is going on?' I asked.

'They are searching for drugs. Hash, opium, heroin.'

'But these searches are totally inefficient, and they're doubling the length of our journey.'

'Don't worry, it isn't this bad when you get further from the border. But here, it's just something you have to put up with.'

'Do a lot of people take drugs in Iran?

'Well, yes . . . It seems that half the drugs that go to Europe from Pakistan and Afghanistan come through our country. With so many drugs available, it would be surprising if people did not use them.'

'You amaze me.'

'Really? You don't have a drug problem in England?'

'Yes, of course we do. But I had never heard that it was a problem here. One has the impression that life in Iran is so, well—'

'So controlled, yes, of course.' He showed me his lovely white teeth in another sardonic grin. 'Life *is* controlled here. But . . .'

'What are the penalties if they catch you with drugs?'

We were next in line to be searched. Alireza winked. 'I'll tell you later.'

Back on board he asked the person alongside me if they could change seats, settled in beside me, and lowered his voice.

'The penalty for drug trafficking is death . . .'

'Ahh.'

'. . . It's rare, though. For possession of less than five grams of hash, it's a fine, but for half a kilo to five kilos you get fifteen years. And there are huge fines. But now people are getting out of prison after three years, instead of fifteen.'

'Drug use has increased in recent years?'

'Before the Revolution it was unknown. But now a lot of young people are smoking hasheesh. Of course, it has been used in this country for many years. You have heard of the hasheesheen?'

'The Old Man of the Mountains. He gave his followers hasheesh and told them they had died and gone to Paradise.'

'And they became fearless assassins. Correct. Well, of course, people have different ideas about the extent to which hasheesh is a bad thing. But there is heroin addiction – that is more worrying.' He grinned again. 'You are shocked. I think there are many realities of Iran not well known outside. Anyway, nowadays, with the ayatollahs, there are strict controls on alcohol. So people have to do something to . . . *unwind*.'

He smiled, relishing the idiomatic phrase.

A dead camel lay by the roadside, its bleached bones almost phosphorescent in the predusk light. The bus was crossing a grey plain. By the roadside there were clusters of inedible wild melons. The only signs of humanity were an occasional discarded tyre, or a shattered windscreen held together by its plastic laminations, embracing a roadside rock like a crystal cobweb. We passed an ancient fort, with a big and obviously manned anti-aircraft gun mounted on one of its towers – here, seemingly in the midst of nothing.

It was still uncomfortably hot. The slowly sinking sun seemed reluctant to depart, burning to the last and warning that it would be back, in force, in a few hours. At last the hateful disc began to set, leaving the air soupily hot and draining. Darkness settled, and stifled people drifted into uncomfortable sleep.

The two black-*chador*'d women in front of us were eating. They glanced at me once or twice, and I knew what was coming: the gift of a chapatti, stuffed with fried spinach, and one for Alireza. It was delicious, and I managed to produce *motashakkeram* – thank you. They laughed, gold teeth flashing. A chink of light in the black tent.

We reached the town of Kerman in the early hours, and Alireza insisted that I go to his home. When I refused, he became almost angry. Iranians, who have a reputation for religious fanaticism, are even more fanatical about hospitality. But I was reluctant to disturb his household in the middle of the night.

The next morning I found a taxi to take me around Kerman. The capital of an arid province, it is an unprosperous city on a mildly fertile plain that is fast filling with ugly houses. We drove up to the brick and adobe fort that overlooks Kerman. Rain is slowly dissolving it like a sand castle, but its erratic skyline of brick towers and mud bulges make it the most exciting building in Kerman. This ancient defence offers fine views over a town whose biggest threat today comes from construction. The local architectural vernacular is an elongated dome, a sort of lozenge made from mud-brick – a form often seen in North Africa. These cool and elegant old buildings are being torn down wholesale. The area around Kerman's central Masjed-e Jame mosque was a demolition site, fringed by broken pillars and arches where an entire quarter of the mud-domed old city was being cleared. New

Kerman has the anonymity of any mushrooming modern town, rambling concrete cells that do not feed the soul. People have to live and work somewhere – but why must ancient, organic buildings be destroyed?

I found Kerman's covered bazaar intact, a tunnel of black shadows, the darkness intensified by pencil-beams of light from lofty glass ports – and by the fact that all the women wore black. The shops were bereft of goods, and many of them were boarded-up, more evidence of Iran's economic depression. I could not reconcile this gloom with the rash of rebuilding going on all around.

One image pierced the gloom. Under an archway two young country women in brilliantly coloured skirts and blouses sat on their haunches. They had swarthy faces, and as they chatted they threw back their heads and laughed, oblivious of Islamic proprieties. I looked eagerly at their faces, and realized I had not seen a woman's naked face since Lahore. One of them had a baby on her knee. She reached inside her blouse and with a forefinger and middle finger tipped a pale breast into the baby's mouth.

'Martin!'

I heard Alireza's voice as I was trying to find the tea-shop where we had arranged to meet. I was surprised to see him with a girl and two other young men.

We climbed down a stone staircase, a passage lined with old tiles of a warm, yolky yellow. The tea-shop was a former bath-house, a long, cool room of turquoise mosaics, where water in white marble pools caught the smoky shafts of light falling from the coloured glazing in the domes.

I almost started when the girl, whom Alireza introduced as 'my girlfriend, Miryam', lit a cigarette. I was further amazed when Alireza leaned towards her and adjusted the scarf she was wearing, pushing it further *off* her forehead, and plumping up the curls. Miryam had a strong, fleshily attractive face (Kermanis are said to be slightly Indian in looks), eager eyes and a wide smile. They made a good-looking couple. Miryam was grinning at me, dragging on her cigarette for all she was worth.

Alireza was the only one of the four who spoke English. I told them that the next day I was planning to go on to Mashhad, in north-eastern Iran – via the desert.

'It is very strange to want to go to the desert now,' said Alireza. 'It's very hot there.'

'It doesn't matter.'

'I have an uncle who lives in Tabas. You have to stay with him.'

CHAPTER SIXTY-FOUR

Happiness

L IKE MANY WELL-WATERED areas on the rims of deserts, the hills west of Kerman had the hot, fertile feel of Provence. The bus passed orchards of oranges and apricots and climbed on, finally emerging onto a scrubby plain, the start of the Lut Desert. I saw twists of smoke like the hearth fires of a scattered hamlet, but they were a troupe of dancing dust-devils.

Alireza was sitting alongside me. He had decided to come along for the ride.

'Don't you have any work to do?' I asked him.

'I am an entrepreneur, I told you – I have work everywhere.'

I was almost starting to get suspicious about him; he had no regular job, yet he was not short of money; there were his film-star looks, and the carefree attitude that made him stand out sharply from every other Iranian I had met. It crossed my mind that he might be a drug dealer, though he seemed too intelligent, too flamboyant – drug dealers are usually cautious to the point of paranoia.

'Listen, Martin, nobody survives in Iran without a struggle. I am struggling, but I am surviving.'

'*Inshaallah.*'

He grinned. 'Yes, *inshaallah*. You think I am a strange Moslem.'

'No, I've met much stranger Moslems than you. You *are* a Moslem, I take it?'

'*Of course*; but not the extremist kind.'

'Are there any other kinds?'

'Are you serious?'

'No, I'm not.'

We had reached a small oasis, and the bus pulled up for lunch beside a public garden. There was only a small kiosk, and I had to drink Coca-Cola. I found the Iranian variety unfamiliar, faintly reminiscent of dates. The other passengers all drifted towards the garden. Iranians, I was starting to discover, adore gardens.

Alireza said, 'Do you really think all Iranians are fanatics?'

'No, Alireza. I suppose it's because you seem . . . well, freer than other Iranians.'

'There is every kind of person and thing in Iran. Don't believe the propaganda. When things are made illegal, they don't stop – they just change, and sometimes they go underground.'

'Tell me – what exactly do you do?'

He laughed. 'What I do is not very romantic. If I told you that I sold potatoes, would you believe me?'

'Of course.'

'I sell potatoes. Also onions. Also sometimes toilet paper, ladies' underwear and T-shirts, but mostly vegetables.'

'And you're successful?'

'I will tell you a true story that will show you what a shithole this country is. My father is a doctor. After the Revolution, when I was studying, he was prevented from working as a doctor – because of politics. He had to take two jobs. In the day he used to work as a hospital porter, then he would work from maybe six until after midnight as a hotel taxi-driver – they earn nothing, nobody has money for taxis. A qualified doctor, doing manual work for sixteen hours a day to bring in just enough food to feed his family.' Alireza's eyes were moist. 'What I do is legal, and I have almost nothing to do with officials. In Iran, that is freedom.'

Across the bus aisle, a young woman was reading. I noticed that she had a sort of hardcover with a picture of a mosque on it, but tucked inside it was a different book. I hoped it was something subversive. Like every other woman, she was wearing one of the bloody black tents. Some of the *chadors* were, I could see, made of hot, artificial fibres – they must be very uncomfortable. I kept wanting to say, 'OK, I get the point; you can take it off now.'

In the afternoon, mid-way across a bleak desert plain, we came to a road block. There were a few buildings and aerials, a clump of trees and a small mosque. We sat in the hot desert wind, and I watched the young soldiers slowly, doggedly working their way through the papers and possessions of sixty passengers. The Khomeini experiment was amazing – so thorough! So oppressive!

One of the soldiers called to two colleagues who were sitting under a tree, and they reluctantly stood up and came to help him. Nothing else stirred on this great plain.

Salt crystals gave way to black rocks, then red ones, then tangerine streaked with dried blood. The hills had eroded into veins like the roots of giant, fossilized trees.

In the afternoon Alireza went to the back of the bus to chat to some young men, and an old lady climbed aboard and sat beside me. She started a conversation, and though I could not understand a word she said, she kept up a stream of high-pitched gossip, much to the amusement of my fellow passengers. She took out a small penknife and began to trim her nails as though she was peeling a potato.

Another old woman made tea, crouching on the floor, pouring hot water from a thermos into a teapot. She wore black socks, a turquoise nylon blouse, then the black tent, and lastly a pale pink floral scarf, darned many times over. As the guest, I was the first to be served tea.

The sun set as we were traversing another plain of salt deposits. It looked exactly like frost, and with a sharp pang I was taken back to a walk I had made with Penny the previous Christmas, beside a frozen Yorkshire river, opaque puddles fracturing underfoot. Now, across the bleached plain, a white sun sank onto a black mountain.

At 9 or 10 p.m., the bus stopped for a dinner break. Searching for the toilets, I bumped into one of the young men Alireza had been talking to at the back of the bus. He gestured for me to follow him, and we walked down a corridor, then out of the back of the building. He led me towards where his friends were smoking a cigarette, and I caught a familiar smell. 'It's hasheesh,' I said, unnecessarily.

'Sure,' said the young man. He spoke English with a vaguely American accent. He proffered the joint.

'No thanks.' There were armed police a few yards away. 'Isn't it dangerous?'

He shrugged, and said something to his friends, who grinned.

Back in the bus, I told Alireza what had happened.

'So you see, people are not so afraid of being caught any more. But *you* should be careful. If the police catch you with drugs, they will lock you up.'

We reached Tabas in the evening, and Alireza led me through dark, scarcely illuminated back streets to his uncle's house. It was late, but we were well received by Mahmoud, a large and enormously hairy man, who put me in mind of an Afghan brigand. He welcomed us in, looked up and down the street, then closed the door and locked and bolted it several times. His wife, Fariba, was a tall, thin woman about ten years his junior. She hurried into the kitchen and began to cook.

Alireza and I sat on cushions on the floor of a well-furnished room, with large, varnished art deco-style furniture, a thick beige rug, and a number of red and gold-embroidered cushions. Mahmoud vanished for a minute, then reappeared from some hiding place with a selection of alcohol. It was the first time I had seen booze in Iran, and as I watched him jovially slosh vodka into a glass, I realized I might not be in a typical household.

It felt good to sit on the carpeted veranda outside the house, with Mahmoud and his wife and Alireza. Breakfast was rose-scented jam, fried eggs, raw tomatoes and onions, *paneer, nan,* and dainty discs of caramelized, toffee-apple sugar that melted in your mouth with the tea. I was astonished when Fariba stood up and her *chador,* her purdah, unravelled to reveal a baggy American T-shirt and fuchsia shorts.

As I flicked through a copy of the *Tehran Times*, I saw a photograph had been printed, without comment, of an Aston Martin convertible. Mahmoud leaned over in a companionable way to look at the photograph.

'It's an English car,' I told him, with unexpected pride, an 'Ast—'

'English cars no good. Germany cars good.'

'*This* car is *very* good,' I insisted, pointing at the technical miracle of sculpted steel.

He looked unconvinced. Then came the inevitable despairing coda to any discussion of national merit.

'Iran no good,' he said, gloomily. 'Iran *kharab*.' He pointed at his head, and with a twist of his finger sketched the turban of an ayatollah. '*Kharab*,' he repeated, gloomily shaking his head.

'You don't like Khomeini?'

Mahmoud simply held his hand in front of me and curled his thumb and forefinger until they made a circle. His grandfather, an old man with warm eyes and a soft, unintelligent smile, came out to join us. Mahmoud said something in which I caught the word 'Khomeini', and the old man made a vigorous reply.

Alireza laughed.

'What did they say?' I asked him.

'Mahmoud asked our grandfather what he thinks of Khomeini. He said he was a good man. Mahmoud replied that he was a . . . I don't know the word.'

'The old man is stupid!' Mahmoud grinned, without malice. 'He is a good Mohammedan, so he thinks Khomeini was a saint. But no, Khomeini was a stupid old man. Look what he has done to Iran. In the time of the Shah, everything was allowed here, gin, cognac, whisky, vodka. Heh, Alireza! We must make a Haj to London – a whisky Haj!' He turned to his grandfather and bawled this blasphemous proposition. The old gent winced, but kept smiling. Mahmoud turned back to me and his face soured.

'When Khomeini got in everyone who disagreed with him was killed.'

'But Khomeini is almost worshipped as a great holy man. Why did he kill so many people?'

Mahmoud shrugged. 'We don't know. The Shah was not popular, but no one had any idea of what Khomeini would do. Khomeini . . . he was mad.'

There was a knock at the door, and a visitor came in. The atmosphere became tense. He was a man they knew slightly, who had brought a message from a mutual friend. There was polite conversation, and the visitor drank some tea, then took his leave.

Alireza had an expression of disgust on his face. 'Look at us: we are afraid of our own shadows.'

'What do you mean?'

'You didn't understand the conversation? Mahmoud asked him if he liked Khomeini, and he said, "Yes". Maybe he means it, or maybe he said so because that is what you're meant to say. He could be a spy – or we could. He thinks we are, we think he is, and so it goes round, in circles of suspicion. You see how Khomeini destroyed this country?'

Mahmoud said to me, 'You know what I was doing when the Revolution happened?'

'No?'

'Guess.' He grinned.

I stared at him, his unshaven chin bright with bristle. Expecting the unexpected, I suggested, 'You were a policeman?'

He laughed. 'Not bad. No. I was in the army.'

'Did you – fight in Iraq?'

'No! I was thrown out before that.'

'Why?'

'Because at the time of the Revolution I would not grow a beard. They told me I had to grow a beard – to be clean-shaven was unislamic. I told them to fuck off, and they threw me out. Maybe it saved my life, God be praised.'

'There is nothing to see here,' Mahmoud had said, 'but Alireza can show it to you anyway.'

I found the arch built from shell cases somewhat dispiriting, but the most remarkable thing about Tabas was its verdure. The dual carriageways running through the town had grass and cypresses in their central reservations, and there were lush public gardens bursting with date palms. The inhabitants of this oasis *loved* trees. I tried to think of somewhere else that made such a fuss of its green amidst a desert – Palm Springs? Tabas was far greener. There were neat arcaded shops in yellow brick and adobe, lawns, fountains and the palms, giving the place an air of gaiety.

The next morning, Alireza walked with me to the bus stop. I was going to continue north to Mashhad, close to the Afghanistan border, then turn east towards Tehran. The bus was not full, and Alireza said he would sit next to me until it was ready to go.

In two years of travelling in remote places, I had often been asked urgent questions about the real or imagined outside world. But the question Alireza asked me was the most urgent and puzzling yet.

'Martin, if I want to be happy, what should I have?'

'Pardon?'

'What do I need if I want to be happy?'

I tried to think of some irreducible condition of happiness. It could, in Islamic Iran, be a test question. I suggested, 'A love of God?'

He smiled. 'I want to know *your* idea.'

'Maybe that is my idea.'

'OK; if I want me – and one day I will marry, so me, my wife, my children – if I want us all to be happy, what do I need?'

'I, er—'

'I don't need money?'

'Well, I have met people who seem to be happy without it.'

'Can one be happy without work?'

'No, I don't think so.'

The driver had started the engine, and looked around enquiringly. Alireza stood up.

'No. A man needs to work. And if you are not going to go mad, what do you need?'

'I don't know. Are you going to tell me?'

'But I want to know the answer myself!'

He climbed down, and waved as the bus pulled out of Tabas.

The bus drove all day. Slowly the road clawed its way out of the desert. The plains became less barren, then studded with small shrubs that caught the sun dazzlingly, yellow or lime green. Then, suddenly, there were fields of raucous yellow hay, and harvested bundles shining metallically in the sunlight. I put my nose to the window and inhaled the scent of the hay. Orchards, and tilled fields where tractor-drawn ploughs lay among the furrows of soft, pink earth. Ranks of sunflowers, and crops of enormous yellow melons. Poplars and cypresses leaning in the wind. I realized that it was much cooler than it had been this time yesterday. We had left the desert, and someone had planted marigolds along the roadside.

CHAPTER SIXTY-FIVE

Tehran

Marg bar Amrika!
Graffito

IN LARGE CITIES, ENCOUNTERS with people are always more febrile than in the countryside, but Tehran had the particular tension of a place where a desperate struggle for material survival is combined with fear of the police state. It was also ugly and polluted. I checked into a hotel whose windows were so thick with smoggy accretions I could barely see out of them. On a wall outside, a street lamp illuminated the words *Marg bar Amrika!* Death to America!

On my last night in Iran, I decided to visit The Resting Place of His Holiness Emam Khomeini, under construction in the southern suburbs. The official cult of the Khomeini personality seemed undiminished, and all over Iran I had seen photographs of him, banners, even murals that plastered his face across eight-storey buildings. Perhaps with time Khomeini would become an irrelevance, and Westernized adolescents would grin with patronizing impatience every time their elders mentioned his name. But in the meantime, everything was being done to turn Khomeini into a holy figure second only to the Prophet himself. In the sacred city of Mashhad I had walked into a shop that sold religious souvenirs, many of them posters, plates and clocks plastered with Khomeini's brooding likeness. (Islam, I reminded myself, was the religion that supposedly banned idolatry.)

In honour of the Ayatollah, Iran was constructing one of the greatest buildings in the Islamic world. A taxi brought me to a massive five-domed structure, with huge cliffs of grey-slab concrete. The domes were being bricked over, later no doubt to be gilded. But the huge central dome was of great beauty, and hovered exquisitely in the night. The complex was surrounded by a gigantic car-park, now barely a twentieth full. A special metroline was also being built to the shrine.

My driver, Ali, was a tubby fellow in his early forties, genuinely delighted that I wanted to see this place. He solicitously ensured the correct removal of my shoes, and hung on to them himself, urging me to go alone across the cavernous concourse.

The floor was of green onyx. Green-painted, five-foot diameter columns rose to a ceiling bristling untraditionally with industrial struts and galvanized cooling vents. The inner dome was immense, and decorated with traditional mirror-work, a gigantic double chandelier cascading from its heights. Beneath it a grilled cage protected the Ayatollah's remains. A largely female crowd of pilgrims clustered around it, mostly looking enthusiastic, some bored, a few, gripping the bars in their fists, desperate. The cage contained a moraine of banknotes, a mound that started at the very perimeter and gradually rose two feet high. Ascending from this paper tribute was the cloth-draped tomb.

As we drove away, Ali asked me to repeat a phrase after him. He said it in small, easy to repeat chunks, giggling to himself. Three times I mumbled after him, '*Ana ashhadu an la illaha illa Allah, wa Mohammed Rasul Allah.*'

He laughed hugely. 'You are a Moslem! You only have to repeat that phrase three times to become a Moslem – you are a Moslem now!'

Not far from the mausoleum was the Behessht-e Zahra, the graveyard of, among others, the Iran-Iraq war-dead. I asked Ali to take me to it. A square white arch bore the words in Persian and English, with here and there a letter missing,

Do not thi.k that those who were slain in the cause of Allah are de.d, they are alive, and well provided for by Allah.

The graveyard was closed, but Ali earnestly petitioned the soldiers on guard to let us in. The corporal in charge was a pleasant-faced young man, and most unofficious in the face of Ali's pious and patriotic onslaught. Smiling, he waved us through.

Ali drove in a circle through a vast necropolis several miles in diameter, intersecting straight, infinite-seeming moon-grey avenues. We ended up in the township of the war-dead, with Khomeini's lovely dome hovering over the trees. Ali rolled his eyes around, then threw out his arms in all directions, row upon row, thousands upon thousands of graves. 'Saddam Hussein,' he said, 'Saddam Hussein, Saddam Hussein.'

'A very bad man,' I agreed.

He looked sharply at me. 'Not only Saddam Hussein. All against Iran.' He counted off countries on his fingers. 'America, *England* . . .'

I nodded grimly. In the belief that Saddam Hussein was the lesser of two evils, the West had provided him with weapons.

'. . . France, Germany . . .' Ali came to his thumb and paused, trying to think of a fifth country. Football came to his aid. '*Brazil!*'

I thought it best to change the subject. 'Are all the Iran-Iraq war-dead buried here?'

'No!' he said. 'There are thousands of them buried where they died, in Khuzestan, Abadan, Khorramshahr.'

Here and there an oil lamp cast a glow that allowed us to pick our way through the tombs. Many of them had shallow glass cases erected over them, containing photographs and mementoes. I went to one case that was illuminated by a flickering night-light. There was a large framed photograph of the dead boy, a studio portrait showing a soft, lightly bearded face. The case was lined with red mirrored wrapping paper, and curtained with lace. There were smaller photographs, too, and bizarrely, two tiny toy tanks. Had they been his childhood playthings? Or had he been in the tank corps? Something in the shadows at the top of the case caught my eye. Pinned above the studio portrait were three snapshots of the dead body as the parents must have received it, half-wrapped in a shroud, blank-faced, punctured, drenched with blood. Suddenly the whole necropolis seemed to fill with weeping mothers.

Ali led me to another tomb, and pointed. 'My brother,' he said, 'look at his face.' I peered at the photograph, trying to see a resemblance. 'He was bigger than me, two metres one. I am fat now. How old do you think I am?'

'Forty?'

'I am forty-two now . . . He was seventeen.' Even in the dim light

418

I could see Ali's eyes were bright with tears. He said, 'I nearly died too, you know,' and seemed almost guilty, as though he felt he should have died. He took my hand, lifted his shirt and rubbed my fingers against his fleshy flank. I felt the lumpy scar of a bullet wound. He moved my hand around to his abdomen. Another. He moved my fingers to the nape of his neck and rubbed them in the sweaty, curly hair. A third lump. 'Three, I took. This one nearly killed me. But Allah be praised, it was not my time.'

'You must hate Khomeini,' I said.

'No,' he said, with transparent sincerity. 'He was a great man. We love him.'

'But so many young men died . . . your own brother. Don't you feel Khomeini caused their deaths?'

'No! It was Saddam Hussein, America and its friends – your government. Khomeini was a saint. You saw his funeral on TV?'

'Yes,' I said. Who could forget the scenes of frantic mourners, so desperate to obtain fragments of Khomeini's shroud that the pale corpse tumbled from its coffin.

'You will see on his birthday. Thousands will come to that place—' he pointed at the levitating dome '—*millions*, you cannot move. You will meet city people who tell you Khomeini was a bad man, but they are the bad people. The country people – the *real* people of Iran – they love Khomeini, because he loved them.'

On our way back into central Tehran, we were pulled over at a road block by a group of young, bearded men carrying sub-machine guns. Ali was questioned aggressively, but, because of a foreigner's presence, they did not detain him for long. We stopped at a small eating house, and I offered to buy him dinner. We sat down at a table with a white plastic tablecloth, and, as is often the case in Iran, were brought iced Coca-Colas before we had even seen a waiter.

'So the Islamic Revolution took place so that innocent taxi-drivers could be harassed by fundamentalist thugs?' I asked.

He shrugged, and grinned without bitterness. Ali was an incorrigible optimist.

CHAPTER SIXTY-SIX

The Empty Quarter

This cruel land can cast a spell
which no temperate clime can match

T. E. Lawrence

THE IDEA OF 'DESERT' exists in the imagination of everyone, evoked by a handful of iconic images and words – Sahara, Gobi, Lawrence of Arabia, date palms, dunes and oil-wells. But the phrase that distills the essence of the sandy desert is 'The Empty Quarter'.

'Last year I could have (written) – "There is but a hand's-breadth we do not know" – thinking of that virgin Rub' al Khali, the last unwritten plot of earth big enough for a sizeable man's turning in twice or thrice about, before he couches. To-day we know the whole earth.' Thus T. E. Lawrence, sorrowfully reflecting on Bertram Thomas's exploration of Arabia's Empty Quarter in 1927-30. The Rub' al Khali was remote, forbidden, seductive, and it captivated the imagination of everyone who came into contact with Arabia. Many men dreamed of stealing that virginity before Thomas achieved it. As I passed along the southern edge of the Great Southern Desert of Arabia, the Empty Quarter, I knew that I was about to cross a line between legend and reality.

It was a fine road, paid for out of oil revenues. The desert shimmered. Occasionally there were roadside halts where huge numbers of Indians, Oman's multitudinous migrant workforce, crowded at formica tables eating curry. I made a detour into the northern highlands and discovered a country of startling physical beauty, prosperity and cleanliness – a sort of Arab Austria. Frequently there were road signs, '*Mosque 2 km*', '*Mosque 1 km*', informing people of a proximate shady mat when the time to pray approached.

In Muscat, the capital of Oman, I had made friends with Ahmed Al-Muhainey, an amateur historian who introduced me to the precious perfume on which for centuries Oman's fame and wealth rested.

'Frankincense is still a very important element in Omani culture. It's thought to get rid of evil spirits, and protect you from, er, malicious hypnotism. It's cleansing – and of course, it smells beautiful. I don't know which of them is at the root of its use in religious ceremonies – maybe all the above. It has been widely used in Europe, of course, in Christianity. It was taken to India, Iran, Iraq, the Queen of Sheba – and the Greeks and Romans bought it by the ton. Hence the wealth of southern Arabia in ancient times. Hence, *Arabia Felix*.'

We drove to groves in the desert outside Muscat where the pale grey, scant-leafed, gnarled trees grew, and Ahmed pointed to the rust-brown wounds where flaps of bark had been sliced off, and the white blood oozed and congealed. In shops where frankincense was burning he would say, 'This is not the best stuff, of course.' He took me to the souk, and showed me small and large and even in-car incense burners, and the stalls where the many grades of frankincense were sold. He scooped his hand into baskets and held up handfuls of the dried, crystalline gum. 'Smell that – now, that's good!' He insisted that I buy a small quantity of the finest grade and try it out: musky, sweet, intoxicating.

I liked Ahmed immensely. He had an amused, faintly satirical view of himself and of the world. He was anglophile, jovial and refined, a slightly theatrical character who always carried a *khaizaran*, a cane swagger-stick that was an essential part of Omani formal dress. He was in his mid-thirties, and a modest savour for the good things in life had grown him a slight paunch. We ate together several evenings in a row – never touching alcohol – at traditional Arab restaurants, where we reclined on mats on the floor behind tall screens, and food was brought on brass trays.

'These restaurants are dying out,' Ahmed told me. 'Don't ask me why. When people go out they want to sit on a chair and eat Indian or Thai or Italian food.'

I asked him what role the desert played in the lives of modern Omanis. His reply was romantic. 'For us, the desert is Mother Earth. You feel a surge of power when you stand with your toes in the sand. We younger Omanis go to the wadis, to unspoiled nature out in the desert, and let go of the complexities of modern life, and let Mother Earth, or mother desert – desert is a feminine article in Arab – refresh our souls . . .'

I teased him. 'Ahmed, you sound like a Westerner. You lot are Bedouin stock, you're meant to *have* the desert in your souls, not to need to go out and get spiritual top-ups!'

'Ah, well. This is a modern country, Martin. We all want to be rich, and that means work, computers, oil, aluminium, and that means stress. What percentage of the population of Oman still lives in the desert, do you think?'

'Ten per cent?'

'Maybe one per cent.'

It took me only a day to reach the far south of Oman, the region of Dhofar, the only part of the Arabian Gulf region that receives the south-west monsoon, and the place where the finest frankincense trees grow. As the road climbed into the Jebel Qara, the hills that form a barrier between the desert and the sea, there was suddenly riotous vegetation, and a violent storm began to lash the windscreen. In darkness I reached Oman's southern capital, Salalah.

The next day I continued south. A recently completed, magnificently sinuous ghat road ensures the Omani military rapid access to their southernmost border, with Yemen. I would not be allowed into Yemen – but then the border is open to no one. Turning back, I stopped at the tomb of Job, the Old Testament prophet who held on to his faith despite the great suffering visited upon him; as Ayoub, he is revered as much by Moslems as by Christians.

I had arranged to meet in Salalah the historian and archaeologist of the Dhofar region, Ali Ahmed al Shahri. The recent history of Dhofar has not been tranquil. A war of secession was fought from 1967 to 1975, the Dhofari rebels receiving assistance from Yemen and the Soviet bloc, the Sultan of Oman receiving military assistance from, in particular, Britain. The failed rebellion still casts a long shadow, and Ali Ahmed, a man in his fifties, drew a veil of discretion over the recent past.

We travelled together to Sunhuram, a few miles south of Salalah, a pre-Islamic port built between AD 50-200 for the export of frankincense. He offered me a story to illustrate how legends live on in the folklore of the Bedouin of southern Arabia.

'Some Bedu told me, "We know a cave where Queen of Sheba and her genie buried her gold; and the Queen left the genie behind to guard it." I asked them where the cave was, and they told me, "No, it's dangerous!" But I said, "I *want* to see it." So one day they took me out into the desert, and led me to the cave. Inside, there was a circle of stones. They were afraid of what I might do, and they said to me, "Look, this is the Queen of Sheba's gold – if anyone touches it, the genie will eat them!" To get rid of them, I said, "You go and search for salt" – desert salt, they use it as a medicine. And left alone, I dug – and just a foot deep, I found a dead body. When they came back, I said, "Look, no gold – just a body." And they were astonished.

'This land is full of such stories. There's a saying that if anyone comes to Dhofar and steals anything, flying snakes will kill him. There are lots of frightful stories about this land. It means that this *is* a frightful land.

'The desert is a very hard place, you see, and the people living there are hard people, but brave, and very generous. The ancient Arabs believed that if you wanted your child to be well-educated and strong and famous, you had to send him to live with Bedu in the desert, and he would come back fair-minded, tough and self-reliant. The prophet Mohammed himself lived in the desert with the Bedu, and he was educated there. He learned their values – he never lied, he never spoke ill of anyone.'

The road I took north towards the Empty Quarter was a former frankincense route, believed to have continued north to the Persian Gulf. At that time, this flat tract of desert was green. The road turned east towards the Gulf, but, according to legend, it first passed through Ubar, the finest city in all Arabia.

T. E. Lawrence said in 1934 that he was convinced that the remains of an ancient civilization were hidden somewhere in the desert. Arabs had described castles seen by wandering tribes, and, Lawrence said, 'There is always some substance in these Arab tales.'

Around AD 200, Alexandrian geographers recorded the existence of a major market town deep inside the Empty Quarter; the Holy Koran tells of desert cities punished for their godless excesses, like Sodom and Gomorrah; the historian Al Hamdani talked of a city called Irem, 'astride the fabled incense routes, with imposing architecture, vast orchards and fabulous wealth', and, again in the

Koran, Irem was a place of lofty or columned buildings 'the like of which has not been created in the land'. Irem and Ubar were believed to be one, and turned by legend into a sort of gold-spired Xanadu.

In 1991, Sir Ranulph Fiennes mounted an expedition to try to find Ubar. He was reluctantly forced to conclude that, although all the folklore places it deep in the Empty Quarter, it was in reality a modest walled settlement on the edge of the desert, known today as Shis'r.

Shis'r is a small town with a mosque, an administrative building and a handful of houses, all smartly painted with white and brown wash. A small sign proclaims, 'The Lost City of Ubar'. As I walked around the site, I could understand Fiennes' and everybody else's disappointment that this tiny dig was fabled Ubar; like Timbuctoo and Agadez, it had signally failed to live up to the hype.

'That's not the lost city of Ubar!' Ali Ahmed told me hotly. 'Fiennes didn't find Ubar; the walls and remains of Shis'r were known long before Fiennes arrived. Nobody believes that's Ubar. If you ask the Bedu in the desert where it is, they say it's about eighty-five kilometres west of Shis'r, in an area of huge dunes. You have to dig in the desert. Ubar is buried under the sands!'

My guide and I drove north across a gravel plain, then past taller and taller sand dunes, towards the border with Saudi Arabia. I had asked for a visa to visit Saudi, and been refused. A few days later I would fly over Saudi Arabia's 864,000 square miles of sand. I wondered whether my inadvertent embrace of Islam in Tehran would qualify me to apply for a pilgrim's visa to visit Mecca. But for now, this was as close to Saudi Arabia as I was going to get.

And so into the Rub' al Khali, the Empty Quarter.

The pyramids of gold sand had become mountainous but well spaced, and the ground at their bases was firm. We drove on among them until the sun began to set, then we stopped and began a long climb. The sand was fine and golden, and still hot beneath our feet. High in the cerulean blue above was a dappled mackerel sky. My guide, who had fathered ten children of whom the youngest was two months old, scampered ahead of me, mocking my gasps. Reaching the top, he squatted and lit a cigarette. He smoked sixty a day, he told me, 'One for every year of my life!'

I joined him at the summit. The sun sank, a rosy disc in the east, and an evening wind blew up, blunting the razor edge into a mist of sand which poured into our mouths and eyes. I did not want to be driven away just yet. We pulled our turbans back around our mouths, turned away from the wind and narrowed our eyes against the sand. The landscape that lay before me was vast – vast and still mysterious, this empty quarter of the Arabian Desert, where even the Bedouin no longer go.

I was glad for the two years I had spent in the desert; it was where I had wanted to be. Each time I left to return briefly to urban life, I felt tugged uncomfortably by hidden threads, like too-tight stitches in a wound. When I went back to the city, I knew I would enjoy its distractions for a while. But each time I left the desert, I left behind something I missed deeply, a version of myself, someone focussed, who seemed to know who and what he was. I remembered George Gurdjieff's depression in the Chinese desert after his bullet wound, his dull sense of an imminent return to daily, self-satisfied life, and his determination to commit himself to some inner revolution. The isolation and unity of the desert make mystics of us all. But whatever it is we gain there we risk losing on our return to the real world. To reap the rewards of whatever it was I thought I had gained from the desert, I would have to take the desert home with me – inside me.

In the course of this journey there had been times of total solitude, and times, like now, of contented appreciation of a scene with another human being. Robert Frost wrote:

> *They cannot scare me with their empty spaces*
> *Between stars – on stars where no human race is.*
> *I have it in me so much nearer home*
> *To scare myself with my own desert places.*

My desert places, my empty quarters, did not scare me as they once had. I had a sense of their geography; I felt they could be crossed.

That night, leaving the Empty Quarter, we reached a small Bedouin encampment at a well. A few men sat around a fire, drinking tea. There were a couple of dozen camels in the shadows, but in an age of oil wealth and motor cars, they served no practical purpose. I asked one of the men, 'Why do you still have camels?'

'It's a hobby. It comes from our culture, our fathers, grandfathers. They used to be very important – for transport, milk, meat; but now modern life has changed everything.'

'But you still keep them?'

He grinned. 'We keep them.'

Camel ownership had always been a sign of wealth and prestige; Ali Ahmed had told me that nowadays people owned so many camels that their grazing was causing desertification in Dhofar.

I asked the man if he ever took his camels into the Empty Quarter.

'Yes, but not very far. In the old times we went in deep, especially in the winter. But sometimes we still go in a little way, searching for grasses.'

The contrast with the harsh, necessity-driven lives of the Tuareg I had met in Niger could not have been greater.

The next day, I pulled over at a Bedouin encampment out in the sands, and an old man welcomed me as a stranger, and called for coffee and dates to be brought, as they always are for the guest in Oman.

As we talked, I asked him if he wanted his children to be educated.

He hesitated. 'God knows, we have everything we need here in the desert. But the young people . . . they only want to know the city. But in the city one sees many bad things. So why go to city? I never go there now. And my mind is very clean!'

I laughed. 'So they should all stay in the desert?'

'Yes,' he said. 'We should be happy with what God gives us.'

In London, I had met Wilfred Thesiger, the British explorer and author whose book *Arabian Sands* describes a journey through the Empty Quarter in 1946-48. With his wistful talk of the loss of 'the spirit of the land and the greatness of the Arabs', Thesiger has done more than any other writer to cast the Empty Quarter as the last and ultimate desert, and the Arab as a noble savage whom oil riches have caused to fall from grace.

'The harder the life,' he told me, 'the more hospitable they were – and they did live a *desperately* hard life. When I went out to live with them, I thought that was going to be the difficulty – the physical hardship – but I was able to do all that. What I found more difficult was to meet them on level terms in their behaviour. They had a quality of nobility which I've met with in no other society. Their generosity, their hospitality, their courage, loyalty, endurance, patience, their love for their camels – all these were qualities which appealed to me.'

I asked Thesiger how he thought this tradition had evolved.

'*It was the hardness of their life*. They lived the life of Bedu because that's the life they wanted to lead. They rejected the easier life of lesser men. Soon after I left, the oil company arrived there with its Land Rovers and helicopters, and the whole structure of life was altered. Enormous wealth was available. Now bin Kabina and bin Ghabaisha drive around everywhere in cars, and Bedu life is finished. If they hadn't got the oil in Arabia, they'd still be riding around on camels.'

The writer and desert explorer (and biographer of Thesiger) Michael Asher is uncomfortable with the view that greater affluence has destroyed the authenticity of Bedu life. 'Thesiger says the Bedu are doomed,' he told me, 'but as long as they preserve their culture, I don't think motor cars or sending their children to school will necessarily destroy it. Nobody *chooses* a harsh, impoverished life.'

'But they can't preserve their culture,' I said. 'It will be replaced by a trashy Western lifestyle, where your life is defined by TV, pop stars and shopping.'

'I think there's something questionable about a rich man telling poor people that they're better off poor, yet remaining rich himself. The Bedu are happy that their children no longer die of diseases for which they have no cure except camels' urine and spells and potions. The changes in Bedu life are only bad from the point of view of Western travellers – they're not so picturesque when they've got Land Cruisers and money, they're not so easy to hire as guides and companions.'

I had already put this point to Thesiger. 'I know you can't keep them locked in the past,' he retorted, 'but all I say is, they were better off in the past. Not materially, but in ways that matter.'

Asher concedes that the modern Bedu have lost something – 'It wasn't the hard life that they lost, it was a sort of spiritual sense of belonging. They used to live in harmony with their environment, because they didn't have rational explanations for anything. If you said to them, "What are the stars?" they would say, "We don't know *what* the stars are – they're the work of God." They had a perspective on the world that we lost long ago.'

PART X

The Christian Desert

CHAPTER SIXTY-SEVEN

The Desert Fathers

What lies behind us and what lies before us are tiny
matters compared to what lies within us.

R.W. Emerson

MY WANDERING IN THE DESERTS was almost at an end.
From the afterdeck of a ship called the Minerva, Penny and
I regarded the meeting place of Saudi Arabia, Jordan and Israel. At this
wide bay, the upthrust eastern finger of the Red Sea parts two
mountain ranges and becomes the Wadi al-Arabah, the river valley
that descends from the Dead Sea, serving as a natural barrier between
Israel and Jordan.

On the road from the airport, we had watched the chain of lights
that circles the neck of the gulf, thin on the Jordanian side, thick on
the Israeli with the luminous holiday resorts of Eilat. Close behind our
moored ship was the small stone fort of Aqaba, and behind it new
Aqaba, a sprawl of incomplete apartment blocks, roundabouts and
newly planted palm trees.

This flat, fresh-watered inlet has been a home to human beings for
millennia. It was King Solomon's port, a Christian bishopric in the
fourth century, and after the rise of Islam, a major way station for
Egyptian *haji* en route to Mecca. The holy city of the Moslems is 600
miles south-east of here, cut off by steep dusky mountains, the
advance guard of the fierce Al Hijaz range.

It was out of these mountains that the Arab force of T. E. Lawrence
– Lawrence of Arabia – came in July 1917, after a celebrated two-
month march. Their crucial battle had been just to the north-east of
Aqaba, in the valley of Aba l-Lassan. A Turkish battalion of just under

500 men was camped by a spring at the base of a natural rock amphitheatre. Lawrence's troops gathered on the rim and exchanged rifle fire with the Turks for hours, to little effect. Lawrence remarked that they seemed to be shooting a lot, but hitting little. Insulted, fifty Howeitat horsemen recklessly charged the Turkish battalion, firing from the saddle. Three of them fell to Turkish fire, but the Turks broke and ran. Now Lawrence and 400 camel-mounted troops swept down onto the fleeing Turks. It was the famous incident in which an excited Lawrence accidentally shot his own camel. Three hundred Turks were killed and 160 captured.

In the moonlit aftermath of the battle, Lawrence was in a melancholy mood. He wrote in *The Seven Pillars of Wisdom*,

> *The corpses seemed flung so pitifully on the ground, huddled anyhow in low heaps. Surely if I straightened them they would be comfortable at last. So I put them all in order, one by one, very wearied myself, and longing to be these quiet ones, not of the restless, noisy, aching mob up the valley, quarrelling over the plunder, boasting of their speed and strength to endure God knew how many toils and pains of this sort; with death, whether we won or lost, waiting to end the history.*

Ahead of Lawrence's party lay three more Turkish posts, all of which surrendered or were already abandoned. They pressed on towards Aqaba, a hamlet of mud-brick buildings scattered around the fort, only to discover that several weeks earlier British gunboats had shelled and destroyed it. Lawrence seized control of a deserted town.

Penny and I sailed south to the mouth of the Gulf of Aqaba, then dog-legged north into that first finger of the Red Sea, the Gulf of Suez. To the east were the mountainous and impenetrable-looking shores of the Sinai Peninsula, its sandy coast stuffed with holiday complexes. The ship navigated its way among the visible, notoriously treacherous sandbanks of the Red Sea, where the eviscerated hulls of merchantmen sat high and dry. Now we glimpsed in the west the scarped Galala plateau in the Sahara esh Sharqiya – the Arabian Desert. Beyond it was the Nile. Standing on deck, Penny and I remembered the time we had spent along the Nile two years earlier. We had travelled to Egypt in the hope, among other things, of seeing the Christian monasteries in the desert.

The man who founded Christian monasticism was one of the religion's earliest monks, a hermit called Antonius (St Anthony), born in Egypt in AD 356. Before the age of twenty, he withdrew to a life of absolute solitude on a mountain by the Nile called Pipsir (today, Dayr al-Maymun). It is said that it was persecution that first drove Antonius and the rest of the so-called Desert Fathers (we hear about them less often, but there were Desert Mothers too) into the desert; but when the persecution of Christians ended in the early fourth century, the monks stayed on in the desert, in a sort of self-imposed persecution. A life of self-abnegation was considered voluntary martyrdom; people spoke of a red martyrdom, where blood was spilled, and white martyrdom, where suffering was voluntary. According to Athanasius, Anthony's first biographer, 'The man who voluntarily inflicts suffering on himself is ranked among the martyrs of God, for his tears are valued as drops of blood.'

There seems to have been another motivation. As newly legalized Christianity settled down, some Christians found that it was becoming increasingly smug; they sought a more fundamental, internal Christianity, beyond the diluting influences of cities.

Anthony spent nearly twenty years on his mountain, fighting a titanic combat with the devil and withstanding a legendary series of temptations. During his fasts, the devil would appear bringing bread; at other times he appeared as a savage animal, or a sexually desirable woman.

The desert was believed to be populated by demons. In Matthew 4:1, 'Jesus was led by the spirit up into the desert, to be tempted by the devil.' St Anthony met a demon who cried out to him, 'This is our place! What are you doing here? Get out!'

Those who witnessed Anthony's visions were convinced that the things he saw were happening. A sceptical modern reader may conclude that they were the almost inevitable psychological manifestations of a life based (presumably due to a too-earnest reading of Christ's teachings) on total self-denial.

Anthony became one of the Middle Ages' most popular saints, and around AD 1100 an order was founded in his name in Grenoble: the

Order of the Hospitallers of St Anthony. It became a pilgrimage centre for sufferers of the disease known as St Anthony's Fire, or ergotism. Eating flour milled from infected rye causes ergotism, whose symptoms include mental disturbance and convulsions. Ergot is also the source of lysergic acid, from which the hallucinogen lysergic acid diethylamide (LSD) is synthesized.

Which brings me back to Aldous Huxley, whose own desert illuminations I described earlier in the book. Huxley was naturally interested in Anthony; he had made a study of mysticism in its many manifestations, and had been one of the first people to experiment with the visionary properties of LSD.

It is too easy to concentrate on the lurid elements of the Desert Fathers' story, for there was a more spiritual aspect to the desert, too. It was seen as a special place where, in the past, the Lord had led His people in order to purify them. This Exodus is seen as being relived in each human breast. The desert was the privileged scene of a mystical encounter with God. Little, unfortunately, has come down to us of these unions. Can such things be communicated? Would we have believed them?

For the twentieth-century theologian and monk Thomas Merton, a monastery is a brick-built, formalized desert; every monk must follow Jesus into the desert:

> Go up, go up! This desert is the door of heaven!
> And it shall prove your frail soul's miracle!

It was only after our arrival – and after two days of travel in the Western Desert – that Penny and I discovered we had come during the one time of year when the monasteries are closed to visitors: Lent. Undiscouraged, we headed south, to the glories of Luxor.

Returning north a week later, we had one more piece of spiritual tourism in mind. In Cairo we had visited the Coptic Museum to examine the fragments of the Nag Hammadi Codices, or Gnostic Gospels, on display there.

The Gnostic Gospels purport to be accounts of the life and teaching of Christ. If they are accurate, they are a remarkable supplement to the Matthew, Mark, Luke and John who blessed the infant bed that many of us lay on. In the four centuries after the death of Jesus, the Christian gospel spread as a sort of house church movement, scattered throughout the Middle East. Isolated sects developed, fostering their own revelations and gospels. These Gnostic or secret gospels portrayed a puzzling, ludic, psychologically edgy Jesus, a man with a secret doctrine for special initiates. Which is the True Jesus? The conventional, loving moralist? Or a mystic who taught about *inwardness,* where the practice of religion is a path of self-scrutiny?

Every mainstream religion has an esoteric branch, but Christianity has an unfortunate tradition of suppressing it. Seen in the context of the time, it may be understandable. Christians were subject to persecution and execution; the church sought homogeneity, mass membership, and political respectability; heretical writings were banned, and ordered to be burned. So we have the image of a monk at a monastery outside Nag Hammadi, kneeling in the sand to bury the sacred doctrines of his cult, where they would lie for 1,600 years.

In December 1945, a farmer called Mohammed Ali was digging with a hoe. With a sudden crack he shattered something glass or ceramic, and the air was filled with fragments of gold – or so he thought for several marvellous moments. But the gold turned out to be dust, a cloud of dried fragments of parchment from manuscripts sealed in clay jars. Mere paper – and as such, he used it to light fires.

But someone said, 'Wait a minute! You know how obsessed these foreigners are with antiquities. Look at that old writing. These scrolls could be worth something!' Mohammed Ali travelled up to Cairo to look for an antique dealer, and some of his parchments duly reached the open market.

When at last they were seen by scholars, they were recognized as extraordinary survivals from the early Christian world, of which so little is known today. These voices from an obscure past threatened to rewrite the story of Christ, even to rock the foundations of Western civilization.

It was not easy to reach Nag Hammadi. Militant Islamicists had been sporadically murdering tourists, and Nag Hammadi is near the university town of Assyut, the nerve centre of Islamic revolt. Penny and I had travelled all day, changing from bus to minibus in a tense town where foreigners were an uncommon sight, and plain-clothes police officers with pistols in their trouser waistbands strutted on every corner. The minibus halted at a police checkpoint supported by an armoured car and two heavily fortified sentry towers, and a plain-clothes man noticed us and questioned our driver. When he gave a rather flippant reply, he was dragged out of his seat and cross-questioned. Suddenly the driver was released, our bags were thrown down, and the minibus departed without us.

The swaggering cops were tough guys. They were fighting a civil war which involved bombs, counter-insurgency and torture. The officer who spoke to us in English had a semi-automatic pistol nestling against his paunch. He was about thirty-five, with a pocked face inset with small teeth, and cold eyes. As he halted a passing taxi and bundled us in, I saw myself: the tourist-stuff on which Egypt's economy depended.

Was ever a name less rich in the promise of pampered luxury than 'The Aluminium Hotel'? The taxi brought us into Nag Hammadi's aluminium factory compound, and we settled into our room as the sunset was turning pink the mist over the Nile.

I had a book which gave the precise location of the discovery; unfortunately, I had left it in Cairo. Our taxi-driver was willing, but had never heard of manuscripts, Gnostics, or even of the Jabal al-Tarif, the hills where I knew the discovery had been made. We realized that Nag Hammadi could simply be the nearest large town to the site, which might be miles away. So the next morning we drove into Nag Hammadi in search of a local museum, library or school. Everything was closed. The police had not heard of Jabal al-Tarif, either.

At last, on the fringes of Nag Hammadi, someone told us, 'They're over there.' In the shimmering heat in the east, beyond the Nile, a range of cliffs reared up like Alcatraz. 'Oh yes, *now* I remember,' said the driver. An iron-girder bridge led over the Nile, and we drove through the fertile green of the flood plain until we reached the foot of the barren cliffs.

We drove on, stopping various people.

Caves, jars, parchments?

'No, no, no.'

At length, we reached a village where the people told us that there were some caves in the hills, though no one had ever heard of anything being found there. They pointed up at black dots on the tan faces of the cliffs. We decided to take a look.

The sun was high as we scrambled up the steep slopes. We had with us, in addition to our protective driver, a dozen ragamuffin children who scampered eagerly ahead, calling out to each other, pointing out the quickest ways up. It took over an hour for our good-natured guides to show us all the caves they knew. In the first, thick clay fragments and handles from large urns lay among a litter of rocks on the floor of the cave. Then there were dark, dangerous caves with sudden potholes in them, networks of fissures filled with musty animal smells, but little else. What did we expect to find? Another buried codex?

Penny and I, our taxi-driver and the dozen village boys, did not find new gospels. Nevertheless, the last cave contained something to astonish us. Its walls were lined with ancient Egyptian hieroglyphs, a delicate tracery of royal blue and gold. Where there were faces, they had been painstakingly scratched out, for the representation of life is inimical to most Islamic traditions. But it was startling and wonderful to find this evidence of – of what? – an Egyptian hermit who lived in a cavern high above the Nile, even before the Christians arrived?

CHAPTER SIXTY-EIGHT

A Grain of Sand

I have written much about many good places.
But the best places of all, I never mentioned.

Edward Abbey

TWO YEARS HAD PASSED since that rather bungled series of would-be pilgrimages (though we had at least managed to get one thing right: a certain proposal, made *à genoux* on a boat in the mid-stream of the Nile, and accepted).

The ship reached Al Suweis – Suez, and the start of the canal. We joined a great queue of vessels – cruise boats, sailing yachts, container vessels and oil tankers – arranged at intervals of several hundred yards, waiting for the south-bound convoy to emerge from the canal so that we might enter.

Penny's father had been stationed on the Suez in the 1950s, and had told us where to look out for his canal-side camp. With the help of our captain's chart, and from the vantage of his bridge, we located it precisely, a clump of buildings near an airstrip in the haze on the western shore.

I have two uncles who served in the British mandates and possessions of the Near East. One fought in the North Africa campaign, and another was stationed in the Middle East at the time of the Suez crisis. Their desert postings are burned into their memories. The evocative names come out of them – Muscat, Aden, Tripoli, Bengazi – and the memories of being wide-eyed young men at war, or doing national service, guarding the interests of a dwindling Empire. They recall long days spent driving across the desert's yawning plains, and the silence of nights under the stars. They remember camaraderie and regimental songs, including songs about the men who fought and died.

> Come listen and I'll tell you a story
> Of a lad who was taken from home
> To fight for his king and his country
> To fight for the old folks at home.
>
> They put him in The Armoured Division
> And sent him to a foreign land
> Where the flies flied around in their millions
> And there was nothing around him but sand.
>
> Early one morning at daybreak
> Beneath the strong Libyan sun
> It was there that this poor British soldier
> Was shot by a big Gerry gun.
>
> Oh bury me deep in the desert
> Under the Libyan sun
> Oh bury me deep in the desert
> My duty for England is done.
>
> So they buried him deep in the desert
> With Allah to watch o'er his grave
> They buried him deep in the desert
> His young life for England he gave.

When I was twenty, in no danger of being dispatched to fight in foreign wars for king and country, I came across the poetry of Keith Douglas. He fought in the North Africa campaign, and recorded it in poetry that seemed to me to have a quality of sophisticated, detached observation, with hints of raw emotion barely beneath the surface. When I think about the North Africa campaign, my images are mostly from Douglas, who died in action in 1944.

Three weeks gone and the combatants gone,
Returning over the nightmare ground
We found the place again, and found
The soldier sprawling in the sun.

The frowning barrel of his gun
Overshadowing. As we came on
That day, he hit my tank with one
Like the entry of a demon.

Look. Here in the gunpit spoil
The dishonoured picture of his girl
Who has put: Steffi. Vergissmeinicht
In a copybook gothic script.

We see him almost with content
Abased, and seeming to have paid
And mocked by his own equipment
That's good and hard when he's decayed.

But she would weep to see to-day
How on his skin the swart flies move;
The dust upon the paper eye
And the burst stomach like a cave.

For here the lover and killer are mingled
Who had one body and one heart.
And death who had the soldier singled
Has done the lover mortal hurt.

Our trip up the Suez was serene, the stately convoy, the monotone brown shores, the placid water. Here and there the *Minerva* scraped by a village where children played in shallows, yards from the continuously dredged big-ship depths. On the Egyptian bank we saw a totalitarian war memorial, a vast, cruel bayonet.

In Cairo we toured the Islamic sights. I was particularly struck by the Sultan Hassan Madersa, one of the finest examples of Marmeluke architecture, a magnificent slab of stone with a superbly poised and harmonious interior; after two years in the desert, I was in awe of a building designed to remain cool throughout the year, with a brilliantly conceived series of perforated walls and ventilation shafts.

Our next port of call was Alexandria. The pleasures of Alexandria are decadent and perverse – the satisfactions of faded grandeur, nostalgia and decay. And we had pagan pilgrimages to perform. We made for the Cavafy Museum, the home of the great poet Constantin Cavafy, born in Alexandria to Greek parents in 1863. Penny had been seasick that morning, and the ship's physician had given her an injection. Under the influence of the drug, she fell asleep on a polished mahogany bench, surrounded by glass-cased Cavafy in multiple translations.

It is a gracious bourgeois apartment, with tall, well-proportioned rooms, shuttered and cool. Cavafy famously asked, 'Where can I live better? Below, the brothel caters for my flesh. And there is the church which forgives sin.' I stood on the balcony, locating the roof of the church and imagining Cavafy receiving absolution. 'And there is the hospital where we die.'

In fact, the brothel did not cater for Cavafy's flesh; being homosexual, he preferred to pick up boys at the Cecil Hotel. The Cecil was our next stop. We walked past second-hand bookstalls, and I imagined I might find Cavafy or a decomposing Faber paperback of Lawrence Durrell's *Alexandria Quartet*. But there were only pulp novels, medical textbooks and computer manuals.

Wandering among the decrepit mansion blocks of Alexandria, I was reminded of another poet. In my early twenties I lived in Bombay, and had known an Englishman marooned on that other island of decaying Victoriana and art deco. Geoffrey had slept by night in God knows what hovels in the slums around Crawford Market. He kept his possessions – mostly reams of poems – bundled up in plastic carrier bags. Geoffrey Hann's life as a servant of empire and artist had taken him from the Levant to Cambridge to India, but the place where it began made his exotic journey seem almost inevitable. He had been born here, in Alexandria.

We walked, with Penny still in a drug-induced stupor, to the Cecil Hotel. They all came here – Churchill, Coward, Forster, Maugham, Durrell; here the heroine of Durrell's *Alexandria Quartet* made her first big entrance, clad 'in a sheath of silver drops'.

The Cecil has retained a discreet charm, a hybrid sense of its occidental roots. 'Would you like some milk tea and English cake, madam?' the waiter asked Penny.

In fact, it was another pilgrimage we had in mind. I have already mentioned the British film called *Ice Cold in Alex*, with its climactic moment of iced beer foaming in tall glasses. I wanted one too.

The menu said 'Foreign Beers'. We asked the waiter if they served Carlsberg. Sure enough, they did.

Back in the street, we hailed a horse-and-trap and trotted around Alexandria's corniche to the butter-coloured Qaytbay Fort, a fifteenth-century sea defence built from the stump of that wonder of the world, the Pharos Lighthouse. Courting couples and ice-cream-licking families dodged the breakers crashing on the seafront.

In Bombay, I used to ask the exiled poet Geoffrey Hann if he thought he would ever see England again. 'No,' he said firmly. He had family somewhere in the English suburbs; he thought he might even own part of a semi. But he was going to die in Bombay. He knew he would never see Alexandria, either. That, however, did not stop him dreaming:

> *I who am rootless as desert air,*
> *Could I put down an archaeological root in Cleopatra's city,*
> *Cavafy's too – Alexandria?*
> *It is fitting that in the coarseness of time*
> *I should exit where I entered*
> *In my case, just a closing of the sand.*

From Alexandria we crossed a corner of the Mediterranean to Haifa in Israel, then made our way to the shore of the River Jordan, where as a boy I knelt with Mum and Dad, collecting water. The shoreline 'near what is believed to be the site of Our Lord's baptism' has been colonized by a souvenir-shop complex of monstrous vulgarity, and was littered with plastic cups. I had read that many pilgrims suffer a mild form of psychosis resulting from the clash between their numinous expectations and the reality of modern Israel; I felt only sympathy for them.

We travelled further south to Jerusalem, arriving on Easter Sunday. The holiest city in the world made me long for the sanity of the wilderness. At the Wailing Wall, a Zionist demonstration was in progress, its leaders threatening to march on, and occupy, the Islamic shrine of the Dome of the Rock. It is a holy site because Mohammed made his night ascent to heaven from here; but simultaneously it is a Jewish sacred site, where the Holy of Holies in the Solomonic Temple stood, therefore where the Ark of the Covenant was kept. There is enough irrationality and hatred in the history of this single lump of rock to turn a person with a sensitive disposition off religion for life.

As we climbed past the Wailing Wall, we marvelled at the bearded, ringletted Hassidic Jews, their eyes bulging, spit and fundamentalist chants erupting from their mouths; the wary-eyed, armed policemen in bullet-proof jackets who frisked us with metal detectors; the TV men with cameras on their shoulders, hoping to capture some violence. The Dome of the Rock with its quiet if sharp-eyed Moslems seemed, after the scenes below, like a vision of civilization.

Jerusalem is crammed full of lumps of holy rock, and supercharged with religious, nationalistic or racial emotion – at once fabulously religious and utterly insane. I had concluded, terminally I thought, that this city gave religion a bad name, when we reached the Church of the Holy Sepulchre. It too is a scene of violent disputes, though as far as I know they are all confined to one religion – the seven or eight mutually antagonistic Christian sects. On Easter Sunday, the church was churning and frothing with pilgrims, with sobbing women sopping the stone on which the slain Jesus was supposedly laid, and mopping up the water to bottle it. Around them swirled a whirlpool of humankind. Most of the visitors were from the corners of Europe where serious Christianity remains alive – Greece, southern Italy, Portugal – and the near-hysterical, mostly feminine devotions seemed harmlessly therapeutic after the scenes of pop-eyed, bare-gummed hatred at the Wailing Wall.

At lunch, in sore need of calm, we met Israeli friends, Haim and Yael. We bathed our raw sensibilities in the balm of their amused moderation.

Haim is an architect. Recently, he told me, he had been offered a job designing a housing settlement for the Bedu. The Israeli state is – like Chile, Peru, Mexico, America, Australia, China, Oman, Niger, Namibia, Botswana (to name a few) – uneasy about tribal peoples. It would like to see them settled, sedentarized, urbanized; tamed. Haim turned down the job.

We went out to the east of Jerusalem. The city stops quite suddenly, in a desert waste that plummets to the shores of the Dead Sea, bleached, mineral-veined and stark. Jesus's wilderness, the first desert I ever visited.

It was the end of my desert circumnavigation.

It had taken me two years, travelling through the twin bands of desert that ring this planet. I remembered tracks littered with the skeletons of camels, a journey through choppy Chinese sand dunes, and Saharan jaunts with rakish Tuareg rebels. Most of all I remembered the desert dwellers. They were people who owned little, but would always invite you into their tent, and feed you tea and sweetmeats, depleting their precious supplies. The desert had produced a mysterious nobility in them. I, a Western writer travelling the deserts because I chose to, always with the choice to leave – I owed them a great deal.

I thought of Jesus baptized by John the Baptist, of desert hermits and oasis whores; the GPS, the sunset, the gazelle and the scorpion, the nomad and the ringletted Jew; the racing meteorites that in their final moments, after light-years in interstellar space, scratch chalk-marks on the desert night; sand, endless sand, worlds, solar systems, eternities glimpsed in a grain of sand.

Acknowledgements

THE JOURNEY DESCRIBED in *Grains of Sand* depended on the support of many people. Principal among them was my wife Penny, whom I married during this journey, who was with me at the book's conception, for three, too-short periods during the journey, and in my heart at all times. Without her, the journey and this book would have remained a dream. My mother, Doris Buckley, also offered unflagging support.

I will be always grateful to Peter Everett, an inspiring boss during my time at the BBC, who generously gave me the opportunity to begin work on *Grains of Sand*. I also thank Brian Barfield, Managing Editor BBC Radio 3, who commissioned the series of radio documentaries that played a vital role in making this journey possible.

I am indebted to a number of people in the world of journalism whose commissions for articles allowed me to finance much of the journey, and to begin to work out on paper what eventually became this book. Simon Calder, travel editor of The Independent, published my first travel article, and encouraged me to believe that a book might be possible. Sarah Spankie commissioned a large article for Condé Nast Traveller. Gill Charlton commissioned my first piece for the Daily Telegraph, starting a relationship without which the book could not have been completed. And I am especially grateful to Tessa Boase, former commissioning editor at the Telegraph, whose idea of a weekly column filed from the deserts underwrote a third of my journey.

The book was magnificently championed by my former agent Jennifer Kavanagh; *Grains of Sand* was the last book she represented before leaving the publishing profession – she assures me that the two events were not connected. Her successor, Ruth Needham at

M.B.A., has also been tireless in her support and friendship. And I am grateful to Paul Sidey, who commissioned the book for Hutchinson; he has been wise, witty and patient before the tribulations of a novice author.

A number of friends gave their time to read early drafts of *Grains of Sand*, especially Alec Charles, Richard Naseby and Tessa Boase. Many other friends and acquaintances offered crucial assistance or encouragement. I would like to thank Mick Brown, Frances Byrnes, Tim Callan, Jennifer Cox, Sky Garner, Emma Hancorn, Mark Jagasia, Jane Jakeman, Jack Klaff, Ilse Lademann, Andy Martin, Jenni Mills, Nadia Mouici, Olivier Noël, Christa Paula, Olivier Roux de Bezieux, Sebastien Ruggiero, Paul Sen, Peter Smith, Sir Roy Strong, Colin Thubron, Richard Johnson, Robin Waterfield, Jules Wilkinson, Colin Wilson and Marie Wynter.

Throughout this journey many people in the world's deserts treated me with remarkable kindness. I was often in the hands of virtual strangers, who treated me as their own.

Fri 27 July

Start of August